APPLICATIONS OF REMOTE SENSING TO AGROMETEOROLOGY

ISPRA
COURSES

ON REMOTE SENSING

A series devoted to the publication of courses and educational seminars given at the Joint Research Centre, Ispra Establishment, as part of its education and training program. Published for the Commission of the European Communities, Directorate-General Telecommunications, Information Industries and Innovation, Scientific and Technical Communications Service.

The publisher will accept continuation orders for this series which may be cancelled at any time and which provide for automatic billing and shipping of each title in the series upon publication. Please write for details.

APPLICATIONS OF REMOTE SENSING TO AGROMETEOROLOGY

Proceedings of a Course held at the Joint Research Centre of the Commission of the European Communities in the Framework of the Ispra-Courses, Ispra, Varese, Italy, 6-10 April 1987

Edited by

F. TOSELLI

Joint Research Centre, Ispra, Italy

KLUWER ACADEMIC PUBLISHERS

DORDRECHT / BOSTON / LONDON

Library of Congress Cataloging in Publication Data

Applications of remote sensing to agrometeorology : proceedings of a course held at the Joint Research Centre of the Commission of the European Communities in the framework of the Ispra courses, Ispra (Va), Italy, 6-10 April 1987 / edited by F. Toselli.
p. cm. -- (Ispra courses on remote sensing)
Includes index.
ISBN 0-7923-0020-3 (U.S.)
1. Meteorology, Agricultural--Remote sensing--Congresses.
2. Meteorology, Agricultural--Remote sensing--Data processing--Congresses. I. Toselli, F. II. Commission of the European Communities. Joint Research Centre. Ispra Establishment.
III. Series.
S600.7.R46A67 1988
630'.2'515--dc19 88-29405
CIP

ISBN 0-7923-0020-3

Commission of the European Communities Joint Research Centre Ispra (Varese), Italy

Publication arrangements by
Commission of the European Communities
Directorate-General Telecommunications, Information Industries and Innovation, Scientific and Technical Communications Service, Luxembourg

EUR 11757 EN

Published by Kluwer Academic Publishers,
P.O. Box 17, 3300 AA Dordrecht, The Netherlands.

Kluwer Academic Publishers incorporates the publishing programmes of
D. Reidel, Martinus Nijhoff, Dr W. Junk and MTP Press.

Sold and distributed in the U.S.A. and Canada
by Kluwer Academic Publishers,
101 Philip Drive, Norwell, MA 02061, U.S.A.

In all other countries, sold and distributed
by Kluwer Academic Publishers Group,
P.O. Box 322, 3300 AH Dordrecht, The Netherlands.

Printed in The Netherlands

CONTENTS

PREFACE

Within the framework of Ispra Courses, a course on "Applications of Remote Sensing to Agrometeorology" was held from April 6th to 10th, 1987 at the Joint Research Centre of the European Communities, Ispra Italy.

The purpose of the course was to familiarize scientists, active in Agrometeorology and related fields, with remote sensing techniques and their potential applications in their respective disciplines. Conventional ground investigations in various fields of natural sciences such as hydrology, pedology and agrometeorology can be supplemented by a range of instruments carried by airborne or earth orbiting platforms.

The last few years, in particular, have seen many developments in this respect and a growing amount of information can now be derived not only from dedicated earth resources satellites such as the LANDSAT and SPOT, but also from other platforms such as METEOSAT and the series of NOAA-TIROS. Future platforms (ERS-1, Space Station, etc.) with their advanced sensors will further broaden the range of applications open to the investigators.

The use of these data sources, together with field investigations, can lead, at a reduced cost, to a better characterization of the spatial and temporal properties of natural systems.

This course was intended to provide information on the state of the art and on prospective developments in this field. The course was divided into three parts:

1) Principles of remote sensing. This section introduces the fundamentals of remote sensing, the basis of detection, the various types of platforms and sensors and the types of data obtained.
2) Data processing. This part is designed for current or future users of aerospace remote sensing data who have limited experience in digital image processing. Properties of digital images, standard CCTs, their format and disk storage systems are reviewed. The principal methods of image processing are then examined, including histogram manipulation, contrast stretching, edge enhancement, filtering techniques, etc. The characteristics and interpretation of radar data are also introduced.
3) Applications. In this section the contribution of remote sensing to the measurement of a number of basic agrometeorological parameters (e.g. solar radiation, albedo, surface temperature, vegetation characteristics and evapotranspiration, etc.) was discussed.

Advantages, limitations and accuracy of remote sensing approaches compared to usual techniques were also presented. To illustrate the usefulness and application of remote sensing techniques to the study of actual problems, examples of work actually performed at various scales, from world-wide to local, were presented.

A series of invited lecturers from universities and institutes in Europe and in the USA, all well-known specialists in their fields, thoroughly illustrated the various parts of the programme.

After the course, with the support of the Directorate-General of the Scientific and Technical Information of the European Communities at Luxembourg and the organization of the Ispra Courses, we asked the authors, who enthusiastically accepted, to take upon themselves the supplementary task to look over and revise their lecture notes in order to make them consistent among them. We would like to thank all these authors for their cooperation.

Our thanks go also to the Directorate-General of the Scientific and Technical Information offices, the Division Ispra Courses and the publisher for their help in the production of this volume. We believe that it reflects the scope and the coverage of the Seminar and we trust that it will constitute a useful tool for all those involved in remote sensing of Agrometeorology and related fields.

Ispra, August 1988

FUNDAMENTALS OF REMOTE SENSING

Ph. Hartl
University of Stuttgart, Institute of Navigation
Keplerstr. 11
D-7000 Stuttgart 1 - Fed. Rep. of Germany

ABSTRACT. This paper defines the purpose and describes the physical background and basic features of remote sensing techniques. It covers the applicable spectral bands of the optical and microwave regions and the basic measurement concepts, which rely on the scattering and/or self-radiating processes of electromagnetic waves and, more generally, on some interaction of electromagnetic energy with matter.

1. DEFINITION

Remote sensing is defined as a means of observing the earth's surface and environment from air or space by means of electromagnetic waves of the optical or microwave range. The restriction to electromagnetic waves is due to the fact that the observation from a spacecraft excludes other possibilities such as sonic waves, which require a medium like air, water or solid earth for propagation. Other means of indirect observation by, for example, stationary magnetic or electric fields are not sensitive enough for high geometric resolution measurements.

Remote sensing is characterized by the fact that it measures qualitatively and quantitatively features of the earth or its atmosphere without any material contact. The instruments applied are called sensors. If they consist of both a transmitter and a receiver, then we are talking about "active sensors". In the case of external signal sources it is possible to use just receivers, i.e. "passive sensors".

The Target's features can be measured indirectly through the interaction with electromagnetic waves. Various methods are applicable. Figures 1 and 2 show the proper arrangements for passive and active sensors, respectively. In the passive case one determines either

- the amount of electromagnetic emission of the target itself, or
- the reflectivity of optical signal power of the sun at the target, or
- the amount of scattered power, if the "target" is, for example, an aerosol of the atmosphere, or
- absorption of solar radiation power in the media to be measured.

F. Toselli (ed.), Applications of Remote Sensing to Agrometeorology, 1–18.

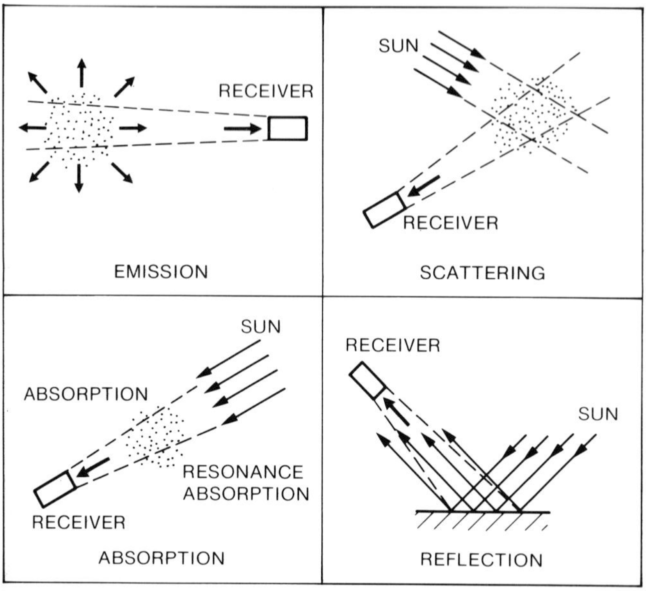

Figure 1. Basic elements of remote sensing techniques with passive sensors.

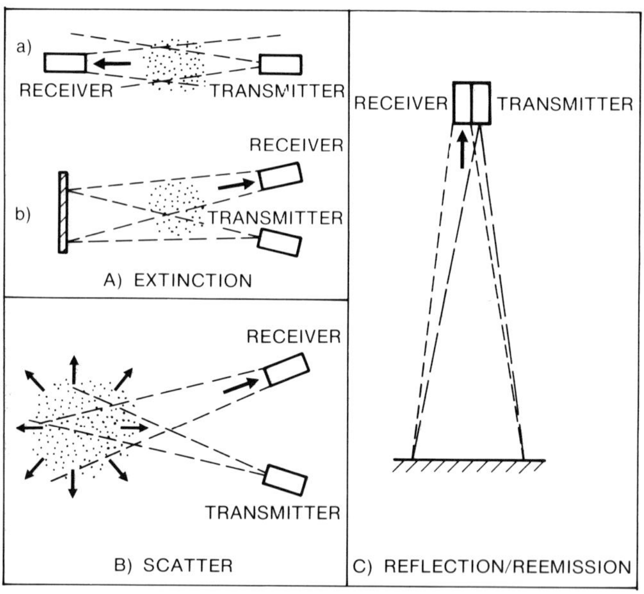

Figure 2. Basic elements of remote sensing techniques with active sensors.

Other methods can also be applied, for example the determination of the amount of signal polarization or depolarization, etc. In all cases it is important to identify functional dependencies of the features measured on frequency or angle of observation.

One can measure "bistatic", if the transmitter and receiver are at different places. In the case of active sensors one usually applies a "monostatic" technique and measures the backscatter function: transmitter and receiver are located on the same platform and the receiver input detects the signal power returned from the target.

The sensors can either "look" vertically down to observe the nadir scene (optical scanner), or sideways (SAR), or horizontally (limb measurements) (see Figures 3 and 4).

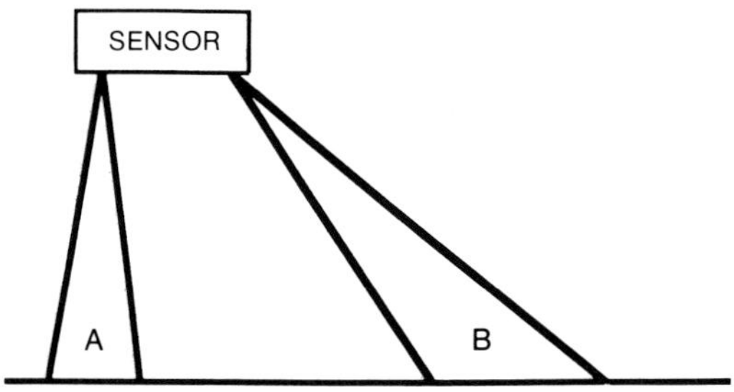

Figure 3. Downward-looking (A) and sideward-looking (B) sensor.

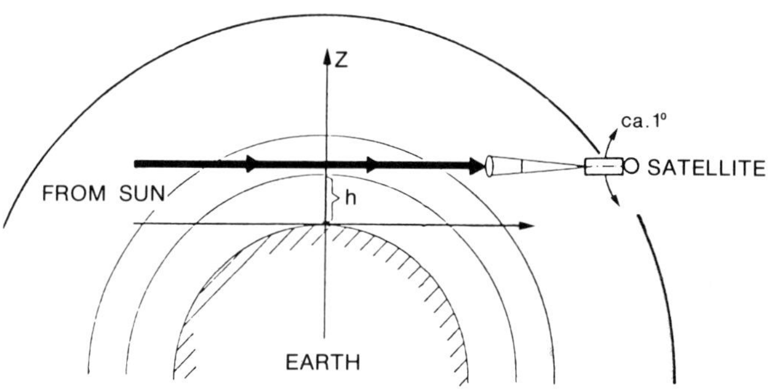

Figure 4. Limb measurement.

The sensor types can be classified in various ways. In Table I the active sensors are categorized into optical and microwave sensors, into imaging and non-imaging sensors, and into instruments, which allow the three-dimensional measurement of target surfaces. In Table II the passive systems are classified as radiometers and interferometers as non-imaging systems, and scanning and framing sensors as imaging systems. Framing systems such as cameras can take a whole image simultaneously, as opposed to scanning systems which produce the image line by line.

TABLE I
Classification of active sensors

Optical sensors		Microwave sensors
	Non-imaging sensors:	
Laser profiler		Altimeter (RA) Scatterometer
	Imaging sensors:	
Scanning laser		SLAR SAR
	Dimensional sensing systems:	
Scanning Lidar		SAR-interferometer Scanning RA

TABLE II
Classification of passive sensors

Non-imaging sensors	Imaging sensors
Radiometers - optical - microwave	Scanning sensors - electro-mechanical - electro-optical - electronic scanning
Interferometers - optical - microwave	Framing systems - photographic systems - electron imagers (RBV)

Framing has the advantage that a misalignment of the platform's orientation affects all pixels in the same way and can be corrected by one transformation. The distortion of a scanned image can vary from line to line or within a line and is more serious and complicated to correct.

2. ELECTROMAGNETIC INTERACTION

Electromagnetic interaction is a function of the physical parameters of the target. It is especially a function of the dielectric properties (permissivity) of the target's material. The material's behaviour against microwaves can be classified into three categories:

- If the material behaves like a metal, then the waves cannot penetrate this matter. Electromagnetic waves irradiating such a metallic surface are almost completely reflected. The type of reflection (specular or diffuse) depends on the roughness of the surface.
- If the material is water, then the incident waves are reflected at the water surface to a large extent. Only some small fractions of the incident energy can enter and penetrate the target and will be

attenuated to almost zero power after a short way, say a few wavelengths or fractions thereof.

- "Dry", non-metallic materials change the wavelength of the incoming signals because the refractive index is larger than in vacuum or air, and the signals split, in accordance with Snell's law, into a transmitted and a reflected component (see Figure 5). With increasing water content of the target the similarity of the material's properties with that of metallic matter under electromagnetic interaction increases.

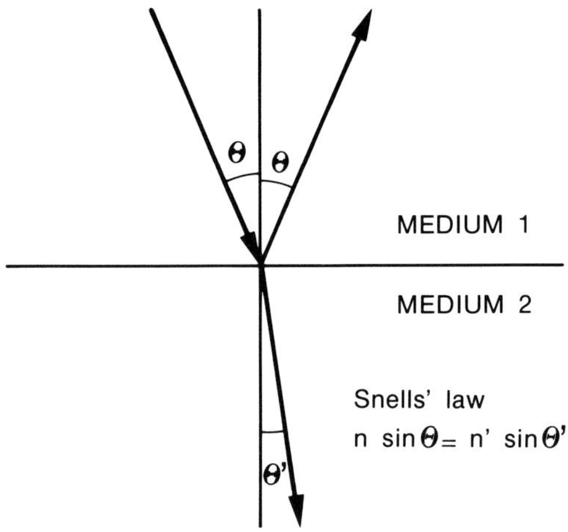

Figure 5. Snell's law of reflection and refraction.

Generally, interaction of the waves with matter and its characteristic features can be described to a large degree by the following factors:
- reflectivity/scatter, backscatter;
- absorptivity, and to some degree, by
- the transmissivity.

2.1 Reflection, Absorption and Backscatter

They define the percentage of signal power reflected, absorbed and scattered by the target. The values of these quantities are functions of the microstructure, the chemical nature and the biological state of the target. As an example: the leaves of plants consist basically of a sponge-like structure, with many randomly distributed holes and droplets of water and other holes filled with some pigments (see Figure 6). They can be modelled simply by a lossy layer of a certain permissivity, where the loss is frequency-dependent. Dispersive multireflection, therefore, occurs in the leaves. The chlorophyll elements absorb radiation energy of the blue and red sections of the light spectrum. The reflected signal lacks these absorbed parts of the solar

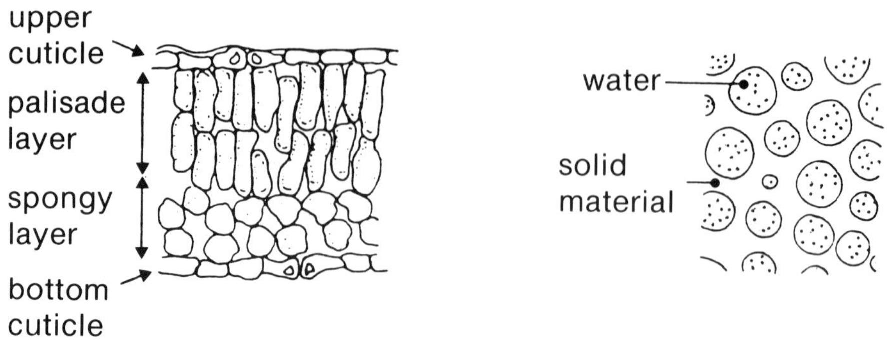

Figure 6. Cross section of a leaf; biological and physical model.

spectrum and, therefore, looks green. In the infrared parts almost all the signal energy is reflected, if the structure is in proper conditions. If, however, the structure is collapsed because of stress conditions, then the reflectivity is affected.

If photosynthesis no longer occurs, then the plant colour changes to yellow, which shows that almost all spectral radiation components of the visible light are reflected. The typical spectral characteristic of healthy leaves is given in Figure 7. It is, of course, also a function of the season, i.e. of the phenological state of the vegetation.

As well as the material, the surface structure and partly the inner structure of the target influence the reflectivity. The "roughness" is a function of the variations of the surface, measured in relation to the wavelength. The "Rayleigh" criterion defines that the surface is rough, if the upper and lower sectors of the surface are $\lambda/8$ apart (Figure 8). Then we have no longer specular reflectivity but reflectivity in accordance with Lambert's law, i.e. the signal is no longer reflected specularly as on a mirror but scattered in all directions according to a cosine law (Figure 9). If the surface has a periodic variation, then a resonant effect occurs (Bragg effect), where in certain directions the signals look extremely bright (Figure 10).

A surface is "rough" if there is a strong interference between the rays reflected by the bottom and top parts of the surface. This happens when the difference of the optical paths of the rays exceeds the value of $\lambda/4$.

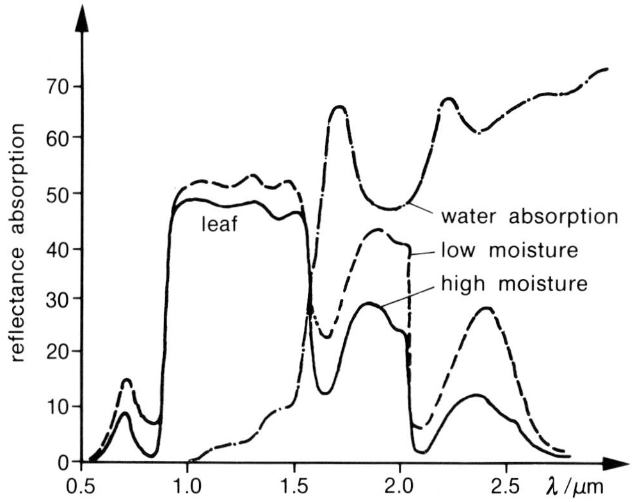

Figure 7. Signatures of leaves and water.

Spectral signatures

<u>Bands of Thematic Mapper</u>:

1) 0.45 - 0.52 μm: Water penetration, differentiation soil/vegetation and deciduous/coniferous flora.

2) 0.52 - 0.60 μm: Peak of "green" reflectance; assessment of vegetation-vigor.

3) 0.63 - 0.69 μm: Vegetation discrimination: chlorophyll absorption.

4) 0.76 - 0.90 μm: Biomass determination.

5) 1.55 - 1.75 μm: Moisture content; soil moisture, snow/clouds discrimination.

6) 10.40 - 12.50 μm: Thermal mapping, vegetation stress analysis, soil moisture discrimination.

7) 2.08 - 2.35 μm: Hydrothermal mapping, rock type discrimination.

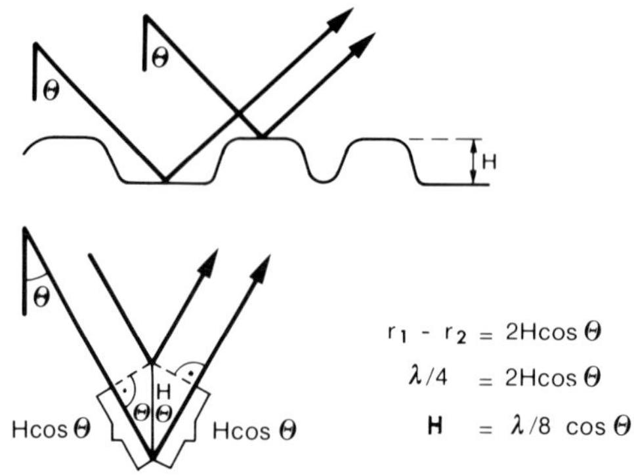

Figure 8. The Rayleigh criterion for the definition of the roughness.

For large θ values (flat incidence), even large H-values can be tolerated to achieve "specular reflection".

It should be remarked that the criteria discussed are scaled by the wavelength. In the microwave region, where the wavelength is 4 to 5 orders of magnitude larger than in the visible region, the structure and roughness must also be 3 to 4 orders of magnitude larger than in the optical region to lead to the same interaction effects, in particular reflection and scatter effects.

Another factor which is to be taken into account is the following: the relative dielectric constant of water is a function of frequency. In the optical region it is in the order of 10, which results an optical refraction index of approximately 3, being interrelated by a square root function. In the microwave region below 3 GHz the dielectric constant is in the order of 80 and the refractive index is, therefore, in

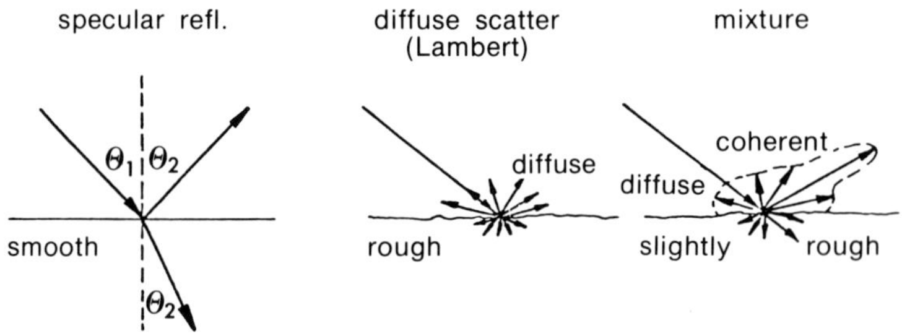

Figure 9. Different types of reflection.

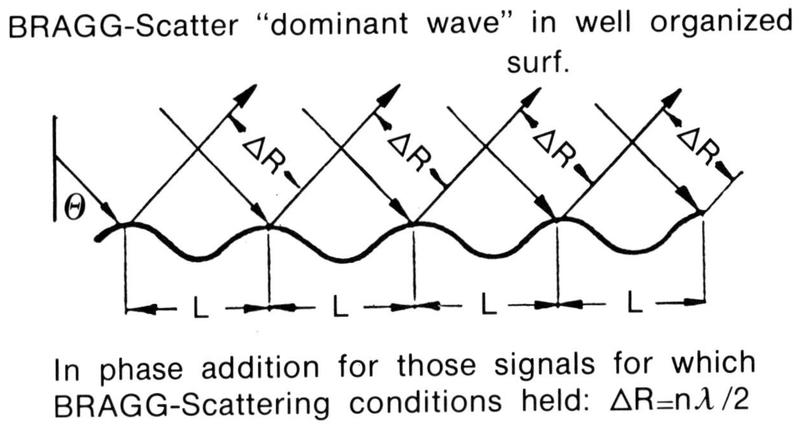

Figure 10. Bragg scattering.

the order of 9. The transition between these values occurs between 3 GHz and 40 GHz. The refractive index n is defined by the ratio of the wavelength in matter as compared to the wavelength in vacuum.

The scatter phenomena are described mathematically by the Mie and Rayleigh functions (Figure 11), as is well known from physics.

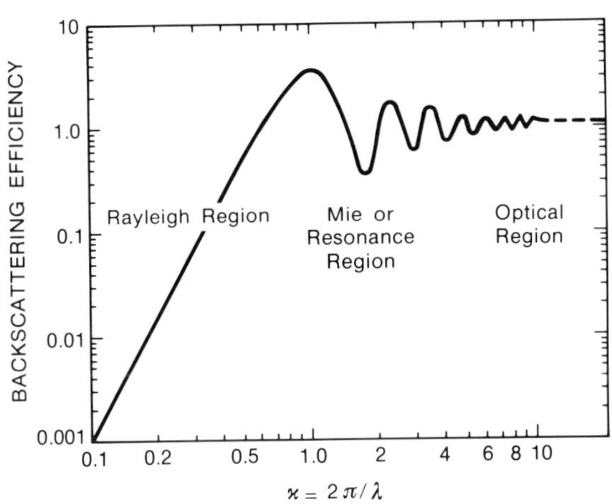

Figure 11. Backscatter characteristic of a sphere as a function of the ratio of its radius to wavelength λ.

As long as the particles are small in comparison with the wavelength the interaction is small. It is largest (resonant), if the particle diameter is about that of the wavelength. If the particle is large in comparison with the wavelength, then we have the case of the geometric optical with "visibility" and "shadowing". This means that the medium is opaque, if there are many such particles in the medium.

By taking into account that raindrops have diameters in the order of a few millimetres or below, one can understand that they inhibit the optical waves, but not the microwaves. Rain only affects the wave propagation for very high microwave frequencies (larger than 10 GHz), with wavelengths smaller than 3 cm. For much higher frequencies (optical area) even fog is not transparent.

Backscatter is a special case of scatter feature, another being reflection. In the "bistatic case" of the optics, where the sun illuminates the target (incident wave) and the sensor receives some scattered signal power, we can expect the received power to be maximum, if the sensor looks towards the imaging point of the sun. A specular reflection means that no signal power is "retransmitted back to the illuminator". In the radar case the illuminator and the receiver are co-located and in the case of a non-zero incident angle the receiver would, therefore, pick up no power at all in the case of an ideally flat boundary surface, for example a metal plate. At the radar image the target would then appear to be completely dark. This is why an ideally flat water surface looks completely dark in the radar image. If the water surface is "rough" due to capillary waves, which are generated by the wind, then the radar image becomes brighter with increasing wind strength. Oil slicks on water cause flattening of the water surface. One can, therefore, detect oil with radar. If the incident angle of the wave is zero and the surface acts like a mirror, then we would receive maximum echo power.

More generally, the situation in the monostatic case is shown in Figure 12. Curve 1 shows the specular case. For an incident angle of zero degree the return power is large, while for other angles it is zero. This is quite obvious.

Almost any surface is more or less rough. Therefore, we can assume that the specular scatter case will rarely occur. It is for this reason that we can assume in case 2 of Figure 12 that the realistic

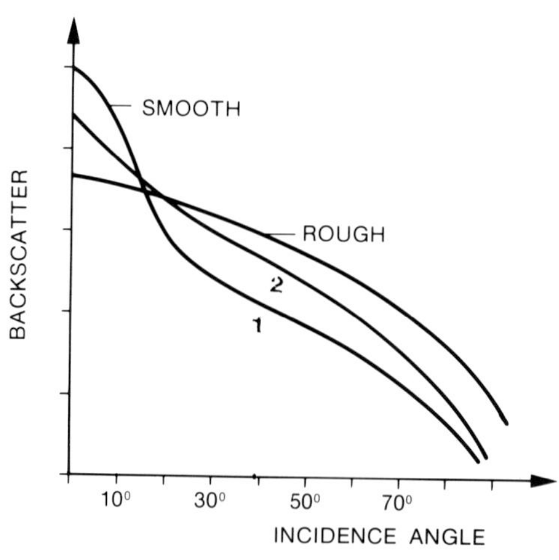

Figure 12. Backscatter as a function of incidence angle θ.

case deviates from case 1. We have generally for small incidence angles α a strong backscatter but it decreases monotonically with increasing α. In accordance with the Rayleigh criteria mentioned above even "rough" surfaces behave like smooth ones for α close to 90°.

For surfaces of modest roughness the backscatter behaviour varies very little with the incidence angle between 30° and 75°. Then the value rapidly becomes smaller due to shadowing effects. It is quite obvious that around $\alpha = 15^o$ the curves for the various degrees of roughness coincide. This is important because, if one requires to obtain by remote sensing information about the type of material and of surface structure, one must be able to separate the roughness parameter from the material information. One case where such information is needed is soil moisture determination. Higher soil moisture means stronger backscatter around 15°.

A practical example is this: assume that there are two surfaces, one with very fine sand and another with coarse stones of diameters at least as large as the radar wavelength. The radar image would then show a dark area in the first region and a bright area in the second one. An analogous case: with airborne SARs with high geometric resolution (2 m) it can happen that an image of a flat roof of a building is represented in an X-band image as being very bright and in an L-band image as being very dark, if the roof is covered by stones of a dimeter between the two wavelengths.

Another interesting case for radar imaging is this: a layer of dry sand with very fine grains, i.e. of particles with diameters much smaller than the wavelength of the electromagnetic signal. Under this surface another layer may exist which consists of larger stones. If they are of sizes large enough (i.e. comparable or larger than the wavelength) to represent large backscatter targets, then one can expect the following: the upper dry layer can be penetrated by the electromagnetic waves. The layer of large stones will scatter them. The radar image can, therefore, in case of very dry matter, show "subsurface phenomena".

Volume scattering is also important for radar scattering and might explain cross-polarized backscatter. This means that some fraction of the backscattered signal power is picked up by a receiver antenna which responds to a polarization orthogonal to that of the transmitting antenna. If the signal was transmitted in horizontal polarization (H) then the received signal is picked up by a vertically polarized (V) antenna (σ_{HV}). If targets have all kinds of orientations (like trees, for example), they can depolarize the backscattered signal to some extent. All such reflecting targets have alternative orientation.

For the measurement of volume scattering targets (canopy) a cross polarization measurement is, therefore, of great importance. But this measurement requires much higher signal power, because σ_{HV} or σ_{VH} are generally lower than σ_{HH} or σ_{VV}.

2.2 Thermal Emission and Brightness Temperature

Let us now look at thermal emission, which is used in remote sensing in the thermal infrared scanner and in the microwave radiometer.

A. Thermal infrared. According to Planck's law any body with a finite temperature T emits electromagnetic signal power at any frequency. This is due to the thermal motion of the charged particles. The higher the temperature, the higher the kinetic energy of the particles and the higher will be the spectral energy of the thermal radiation. For the "blackbody" the spectral radiant exitance is a maximum at λ = 2898 μm K/T (Wien's Displacement Law) (Figure 13). For "normal" temperatures of T = 290 - 300 K this maximum is at about 10 μm. For the

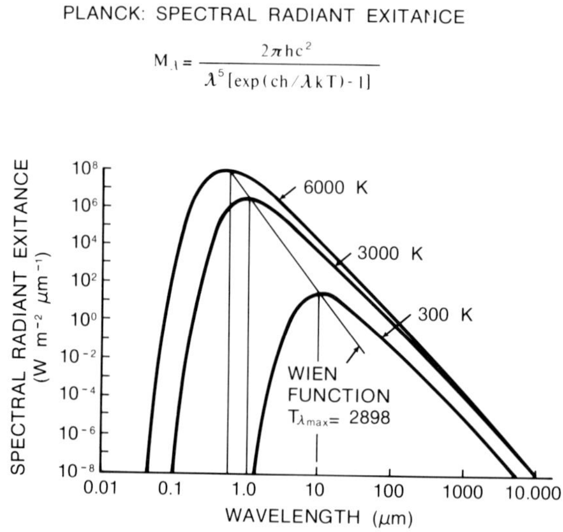

Figure 13. Planck's law.

sun's surface the "brightness" temperature is in the order of 6000 K and the radiation maximum is at about 0.6 μm, i.e. in the "yellow part" of the physical spectrum. The human eye is also the most sensitive to this part of the spectrum. The human sensor is optimized in this sense to the sun's radiation! For the earth's surface the actual situation of spectral radiant emittance is shown in Figure 14, which can be understood by remembering that in the reflective part of the optical spectrum it is the solar radiation reflected on the earth which will be observed by means of the nadir-looking sensors. In the thermal region, i.e. the emissive part of the spectrum, the radiation emitted by the earth's surface is observed directly. During daylight one would receive the mixture of both types in the wavelengths around 4 to 6 microns, whereas in the 11 to 14 micron region it is in any case the surface emitting radiation which dominates. Here, we can see that up to about 5 μm the reflected solar radiation will be dominant, but for

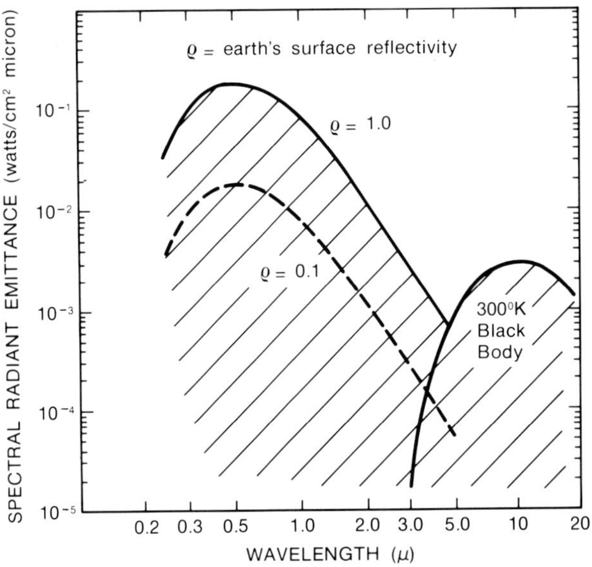

Figure 14. Energy available for remote sensing. ρ = reflectivity of the earth's surface.

larger wavelengths the thermal radiation effect becomes larger. Two facts should be considered in this context:

- The brightness temperature is proportional to the product of the "sensible" or physical temperature (which would also be measured by means of a thermometer) and the emissivity. For thermal infrared most materials have almost the same emissivity of about 0.9. It is, therefore, relatively easy to determine surface temperatures of regions of unknown surface material. In the microwave region the emissivity values of the various materials vary considerably and there can, thus, be confusion about the physical surface temperature.
- The emissivity of a material generally depends on the spectral range. The spectral emissivity and, therefore, also the absorptivity can be very different in the thermal IR as compared to the visible part of the spectrum. This is true, for example, for glass and for the CO_2 in the atmosphere. Glass is transparent for the visible part of the spectrum, but reflects the thermal IR part (greenhouse effect). The same problem might occur in our atmosphere from an increasing CO_2 content of the atmosphere: the energy of the solar radiation is absorbed by the earth's surface and causes its average temperature of about $20^{o}C$. At present the radiation balance is achieved by reradiating back to the sky in the thermal infrared. If the CO_2 atmospheric content is in the future much greater than now, the atmosphere will then not be as transparent to the thermal IR radiation. This would lead to a heating up of our globe.

B. Microwaves. Now let us come back to the brightness temperature. Microwaves have the advantage of being able to penetrate to some extent into the vegetation canopy. For longer waves and/or small incidence angles one can, therefore, obtain radiometric signals about the soil temperature and soil moisture, even through clouds. From this point of view one can see that microwave radiometry has considerable benefits. However, there are two disadvantages, when comparing the features of this spectral region with those of the thermal IR region: different materials have emissivity in the microwave region which differs considerably and, as has already been mentioned, the geometrical resolution of passive microwave systems for reasonable sizes of sensor antennas is very low.

One might wonder why it is really possible to measure surface temperatures remotely so accurately, namely with a resolution of about 0.1°C, almost independent of the distance. This can be explained easily by the following example: assume 3 heated plates; one of them might be as hot as, say 1300°C and is, therefore, glowing red. The second plate should be "cold" and show, therefore, the natural colour of the material (say grey). The third plate may also be cold, but painted in a red colour. From any distance, where the plates and their colours can still be seen, one could easily identify the grey plate as being a cold one. But for the two other plates one must see whether the red colour is due to paint or to temperature, i.e. if the colour observed is due to "reflected light" or "emitted light". Additional information is needed. In the example given it would suffice to observe them in the darkness. It is also obvious that the temperature, which is identified by the colour of the emitted light, can be recognized remotely: the colour is not changed by the medium between the target and the observer! Instead of the colour a real radiometer measures the spectral radiation power of the target quantitatively at several spectral bands and can, therefore, determine to which part of a Planck's curve it fits.

3. ELECTROMAGNETIC SPECTRUM

We are now considering the electromagnetic spectrum in general. As shown in Figure 15 there are various mechanisms which affect the interaction between waves and material, such as dissociation, photo-ionization, molecular vibration and rotation and finally, in the microwave region, the electric currents, and the magnetic effects. Due to Planck's law $E = h \cdot f$ we can see that the energy increases proportionally to the frequency f and, therefore, inversely to the wavelength. The interaction of the electromagnetic waves with the components of the atmosphere affects their porpagation. Only those parts of the spectrum for which the atmosphere is transparent can be used to sense the earth's surface. Figure 16 shows the transparency of the atmosphere for the optical range. Here we have also indicated the bands of the LANDSAT MSS and the LANDSAT TM sensors. They are all located in windows of the atmosphere. As shown in Figure 17, the atmospheric attenuation for the microwaves increases towards lower frequencies due to the interaction

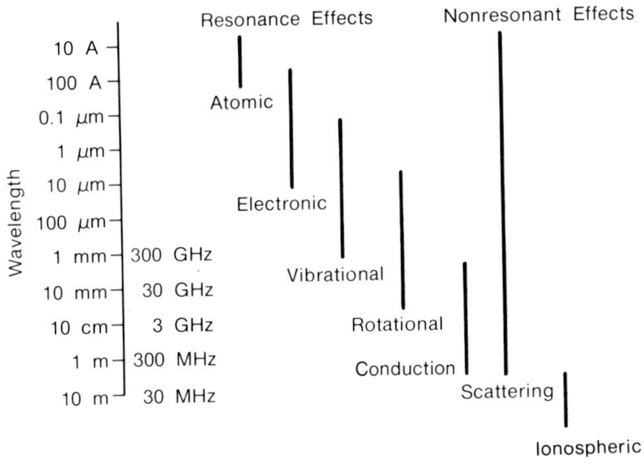

Figure 15. Interaction mechanism as a function of λ.

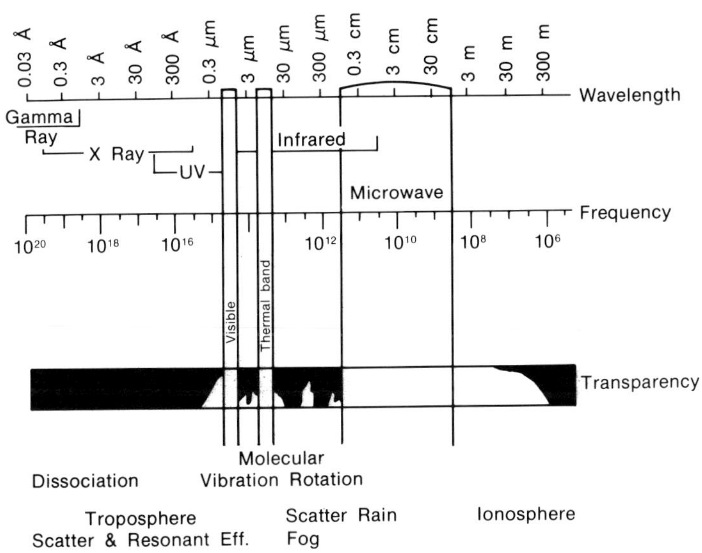

Figure 16. Transparency in percent of the atmosphere for electromagnetic waves.

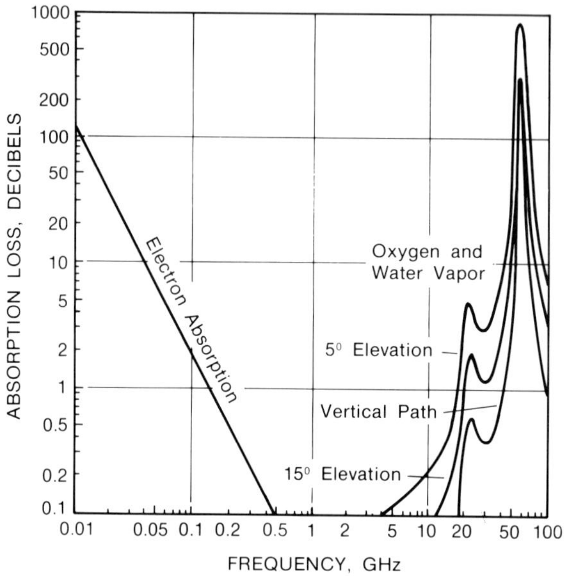

Figure 17. Absorption in the atmosphere by free electron and combined oxygen and water vapour (radar at sea level).

with plasma in the ionosphere and, in the higher spectrum, due to the attenuation of the troposphere. In the visible part, the atmospheric attenuation can be easily identified in the solar spectrum. The overall radiation follows the Planck curve for about 6000 K. There are many absorption lines produced by the constituents of the earth's

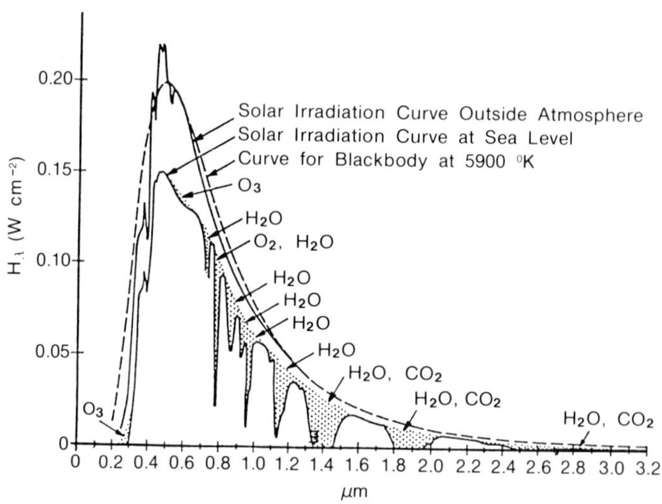

Figure 18. Spectral signal power of the sun measured on earth. There are various absorption lines some of which are caused by the resonance with atmospheric constituents.

or the sun's corona (Figure 18); the spectral ranges are coded in Table III.

TABLE III
Nomenclature of spectral bands

Optical bands:	
UV and visible	0.30 - 0.70 µm
Near infrared NIR	0.70 - 3.00 µm
Middle infrared MIR	3.00 - 6.00 µm
Thermal infrared TIR	6.00 -20.00 µm
Far infrared FIR	20.00 - microwaves

Microwave bands:

	f/GHz	λ/cm
P-band	0.225 - 0.390	133 - 76.9
L-band	0.390 - 1.550	76.9 - 19.4
S-band	1.550 - 3.9	19.4 - 7.69
C-band	3.9 - 5.75	7.69 - 5.21
X-band	5.75 - 10.90	5.21 - 2.75
K-band	10.90 - 36.0	2.75 - 0.83
Q-band	36.0 - 47.0	0.83 - 0.063
V-band	46.0 - 56.0	0.063 - 0.053
W-band	56.0 - 100	0.053 - 0.3
KU	10.9 - 18	2.75 - 1.67
K	18 - 26.5	1.67 - 1.13
KA	26.5 - 36.6	1.13 - 0.83

5. REMOTE SENSING INFORMATION

The tasks of remote sensing can be considered to be

- cartographic mapping of regions;
- thematic mapping;
- identification of substances, structures and regional interrelationships;
- recognition of processes and trends;
- monitoring of regional, seasonal developments;
- etc.

So, it is partly the geometric and partly the radiometric information which is of interest. The combined geometric and mapping information of a B/W or a colour photograph is evident. As regards the radiometric part, the information is contained in

- spectral reflectance;
- scatter/backscatter varieties;
- emissivity as a function of materials and substances;
- temporal variations;
- structural/textural features;
- etc.

They affect variations of the radiation quantities. It is the task of the discipline expert to recognize the variations and to draw conclusions by applying his special knowledge to interpreting the radiometric and other data with respect to the physical/chemical, biological, ecological reality. Some of this knowledge is or will be collected systematically in expert systems.

6. CONCLUSIONS

Remote sensing is based on the interaction with matter and the propagation phenomena of electromagnetic waves. There are various constraints as regards the applicable frequency bands, the sensor capabilities, the transparency of the atmosphere, etc. In general, one needs multitemporal and multisensoral data to extract the maximum information.

REFERENCES

Sabins, F., 1978: Remote Sensing: Principles and Interpretation, Freeman, San Francisco.

Elachi, Ch., 1987: Introduction to the Physics and Techniques of Remote Sensing, John Wiley & Sons, New York.

Colwell, R.N. (Ed.), 1983: 'Manual of remote sensing', I,II, Am. Soc. of Photogrammetry, Falls Church, Va.

Kronberg, P., 1985: Fernerkundung der Erde, F. Enke, Stuttgart.

Slater, P.N., Remote Sensing, Optics and Optical Systems, Addison-Wesley, Reading, Mass.

Ulaby, F.T., Moore, R.K. and Fung, AK.K, 1981: Vol.I, Microwave Remote Sensing Fundamentals and Radiometry. Vol.II, Radar Remote Sensing and Surface Scattering and Emission theory (1982). Vol.III, From Theory to Applications (1986) Artech House, Dedham, Ma.

SENSORS

Ph. Hartl
University of Stuttgart, Institute of Navigation
Keplerstr. 11
D-7000 Stuttgart 1 - Fed. Rep. of Germany

ABSTRACT. This paper reviews the various types of sensors applied in optical and microwave remote sensing from satellites, in particular over land. It covers active and passive sensors, imaging and non-imaging sensors, the basic concepts and some types of instruments used.

1. INTRODUCTION

For the purpose of this paper we define as sensors instruments which can

- collect electromagnetic energy from targets, and
- extract some information from these signals about the features of the observed object.

Sensors can be classified in several ways. We divide them into active and passive sensors. The passive sensors receive, but do not transmit any signals. We have radiometers (one-dimensional mapping geometry), imaging (two-dimensional mapping geometry), and stereo type (three-dimensional mapping geometry) sensors.

The active sensors have, in addition to the "receiving unit", a transmitter which emits the signal, part of which will be reflected by the target and received in the form of "backscattered signal power".

Most important for the present applications in the optical part of the electromagnetic spectrum are the "multi-spectral scanners", which are passive instruments, and in the microwave region the radar instruments which are active sensors.

2. PASSIVE SENSORS

Passive sensors make use of the electromagnetic radiation of either the sun or that emitted by the targets themselves. The solar radiation of interest is reflected or scattered and/or attenuated at the objects.

F. Toselli (ed.), Applications of Remote Sensing to Agrometeorology, 19–34.

2.1 Radiometers

Measurement of radiation intensity from a fixed direction, for example the sub-satellite point (nadir point), can be performed by means of a radiometer. This requires several instrument features: directivity to aim at the object of interest, spectral filtering to extract the proper spectral range of the signal, detection and amplification of the signal received, and some signal power meter to measure quantitatively the collected and filtered signal intensity.

The basic concept is shown in Figure 1. This block diagram is valid for both microwave and optical radiometers. Directivity and signal pickup are performed by a telescope (a lens or a mirror telescope in the optical case, an antenna in the microwave region). The

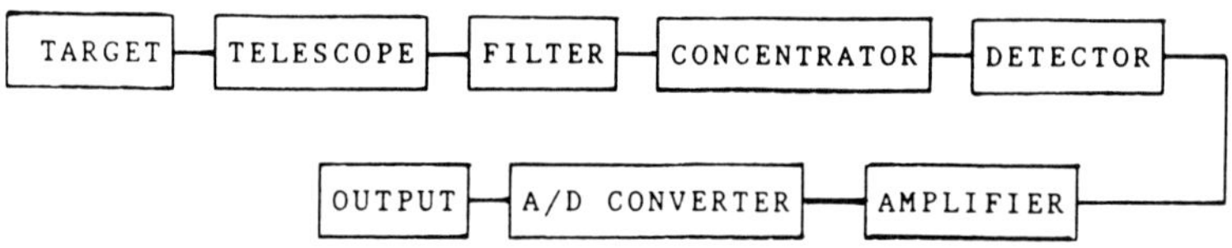

Figure 1. Block diagram of the radiometer measurement concept.

selected spectral range is chosen by the filter. The collected signal power is concentrated (optical case) and/or amplified and filtered (microwave case) and then detected, i.e. converted to some electric signal. This is then amplified and measured by means of a voltage, current, or power meter or an impulse counter, etc.

The radiometer is designed to operate very stably and precisely over a wide range of input signal intensities. It delivers calibrated intensity values, i.e. it determines the signal levels in absolute values (in Watts, for example) and not just in relative values. It allows the measurement result to be compared with and related to other remote sensing values collected by different instruments at different times, locations or measurement conditions.

This absolute measurement requires the comparison of the input signal with some stable and well-known reference signal source which can either be the sun itself or some internal noise generator.

There exist various types of radiometers. We shall consider only a few typical examples.

2.1.1 Optical systems.

Figure 2 shows a radiometer for simultaneous use in three optical ranges, namely in the area of short waves (SW), visible waves (V), and long waves (LW). A common telescope is oriented towards the target. The collected signal power is first split by a half-silvered plane mirror into two parts, one part (LW) of which passes through the mirror, the other part (V,SW) being refracted. The LW light is also refracted at another opaque mirror. Both rays pass a field stop to exclude stray light and a chopper to transform the received light intensity into a periodically varying (chopped) signal, which leads to an alternating current after the detection process.

This is important because the amplification can then be done by ac-amplifiers instead of the more complicated dc-amplifiers. The (SW,V) ray is afterwards split by another opaque mirror and then passes through the optical set-up and filter. Each of the three signals is focussed on a signal detector. Each one is optimized to the appropriate waveband and delivers some electrical signal.

Instead of this parallel channel technique one can also use a one channel technique, where a rotating wheel is placed in front of the aperture, which allows some part of the incoming spectrum to pass sequentially. This wheel has segments of different colours and is used instead of a chopper wheel. Each element of the sensor is unique and is used by all spectral parts. The output of the sensor delivers the radiometric data as a periodical time sequence. This instrument is very simple, but for N spectral channels the energy of each output is 1/N that of the corresponding N channel parallel radiometer. Moreover, it is impossible to obtain an optimum adjustment of the individual bands.

2.1.1.1 Optical detectors. The conversion of the optical signal power into electric current, voltage, or power can be performed by the photon effect or some thermal effect:

- The photon effect can be used to generate a current proportional to the number of photons illuminating the detector. This process is well known from solar cells. Generally the use of quantum detectors is based on the interaction of the photon flux with the electron energy state: charge carriers are released or transferred if photons are absorbed. This effect depends on the solid state material of the detector and is effective only up to a certain frequency (Figure 3).

  The characteristics of the detector are defined by various parameters:

  . The "noise equivalent power" (NEP) is a measure of the minimum signal power which can be detected. In a one Hertz bandwidth it is the signal power necessary to have an output signal-to-noise ratio (S/N) of one.

    The following equation can be used for the determination of the sensor performance:

    $$NEP = (E_d \cdot F \cdot N)/(S \cdot \Delta f^{\frac{1}{2}})$$

    where:

    E_d is the surface radiation at the detector surface, in Watt/cm^2;
    F is the detector sensitive surface in cm^2;
    S is the signal voltage developed by the detector in the radiation field;
    Δf is the noise bandwidth.

  . The "detectivity" D^* is a "figure of merit" used to characterize the detector, excluding the geometric dimensions' dependency of the NEP-value:

$$D^* = F^{\frac{1}{2}}/NEP, \text{ measured in } cm \cdot Hz^{\frac{1}{2}}/Watt.$$

For a given detector surface the NEP should be low, D^* should be high.

Apart from the photon effect one can make use of the thermal effect in the detection process, as was stated above.

- The thermal effect results from the heating of the element due to absorption of radiation. This can change the resistance of the electric circuit (thermistor, bolometer). The temperature variation can be used in a thermocouple to generate a voltage. Alternatively, it can also cause a pyroelectric effect: a slab of pyroelectric material has conducting layers at two opposite face areas. A change of the temperature of one face due to the incoming radiation generates a voltage change proportional to the temperature increase.

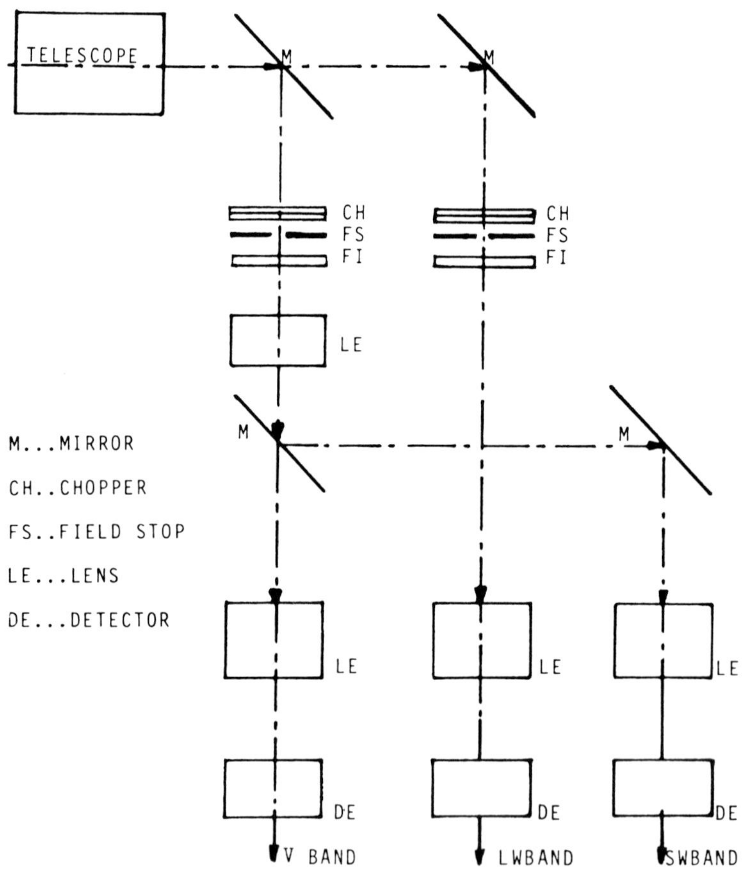

Figure 2. Principle of a 3-band radiometer.

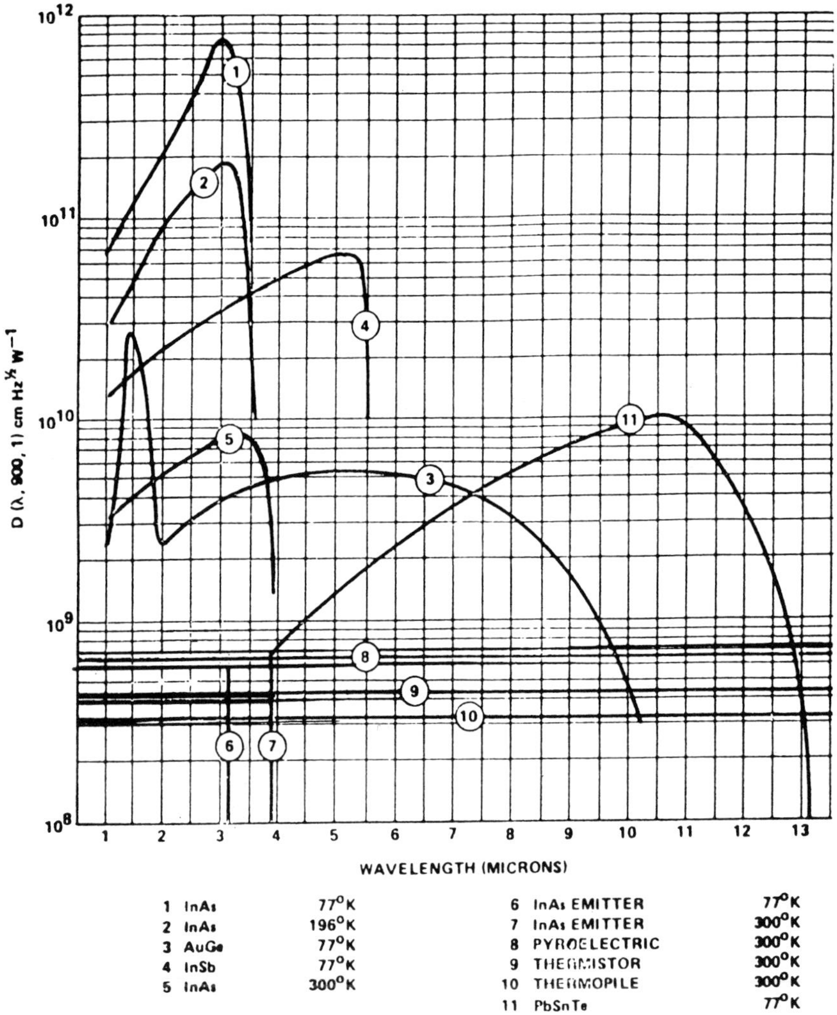

Figure 3. Spectral response of detectors.

2.1.1.2 Telescope. The optical telescope consists either of a lens system or a reflective mirror system. The main parameters which describe its features are the following:

- The aperture A. It defines the geometric size of the light collecting input of the optic. If a lens or mirror is circular and of diameter D, then the aperture is πD^2 times its efficiency, which is smaller than 1.
- The "instanteneous field of view" IFOV is about λ/D in rad.
- The "ground resolution element" (GRE) is the smallest dimension of

the object which can be identified on the surface of the Earth, if the radiometer is looking towards it from space. If the satellite height is H, then

$$GRE = H\lambda/D.$$

The equivalent focal length f of the lens is related to the size GRE of the object and the image d by the following "lens maker's formula":

$$\frac{1}{f} = \frac{1}{a} + \frac{1}{b}$$

b = distance object to lens;
a = distance image to lens;

and with $b = H \gg a$, $f = a$;

$$D/d = H/a$$

If the diameter D is increased, then more light can be collected. This increases the radiometric resolution. The geometric resolution is then improved for a given size of the detector. One can improve the situation by applying many detectors and using glass fibres and filters (see Figure 4).

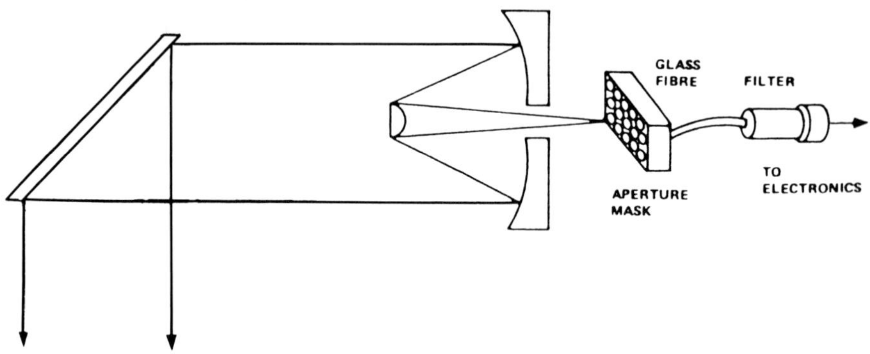

Figure 4. Landsat-MSS with a set of detectors and glass fibres.

If the filters of each line are identical, but different from line to line, then it is the spectral resolution which is increased in the vertical direction. In the horizontal direction we have then improved the spectral resolution. If all glass fibres have the same filters then the geometric resolution is improved in both dimensions. Instead of glass fibres one can, of course, immediately use arrays of electrooptical detector elements.

2.1.1.3 Filters. There are various ways of applying spectral filter techniques. One early way was the application of a filter wheel, as mentioned above.

Interference filters are especially applied if good spectral resolution is required.

Gratings are used for spectrometry with very many channels.

2.1.2 Microwave radiometers. In the microwave region the radiometer consists of an antenna, preamplifier (up to frequencies where they are still available), detectors, Dicke switch, etc., as shown in Figure 5.

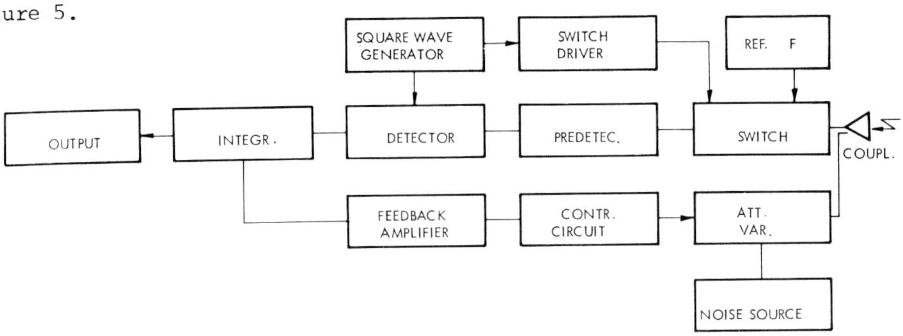

Figure 5. Dicke-microwave radiometer.

The aperture A is an antenna, for example a parabolic reflector or a horn antenna, an array of dipoles, etc. The amplifier is a microwave receiver, which might have one, two or even three different intermediate frequencies (and corresponding frequency converters) before the final detection process.

Microwave radiometers have the problem of geometric resolution. Because these waves are 4 to 5 orders of magnitude longer than the optical waves it would be necessary to have the same factor in the size of the antenna to achieve the same resolution. This is, of course, not possible.

2.2 Spectrometers

Very often the radiometer technique is used for spectral power distribution measurements. So, the incoming signal power must be split into the various spectral bands in order to allow measurement of the spectral power either in parallel by several radiometer set-ups or in sequence.

The spectral filtering can be done by the following methods: one can use a filter wheel in front of the aperture, i.e. by replacing the chopper wheel mentioned above for the radiometer. This wheel consists in this case of various segments transparent to the "colours" to be measured (passband filters). By rotation, each segment appears periodically in front of the aperture, thus filtering the signal intensity. So, one can measure the spectral power periodically as the output of the radiometer. The disadvantages of this filter wheel solution are its mechanical parts and limited use of the signal power because only during a fraction of the whole time can one measure the signal power of a given (filtered) spectral part. As this reduction of available spectral energy interferes with the number N of spectral partitions, it can only be used for a small number N of channels, such as 2, 3 or 4.

For many spectral bands one applies dispersive systems, in particular gratings, which split up the spectrum spatially. Then one can use a separate detector for each portion and measure continuously.

Radiometers are non-imaging systems. They deliver only one-dimensional radiation data as they look in one direction, i.e. towards the subsatellite point. The time-varying radiation then corresponds to the radiation originating from the subsatellite track point.

2.3 Scanning Radiometers/Spectrometers

In order to achieve a two-dimensional image one must arrange for some scanning either of the instrument's aperture itself or of some mirror in front of it. Two interesting mechanical scanning mechanisms and an electro-optical technique are used.

2.3.1 Mechanical linescanners. Mechanical linescanning is performed across the satellite's track. Here one can achieve a swath width (SW) as a function of the scanning angle. As the satellite moves along, one can automatically obtain a line by line imaging of the earth's surface and finally have an image similar to the line characterized image of a TV-screen.

The geometric conditions are given in Figure 6 for the LANDSAT 1, 2,3 MSS and the LANDSAT 4,5 MSS and TM in connection with the repetition rate. The scanning is done by means of a vibrating mirror.

The advantages of mechanical linescanners are the following: one detector is used for the whole image in a given spectral band. The system is, therefore, very precise in the radiometric calibration. The geometrical accuracy is also very good, if one applies precise mechanical scanning devices and very constant rotation or vibration by means of a good scanning motor.

In addition, one can split the beam, and map in parallel a few lines with separate detectors. Finally, one can adapt the choice of the detector to the spectral band, i.e. one can apply different types of detectors in the different spectral bands.

The disadvantages are: as the line scanner maps sequentially across the subsatellite's track, a very limited fraction of the sweep period per target point is used as observation (dwell) time. This affects the signal/noise ratio. In addition, moving mechanical parts are not as reliable and wear out. They influence the satellite's attitude control performance. Variations in the satellite's orientation lead to geometric imaging errors along and across the track. Instead of a straight scanline it might then be curved.

2.3.2 Conical scanning. Another method of scanning is the conical scan method (see Figure 7). Here the rotation of the mirror or aperture is such that the footprint is a circle. Compared with the line

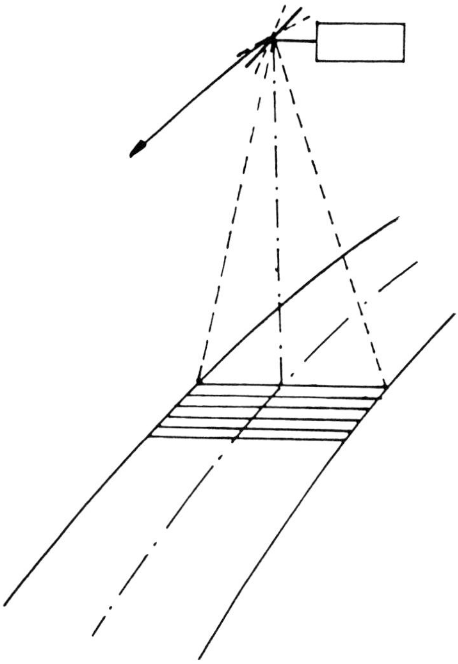

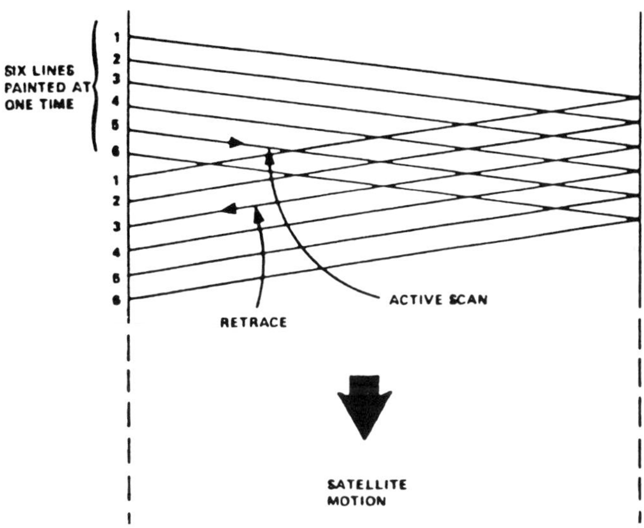

Figure 6. Mechanical linescanning; example LANDSAT MSS.

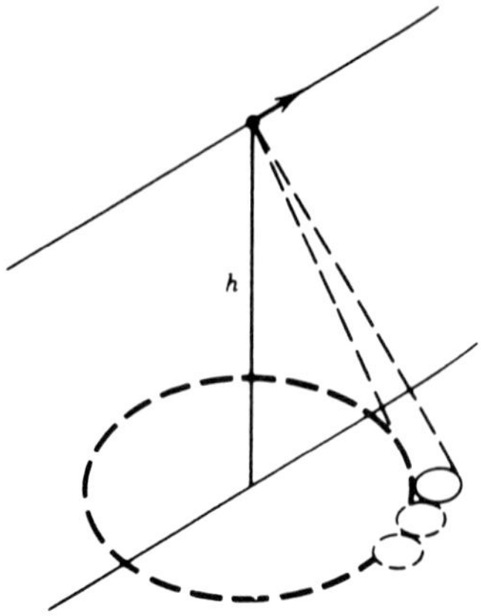

Figure 7. Conical scanning.

scanning mechanism we have various differences. Each point of the image surface is mapped twice, once by looking at it forward and once aft. This allows a stereomapping. In addition, one has the advantage of looking at the object with the same incident angle, whereas the linescanner looks at objects far from the subsatellite point with large incident angle and at objects near it with low incident angle.

2.3.3 Pushbroom scanner. So far most of the mechanical scanners are linescanners. In the future one can expect that many electro-optical scanners (pushbroom scanners) will be used. Examples of this type are the SPOT and the MOS scanning systems and the MOMS system (Figure 8).

This type of scanner is, in principle, a linescanner. The advantages of the pushbroom scanners are the following: they do not have rotating mechanical parts and they associate a single pixel to a single detector per line. Therefore, the dwell time for this detector to a single point of the surface is equivalent to the time which was allowed by the mechanical scanner to the whole line. As the number of pixels per line goes into the thousands, this improvement factor does also. More precisely: the signal to noise ratio varies as the square root of time. So, if the dwell time in the case of a mechanical scanner was 1:1.600 for the pushbroom case, then the signal to noise ratio improves by a factor of 40. This improvement can be used, for example, to increase the geometric resolution. There is still a disadvantage at present: the number of elements per array is still limited and the arrays are not available yet for the whole spectral band from the visible up to the thermal infrared. But with the dramatic progress in integrated electro-optics one can hope that this disadvantage will no longer exist in the next decade.

2.4 Imaging Spectrometry

The pushbroom scanners have, so far, very limited numbers of spectral channels. SPOT has 4 channels, MOMS was flown in the Shuttle experiment with only two channels. By using many detector arrays in parallel,

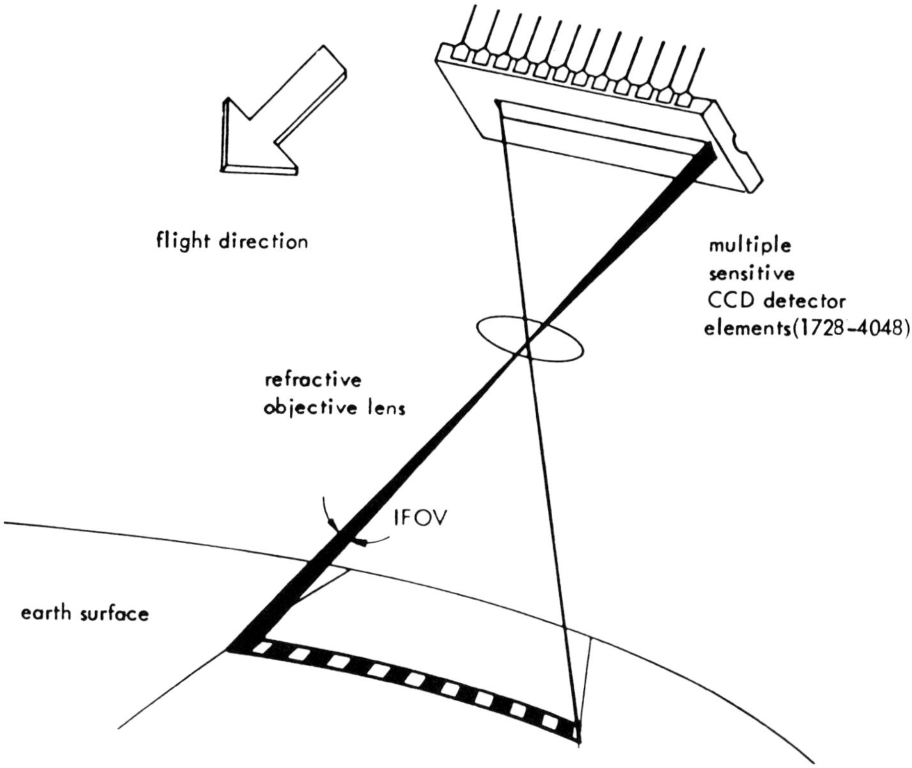

Figure 8. Pushbroom scanning.

each associated with a very narrow spectral band, one can avoid the disadvantages which exist in today's optical remote sensing; generally, the spectral signatures of the objects are undersampled. One cannot discriminate as finely as is possible in laboratory spectrometry. Practical examples have shown, however, the great benefits which imaging spectrometry might have. In the future one will use this technique intensively in connection with the polar platform. ESA concepts as well as NASA concepts exist (Table I). One very important problem is then, however, data storage, transmission and processing. Practical solutions are the preselection of a subset of spectral bands, specifically adjusted to the dedicated mission objectives over a limited period of the mission duration. One could then choose the proper band combinations sequentially. As regards data handling it is quite obvious that one must create expert systems and apply artificial intelligence procedures to make the best use of the vast amount of information offered.

TABLE I
Typical parameters of an imaging spectrometer

Parameter	Value
Field of view (FOV)	15°
Swath width	183 km
Instantaneous FOV	15 m
Spectral coverage: visible	0.4 - 1.0 m
IR	1.0 - 2.5 m
Spectral resolution: visible	64 channels
IR	64 channels
Aperture	52 cm
Data encoding	8 bits/sec
Raw data rate	

2.5 Framing Cameras

The most familiar optical sensor is the camera. The specific device for aerial and space photography is the photogrammetric camera which is designed to give geometrically extremely precise images of large areas with high geometric resolution. These cameras use photosensitive films either in the visible or the near-infrared part of the spectrum, or so-called False Colour Film, which is a colour film with a sensitivity spectrum shifted towards the near-infrared.

The advantages of a photogrammetric camera are:
- very high geometric resolution for large formats;
- relatively simple instrumentation and geometric data evaluation;
- very high geometric accuracy;
- with a large overlap one can derive topographic information by "stereo imaging".

The disadvantages are:
- limited spectral resolution;
- limitations to the visible and near-infrared spectrum;
- non-linear radiometric effects;
- use of films and, therefore, limited recovery capability.

With advances in electro-optical techniques one can replace the photocamera by the electro-optical camera, i.e. the film by one and two-dimensional detector arrays.

3. ACTIVE SENSORS

The active sensors have both a transmitter and a receiver. The transmission allows operation independent of daytime and season. It leads to a "backscatter mechanism". It also allows the signal propagation time to be employed as an additional measurement parameter. So, it is the radar principle which is used. In the optical part of the spectrum it is the LIDAR, in the microwave part it is the radar altimeter (which is not described here because it is almost exclusively used for maritime applications), and the SAR principle.

3.1 Lidar

The basic concept is very simple (see Figure 9). Pulses of coherent optical waves are emitted in the visible, the UV or the IR part of the spectrum. The signals have a narrow beam and the bandwidth is very sharp, but to some extent adjustable.

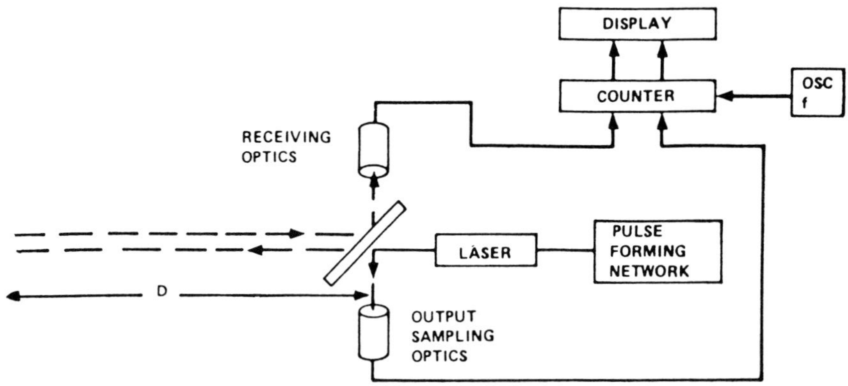

Figure 9. Laser ranging concept.

The emitted signals are backscattered by the targets, i.e. by the scattering aerosol layers of the atmosphere (to detect clouds), by dust, and by the Earth's surface. The echos are received by the sensor and the signal intensity and signal delay are both measured. The delay gives the distance, while the backscattered intensity provides the information about the size and the physical and the chemical compositions of the target. If the laser frequency is adjusted to the fluorescence, then we can identify the chemical substances of the target, such as chlorophyll, oil types, etc.

3.2 Sideways-looking Radar (SLR)

Sideways-looking radar systems are active, imaging microwave sensors which apply ranging techniques across the track in order to achieve good geometric resolution in this direction. Moreover, the scanning mechanism of "mechanical scanners" can be avoided. The pulse duration determines the across-track resolution. The basic concept is shown in Figure 10 for the "sideways-looking radar". In this case a "real aperture" antenna is used for the along-track resolution. It is a long, slender antenna of length L, which then has an angular resolution of λ/L. This value must be multiplied by the satellite height H to determine roughly the along-track resolution. The across-track resolution can be calculated as in the figure.

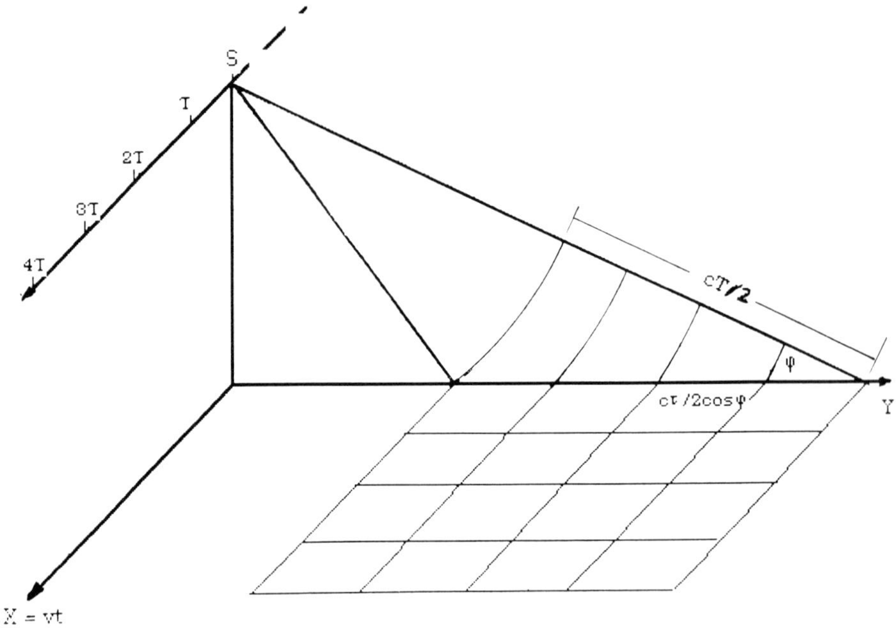

Figure 10. Sideways-looking radar.

The resolution would be too coarse for satellite application. One must, therefore, use a modified and more complex technique which makes use of modern signal and data processing. This is done in the "Synthetic Aperture Radar" technique:

- In order to achieve very high across-track resolution one applies the so-called "chirp technique", which makes use of the fact that correlation processes and/or signal processes with "matched filters" can allow the pulses to be made effectively extremely short. As this is only a well-known technique to replace the extremely short pulse generation/detection process for technological reasons, it is not necessary to describe it here.
- The "synthetic aperture" concept itself is used in the along-track dimension. Here one makes use of the fact that one can also generate the effect of a lens by appropriate signal processing. In order to understand this, let us first consider the optical lens (Figure 11). The focusing effect is gained by the fact that the refractive index of glass differs from that of a vacuum. Parallel and phase coherent light illuminating a lens thus has the same optical length, i.e. path, measured in multiples of the wavelength for all the lines shown. The waves in the glass of the lens are shorter than outside. This compensates for the slant range of the "edge lines". All rays arrive phase coherent at the focal point.

The "synthesized" aperture of a radar also applies various rays which are sent out at different locations of the satellite's orbit.

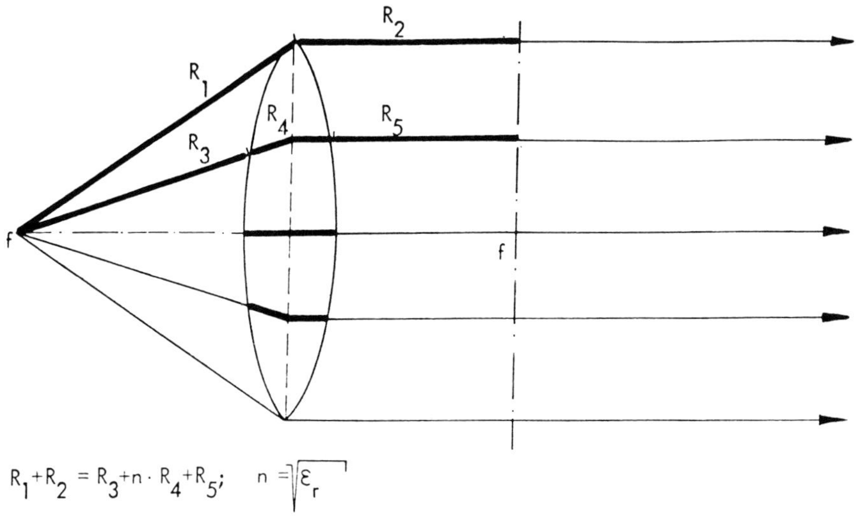

Figure 11. Focusing feature of a real lens.

As long as the real antenna of length L along track can illuminate the target with its beam, which is of width λ/L, it is possible to perform the following:

One periodically sends impulses and measures both the intensity and the phase of the reflected signals. These values are stored. At the end of the illumination these signal echos are compensated for the different distances as a function of time and then added. The echo waves then add "in phase", similar to the optical case. This has the following effect: the synthesized antenna aperture S is given by the "real antenna length" L and by the following condition of Figure 12: the beamwidth of the real aperture antenna of length L is λ/L, which corresponds, due to the height H of the satellite, to a ground length $(\lambda/L)\cdot H$. This is also the length over which the synthetic aperture procedure can be carried out because the target is only illuminated within this distance. So, in principle, the synthetic aperture length S is given by this same value. As the resolution of an antenna of aperture S is of value λ/S, or on the ground H times this value, one can expect that this would correspond to

$$\lambda \cdot H/\lambda \cdot H/L = L$$

In reality it is even better by a factor of 2, because the phase shift is twice as much in the synthetic case than in the optical case because of the two-way measurement.

The resolution of the synthetic aperture radar is, therefore, L/2, along the track, which means that with a satellite radar antenna of length 1 m along track, one can achieve a resolution of 0.5 m, independent of height.

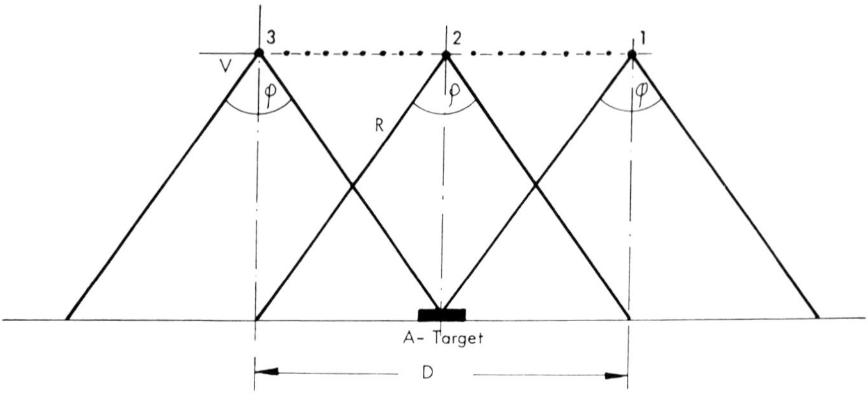

$\varphi = \dfrac{\lambda}{L}$; L = Real aperture = length of antenna

Synth. aperture of length D;

$D = R \cdot \varphi = \dfrac{R \cdot \lambda}{L}$;

$\varphi_1 = \dfrac{R \cdot \lambda}{2\,D}$

$\varphi_1 = \dfrac{R \cdot \lambda}{2} \cdot \dfrac{L}{R \cdot \lambda} = \dfrac{L}{2}$

Figure 12. Basic concept of the synthetic aperture.

In reality one must remember that this arrangement requires considerable signal processing and signal power.

REFERENCES

Elachi, Ch., 1987: 'Introduction to the physics and techniques of remote sensing', John Wiley and Sons, New York.

Swain, P.H. and Davis, S.M., 1978: 'Remote sensing: the quantitative approach', McGraw-Hill, New York.

Short, N., 1982: 'The Landsat tutorial workbook: basics of remote sensing', NASA Ref. Publ. 1078, U.S. Government Printing Office, Washington, D.C.

Goetz, A.G., Vane, J. Solomon and Rock, B., 1985: 'Imaging spectrometry for Earth remote sensing', *Science*, **228**, 1147/1157.

Ulaby, F.T., Moore, R.K. and Fung, A.K., 1981: Vol.I, *Microwave Remote Sensing Fundamentals and Radiometry*. Vol.II, *Radar Remote Sensing and Surface Scattering and Emission Theory* (1982). Vol.III, *From Theory to Applications* (1986), Artech House, Dedham, Ma.

Slater, P.N., *Remote Sensing, Optics and Optical Systems*, Addison-Wesley, Reading, Mass.

PLATFORMS

G. Fraysse
Former Scientist of the Joint Research Centre of the European Communities
21020 Ispra (Va) - Italy

ABSTRACT. This chapter gives a clear view of the various types of platform used in remote sensing: low, mid and high altitude aircraft, orbiting and geostationary satellites, first generation, present and future satellites. Types, altitude, period and inclination of the orbits together with the characteristics of the various types of sensor embarked, data transmission, reception and preprocessing will be treated exhaustively.

1. INTRODUCTION: GENERAL ASPECTS

Remote observation of agrometeorological variables at the earth's surface from aircraft and from satellites is based on the analysis of the electromagnetic radiation reflected or emitted from the earth towards the sensor. Agrometeorology consists in measuring the primary variables affecting plant growth: energy (insolation), temperature (air and soil), moisture (air and soil, due mainly to precipitation; it is generally easier to measure precipitation than soil moisture. This is why precipitation (rainfall) often appears as a primary variable;), and wind. They are not generally measurable directly by remote methods, but they are not totally independent and can be related to other parameters or some indices which can be determined more easily by appropriate sensors.

An extensive list of measurable parameters is: albedo, vegetation index, vegetation stress, cloud cover, bare soil or canopy temperature, precipitation, earth radiation budget, soil moisture or retention and, using adequate theoretical or empirical models, evapotranspiration, stress-degree-day, hygrometry and wind (in coastal areas).

Any measurement can be described by: its accuracy (reproducibility and sensitivity), its repetitivity and its geographical location. These specifications apply particularly to remote observations from air and space.

Accuracy (reproducibility and sensitivity), are characteristics of the sensor being operated while the repetitivity and the geographical

F. Toselli (ed.), Applications of Remote Sensing to Agrometeorology, 35–55.

location are platform-dependent.

Repetitivity depends on the type of satellite orbit or of sampling rate for geostationary orbits. But it is also linked to the observed swathe, which is dependent on the sensor view angle.

The accuracy of the geographical location depends on that of the satellite location at the time of measurement and on the instantaneous field of view which permits observation of a pixel (averaging) and not of a point.

Resolving power is the ability to discriminate fine detail.

One fundamental limitation depends on the ratio of the wavelength of the radiation to the receiving optical aperture of the sensing system:

$$\alpha = K \frac{\lambda}{d}$$

α = smallest angle which can be resolved by a perfect or ideal sensor;
λ = wavelength;
d = optical aperture.

In the IR region, this wavelength-to-aperture relation imposes practical limitations because of sensor size. This region is important in agrometeorology because the maximum radiation power emitted by land surface occurs in it.

Thermal radiation from the earth in the microwave region is very low indeed, but the receivers are very sensitive. On the other hand, the λ/d ratio becomes unfavourable to the discrimination of small details. In contrast to passive microwave sensing, an active microwave radar system can provide coherent radiation. This permits the synthe sizing of an effective aperture along the flight path which is much larger than the dimensions of the vehicle carrying the antenna.

With advanced sensors, another practical limitation comes from the transmission capability of the radiofrequency channel from the satellite to the earth. In the X-band link used for LANDSAT and SPOT satellites (frequency $\sim$ 8 GHZ, wavelength $\sim$ 3 cm), one is limited to a bandwidth of a few hundred megabits/second.

Even with advanced ground receiving and processing systems which can accept a higher flow of data, it would be unpractical to combine large swaths and a small IFOV producing too much data.

The accuracy with which the coordinates of the centre of a given pixel can be determined depends on the stabilization of the satellite, and the telemetry accuracy for the measurement of its attitude and position. In the case of SPOT, the rotation of the satellite around the axis orthogonal to the orbit plane is 1 rotation/orbit, and spurious rotation is kept to $\sim 10^{-3}$ degree/second. The pointing accuracy of the HRV sensor is $\sim 0.15^{o}$. The attitude and orbit control, measuring yaw and pitch movements, combines gyrometers, earth sensors and sun sensors connected to the on-board computer processing the successive correction instructions to reaction wheels and to magnetic coils. Using in addition ground control points, an accuracy of 0.5 pixel can be reached in some cases.

Finally, how long does an earth observation satellite live? The operational lifetime is the period during which it performs its planned mission; it is essentially, on the one hand, the life during which the hardware operates above minimum nominal performance (batteries, solar cells, detectors), and, on the other hand, is governed by the consumption of the propellant necessary to maintain the proper orbit and altitude.

In the case of most satellites, operational lifetimes have been of the order of 2 to 4 years. Technological progress will increase these lifetimes. It should be noted that the revisit capability of the future space stations will modify the actual concept, since most parts (conceived as modular) will be interchangeable, repairable or refurnishable.

The orbital lifetime is the time until the satellite burns up in the earth's atmosphere. This lifetime is much longer than the operational one for civil application satellites, because the air drag, which is the only factor responsible for the decay of a satellite, is very small at altitudes generally above 500 km. For certain military observation satellites, this lifetime can be very short (many COSMOS satellites operate in orbits having a perigee of $\sim$ 150 km, where air drag is dominant).

Satellites in geostationary orbits can have very long lifetimes, of the order of 10^7 years. The geostationary orbit being densely occupied, it is necessary to avoid the danger of collisions and in order to reduce the number of dead objects sited along the geostationary ring, space agencies manoeuvere satellites reaching the end of their operational lifetime to a position above, freeing the location for another active satellite. Obviously, space debris cannot be manoeuvered and it is increasing continuously.

2. PLATFORMS

2.1 Low and Medium Altitude Aircraft

Aircraft are well-suited to experimental missions or to investigations which do not need a high repetitivity. When operated at low altitude (1000-2000 m), atmospheric effects are reduced. In addition, flights can be done when cloud cover conditions are good and at the right time when observation of particular phenomena needs to be done (crop phenological states, crop diseases, natural disasters).

Disadvantages are poor stability (pitch, yaw, roll), considerable scale variations due to altitude variations and relief, and large viewing angles (MSS) over the sides of the swath, requiring the introduction of important radiometric corrections.

Two categories of aircraft are being operated:

- propeller aircraft flying at altitudes generally lower than 7000 m (Dornier 228 used by DFVLR, B 17 until recently used by IGN, Convair 580 used by CCRS, etc.);
- jet aircraft flying above 10,000 m (Mystère 20 used by IGN-GDTA).

Commercial MSS have an instantaneous field of view of 2.5 mrad, corresponding to a ground resolution of 12.5 m at 5,000 m altitude (to be compared with 20 m of SPOT).

2.2 High Altitude Aircraft

When large areas are to be covered, high altitude aircraft are necessary. For high altitude measurements, NASA uses the famous U-2 which can fly above 20,000 m. A modern version of the U-2, the TR-1, is now used by the USAF and NASA-AMES Research Centre operates a high altitude heavy-winged ER-2. In the USA, high altitude photographs are available for civil applications. Modern cameras such as the European Metric Camera or the U.S. Large Format Camera, carried by the Shuttle, have provided photographs of similar characteristics. One can consider Shuttle missions as a transition between high altitude research aircraft and automatic long-lived satellites.

2.3 Satellite Orbits

2.3.1 Orbiting satellites. Orbiting satellites generally have orbits at altitudes lower than 2000 km. Altitude and period are linked, and such altitudes correspond to periods smaller than 2 hours (Figure 1). The inclination of the orbit determines the latitude range coverage. Only orbits having more than 90° inclination can cover the whole earth (Figures 2 and 3). Sun-synchronous orbits are used when the satellite must observe the same area at the same hour (Figure 4). Polar orbits permit the total coverage of the earth.

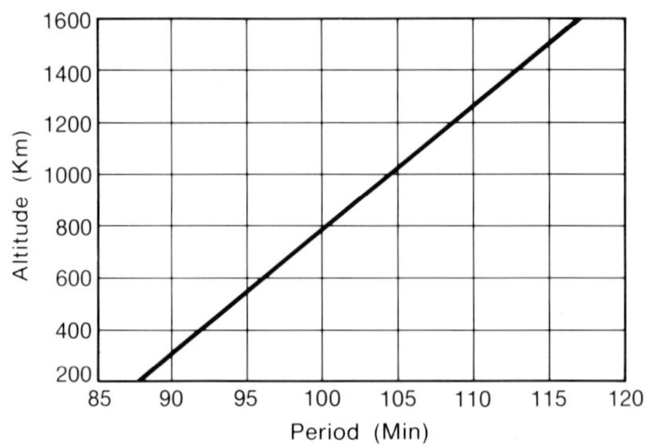

Figure 1. Relationship altitude/period.

2.3.2 Geostationary satellites. When the angular velocity of the satellite is equal to the rotation of the earth, the satellite appears stable at 36,000 km above the equator (Figure 5). The GARP meteorological satellites (GOES, METEOSAT) and telecommunication satellites (ECS-EUTELSAT, INTELSAT, TELECOM) belong to this category. Figure 6 gives the attitude of some satellites.

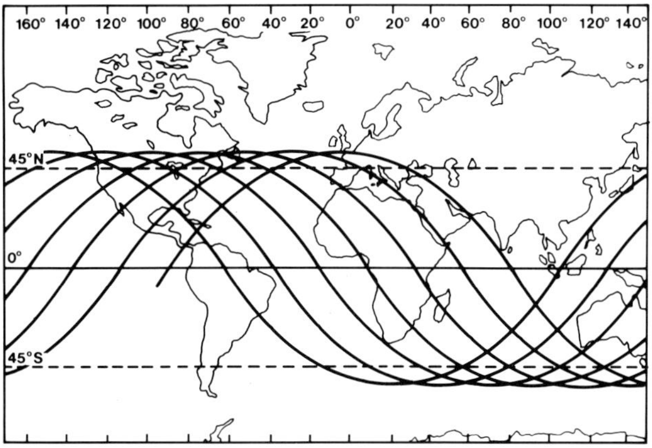

Figure 2. Area covered by SKYLAB (inclination 50°).

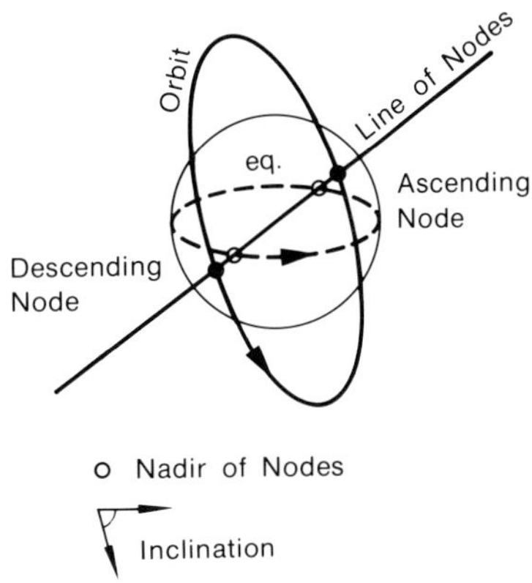

Figure 3. Nodes and inclination of a satellite.

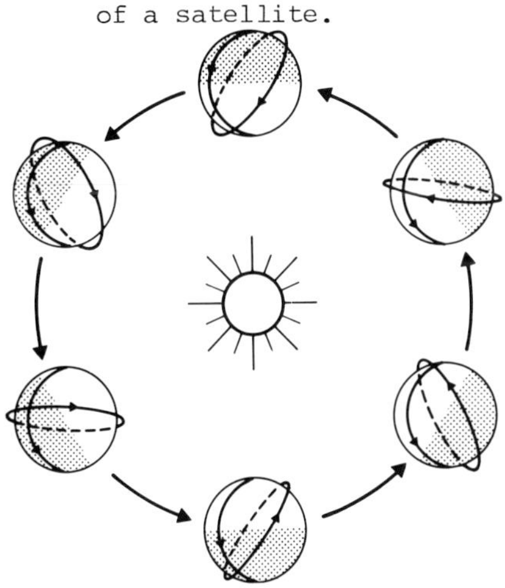

Figure 4. Sun-synchronous orbit (angle orbit plane/direction earth-sun = constant).

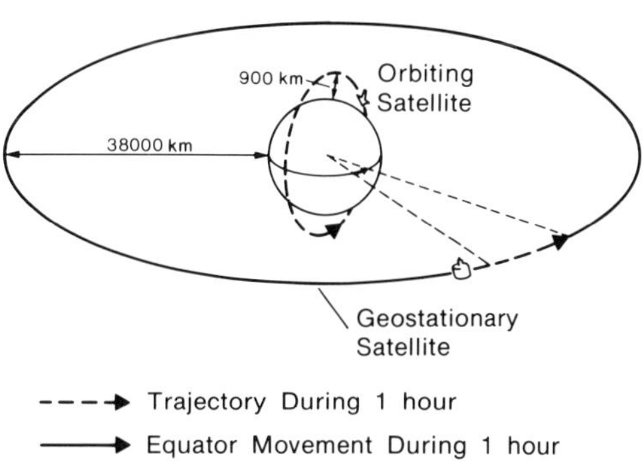

Figure 5. Orbit of a geostationary satellite.

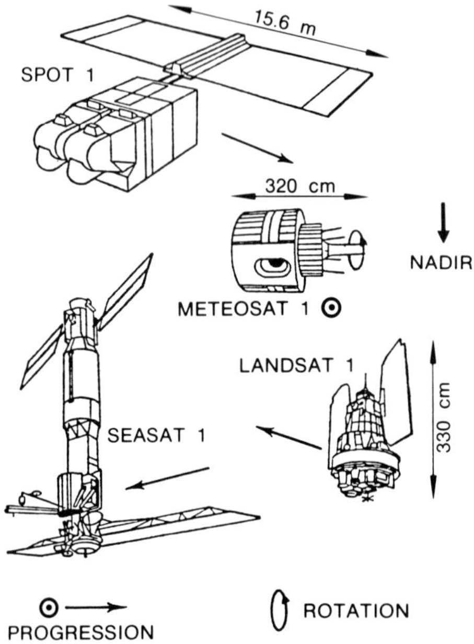

Figure 6. Attitude of some satellites.

3. SATELLITES

3.1 First Generation Satellites

Table I indicates the main characteristics of the geostationary observation satellites. For European countries, the ESA-METEOSAT-2 satellite payload includes a 3-band spin-scanning radiometer (the visible range: 0.3-1.1 µm; the water vapour absorption band in the infrared range: 5.7-7.1 µm; the far-thermal infrared: 10.5-12.5 µm). (In orbiting satellites, scanning along-track is derived from the satellite movement around the earth. This movement does not exist in a geostationary satellite; it is substituted by a rotation of the scanner (spin-scanning)).

The ground resolution at nadir is 2.5 km in the VIS and 5 km in the TIR. A data collection system permits the operation of several hundred data collection platforms (DCP). Table II shows the characteristics of orbiting satellites. While LANDSAT operates in the VIS and IR ranges, SEASAT operates in the microwave range, and HCMM in the TIR range. The NOAA TIROS-N series (NOAA-9) incorporate an Advanced Very High Resolution radiometer (AVHRR-2) with 5 channels. The thermal channel 10.5-11.5 µm is divided into 2 channels: 10.3-11.3 µm and 11.5-12.5 µm, allowing temperature correction to be made. The high repetitivity (2/day) of these satellites makes them well-suited to agrometeorology, but their ground resolution of 900 m limits their use to regional

TABLE I - Geostationary satellites (GARP)

Name	Country	Launched	Mass (kg)	Altitude min (km)	Altitude max (km)	Period min	Inclination (degrees)	Longitude
METEOSAT 1	Europe	23.11.77	697	34 912	35 692	1 411.5	0.7	0°
GOES 2 (1) = SMS1 (2)	USA	17.05.74	286	35 729	35 840	1 436	2	60°W
GOES 3	USA	16.06.78	624	35 769	35 805	1 436.1	0.9	135°W
HIMAWARI = GMS (3)	Japan	14.07.77	280	35 771	35 804	1 436.2	1.1	140°E
GOES 1	USA	16.10.75	295	35 759	35 808	1 435.9	0.5	60°E
METEOSAT 2	Europe	19.06.81	idem METEOSAT 1					

(1) GOES: Geostationary Operational Environmental Satellite
(2) SMS : Synchronous Meteorological Satellite
(3) GMS : Geostationary Meteorological Satellite.

TABLE II
Characteristics of Earth orbiting observation satellites (NASA documentation)

Name	Scientific name	Country	Date of launch	Weight (kg)	Low altitude (km)	High altitude (km)	Incli-nation (°)	Local time at descending node	Ob-jective (*)
Apollo 9 CM (1)	1969-018A	USA	03.03.69	5627	191	511	91.46	variable	(3)
Landsat 1 = Erts-1 (2)	1972-058A	USA	23.07.72	891	898	916	103	9h30	E R
Noaa 2 (4)	1972-082A	USA	13.10.72	312	1451	1457	114.9	8h50	M
Skylab	1973-027A	USA	14.05.73	90607	423	440	93.2	variable	E Ob
Noaa 3 (4)	1973-086A	USA	06.11.73	409	1500	1509	116.1	8h30	T
Noaa 4 (4)	1974-089A	USA	15.11.74	409	1447	1461	114.9	8h30	T
Landsat 2	1975-004A	USA	22.01.75	900	903	914	103.2	9h30	E R
China 3	1975-070A	China	26.07.75	-	183	455	90.9	variable	E R
Noaa 5 (4)	1976-077A	USA	29.07.76	409	1507	1522	116.2	8h30	T
Soyouz 22	1976-096A	USSR	15.09.76	6000	251	257	85.6	variable	E R
Landsat 3	1978-026A	USA	05.03.78	960	900	918	103.2	9h30	E R
HCMM (5)	1978-041A	USA	26.04.78	134	550	642	97.2	2h00	T
Seasat 1	1978-064A	USA	26.06.78	2000	770	798	100.75	variable	O
Cosmos 1043	1978-094A	USSR	10.10.78	-	625	650	97.3	variable	E R
Tiros N	1978-096A	USA	13.10.78	734	849	864	102.1	15h00	M En
Nimbus 7	1978-098A	USA	24.10.78	715	938	953	104.09	23h50	M O
Meteor 1	1979-005A	USSR	25.01.79	2750	617	642	97.2	constant	M N R
Baskhara	1979-051A	India	07.06.79	444	512	557	95.2	variable	E R
Noaa 6	1979-057A	USA	27.06.79	729	807	824	102.2	7h30	En
Cosmos 1121	1979-074A	USSR	14.08.79	-	180	375	89.7	variable	E R
Spacelab		Europe		5000				variable	C
			1982	9000	250	89			
Landsat (7) 4,5		USA		1630				9h30	E R
			1982,84		705		98		
Spot (6,7)		France	1986	1000	830		98.7	10h30	E R

(1) This indication concerns the command module. (2) Earth resources and preparation of the lunar flight. (3) ERTS: Earth Resources Technology Satellite. (4) Noaa: National Oceanographic and Atmospheric Administration. (5) HCMM: Heat Capacity Mapping Mission. (6) Spot: Satellite Pour l'Observation de la Terre. (7) Second generation SGEOS.

(*) C: cartography; E: Earth; En: environment; M: meteorology; N: natural; O: oceanography; Ob: observation; R: resources; T: temperature.

LANDSAT-5

Inclination = 98.2°

Altitude = 705 km (Nominal)

Time of Day = 9.45 AM (Local)

Orbit period = 98.9 Minutes

Repetitivity = 16 Days

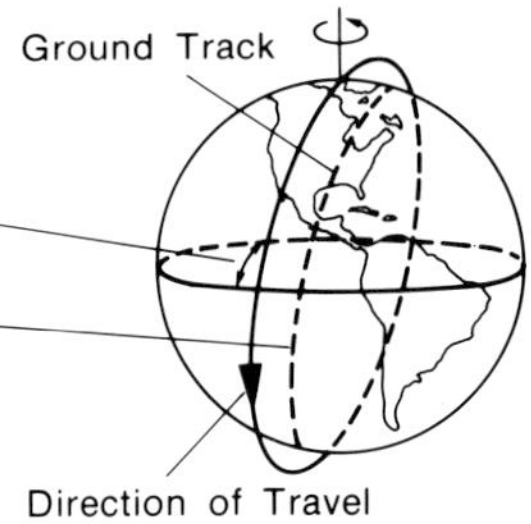

AVHRR NOAA-7

Inclination = 102°

Altitude = 850 km (Nominal)

Ascending Node
07.30/19.30 (NOAA-6)
14.30/02.30 (NOAA-7)
} 2 SATELLITE SYSTEM

Orbit Period = 102 Minutes

Repetitivity = 9.2 Days

SPOT

Inclination = 98.7°

Altitude = 832 km (Nominal)

Descending Node
10.30

Repetitivity = 26 Days

Figure 7. Orbital parameters.

surveys. Figure 7 summarizes the most useful orbital parameters; Table III gives some details on the payload of METEOR satellites.

The LANDSAT satellites have been extensively used for many land disciplines (geology, hydrology, agriculture, forestry), but their repetitivity (18 days) and their ground resolution (80 m for the MSS, 40 m for the RBV) is much better adapted to USA conditions than to European conditions. Table IV gives their sensor characteristics and Figure 8 shows the LANDSAT-3 configuration.

3.2 Second Generation of Earth Observation Satellites

The NASA-LANDSAT-D with its Thematic Mapper and the French SPOT are the precursors of the future satellites. Table V shows the comparison between the Thematic Mapper and the MSS. For vegetation studies, the number of the bands and their smaller width represent a great

TABLE III
Experimental instrumentation payload of METEOR satellites (NASA documentation)

No.	Instruments and characteristics	Application
1	Spectrometric IR instruments, 10-17 μm range, angular resolution 6x1.5^{o}.	Determination of vertical atmospheric temperature profiles.
2	Spectrometer-interferometer developed in the DDR, 6.25-25 μm range, angular resolution 2°.	Determination as above and atmospheric water vapour and ozone content.
3	Microwave 0.8 cm polarization radiometer, two orthogonal polarizations, angular resolution 2.5^{o}.	Tracking of falling precipitation areas, determination of cloud water content and phase composition.
4	Three-channel microwave radiometer (scanning in 1000 km band), 0.8, 1.35 and 8.5 cm wavelengths, resolution (at a height of 900 km) 24x30 km, 90x90 km and 100x100 km on the ground, respectively.	Determination of the total atmospheric water and cloud water content, sea-surface temperature; tracking of falling precipitation areas and ice-cover boundaries.
5	Scanning IR polarimeter, 1.5-1.9 and 2.1-2.5 μm ranges, field of view (at a height of 900 km) 2200 km, angular resolution 3°	Determination of cloud phase composition.
6	Instruments for studying sun-atmosphere relationships, including: - 4-channel corpuscular spectrometer, 0.3-30 keV range; - scanning IR slant sounding radiometer (0.3-30 μm).	Obtaining information on corpuscular radiation affecting the upper atmosphere, intensity of TIR radiation of the upper atmosphere, which is one of the agents of energy transfer from the upper to the lower atmosphere.
7	Scanning 4-channel TV-type instruments (0.5-0.6, 0.6-0.7, 0.7-0.8 and 0.8-1.1 μm), field of view 1930 km on the ground, resolution 1000x1700 m in nadir.	Obtaining overlapping images of earth's surface and clouds in various ranges of the visible and near-infrared portion of the spectrum.
8	Scanning 2-channel TV-type instrumentation (0.5-0.7 and 0.7-1.1 μm), field of view 1380 km on the ground, resolution 240 m in nadir.	Idem.

TABLE IV
Spectral bands of some satellite sensors (NASA documentation)

Satellite	Sensor	WINDOWS A	B&C	D	F	Outside window
1) Photographic cameras						
Apollo 9	S 065	0.51-0.89				
		0.47-0.51				
		0.59-0.715				
		0.68-0.89				
Skylab	S 190	0.50-0.60				
		0.60-0.70				
		0.70-0.80				
		0.80-0.90				
		0.50-0.88				
		0.40-0.70				
Soyouz 22	MKF 6	0.45-0.53				
		0.50-0.59				
		0.56-0.64				
		0.62-0.70				
		0.67-0.76				
		0.76-0.89				
2) Scanners and RV cameras						
Skylab	S 192	0.41-0.46	1.55-1.75		10.20-10.50	
		0.46-0.51	2.10-2.35			
		0.52-0.56				
		0.56-0.61				
		0.62-0.67				
		0.68-0.76				
		0.78-0.88				
		0.98-1.08				
		1.09-1.19				
		1.20-1.30				
Landsat 1, 2 and 3	RBV Landsat 1 and 2	0.475-0.575				
		0.580-0.680				
		0.690-0.830				
	Landsat 3	0.510-0.750				
	MSS	0.5-0.6			Landsat 2 only 10.4-12.6	
		0.6-0.7				
		0.7-0.8				
		0.8-1.1				
Landsat D	TM	0.45-0.52	1.55-1.75		10.4-12.5	
		0.52-0.60	2.08-2.35			
		0.63-0.69				
		0.76-0.90				
Meteosat	VIS	0.4-1.1				
	IR					5.7-7.1
	TIR				10.0-12.5	
HCMM	HCMR	0.5-1.1			10.5-12.5	
Noaa 3-4-5	VHRR	0.6-0.7			10.5-12.5	
Noaa 6	AVHRR	0.58-0.68		3.55-3.93	10.5-11.5	
		0.725-1.1				
Tiros N	AVHRR	0.55-0.90		3.55-3.93	10.5-11.5	
		0.725-1.1				
Nimbus 7	CZCS	0.433-0.453			10.5-12.5	
		0.510-0.530				
		0.540-0.560				
		0.660-0.680				
		0.700-0.800				
Spot	HRV	0.50-0.59				
		0.61-0.68				
		0.79-0.89				
		0.51-0.73 (panchromatic ch.)				
Noaa 7-9	AVHRR	0.58-0.68		3.55-3.93	10.3-11.3	
		0.725-1.1			11.5-12.5	

Atmospheric windows: A = 0.3-1.3 μm; B = 1.5-1.8 μm; C = 2.0-2.4 μm; D = 3.5-4.2 μm; E = narrow at 4.9 μm; F = 8-13 μm with absorption at 9.6 μm.

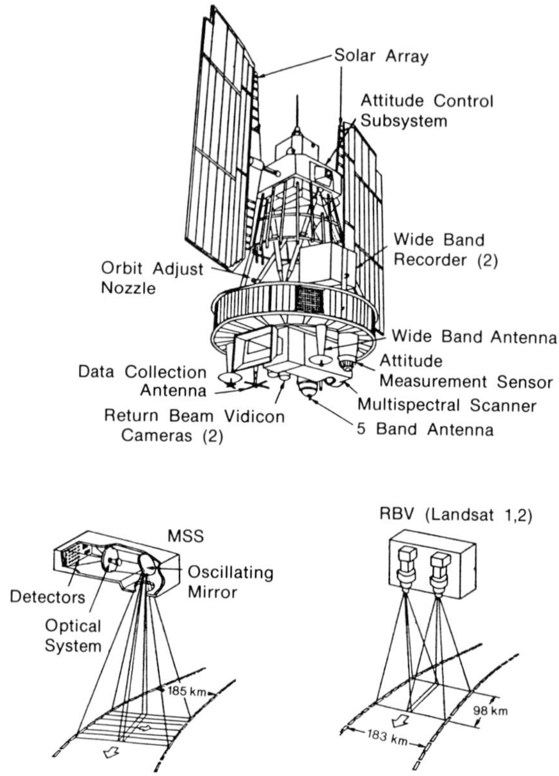

Figure 8. LANDSAT-3 configuration.

improvement. There is a great improvement in the ground resolution which permits more detailed investigations in European conditions. In addition, the thermal channel (band 7) with its 120 m IFOV extends use towards the observation of agricultural conditions and the middle IR channel (band 6) permits observation of vegetation stress.

The French SPOT (February 1986*) with its 2 steerable HRV (high resolution visible) represents a further step in resolution improvement (Table VI). The revisit and stereoscopic viewing capabilities are also

---

*SPOT 2, identical to SPOT 1, is ready for launch at the end of 1987, to be launched in 1988 or 1989.
SPOT 4 will include a number of changes to increase performances (lifetime 4 years instead of 2, substitution of the panchromatic 10 m band by a 10 m band 0.61-0.68 μm, addition of another band around 1.7 μm, incorporation of a low resolution instrument called "vegetation" with IFOV ∿ 1 km and same spectral bands as the HRVs).

TABLE V
Comparison between LANDSAT-D TM and LANDSAT-C MSS (NASA documentation)

	LANDSAT-D TM		LANDSAT-C MSS	
Spectral band	μm	Radiometric sensitivity (NEΔρ)	μm	Radiometric sensitivity (NEΔρ)
1	0.45-0.52	0.8%	0.5-0.6	.57%
2	0.52-0.60	0.5%	0.6-0.7	.57%
3	0.63-0.69	0.5%	0.7-0.8	.65%
4	0.76-0.90	0.5%	0.8-1.1	.70%
5	1.55-1.75	1.0%	-	-
6	2.08-2.35	2.3%	-	-
7	10.40-12.50	0.5K (NEΔT)	10.40-12.50	1.4K (NEΔT)
Ground IFOV		30m (bands 1-5) 120m (band 6)	79m (bands 1-4) 237m (band 5)	
Data rate		83 MB/s	15 MB/s	
Quantization levels		256	64	
Weight		268 kg	75 kg	
Size		1.0x1.0x1.8M	0.35x0.4x0.9M	
Power		250 Watts	42 Watts	
Altitude		715 km	917 km	
Repeat cycle		16 days	18 days	

NASA HQ ER78.346
Rev. 2-22-78

TABLE VI
SPOT characteristics

Characteristics of the HRV instrument	Multispectral mode	Panchromatic mode
Spectral bands	0.50-0.59 μm 0.61-0.68 μm 0.79-0.89 μm	0.51-0.73 μm
Instrument field of view	4.13°	4.13°
Ground sampling interval (nadir viewing)	20 m x 20 m	10 m x 10 m
Number of pixels per line	3000	6000
Ground swath width (nadir viewing)	60 km	60 km
Pixel coding format	3x8 bits	6 bits DPCM*
Image data bit rate	25 M bits/s	25 M bits/s

*DPCM (digital pulse code modulation) is a mode of data compression that does not degrade the radiometric accuracy of the image data (256 grey levels).

important improvements for agriculture and forestry (Figures 9-11). A selected site can be observed with a repetitivity of a few days, much smaller than the "nadir repetitivity". This advantage will become particularly important with SPOT-4 in agrometeorology when detailed highly repetitive observations of vegetation stress will be possible.

3.3 Satellite Data Requirements for Agrometeorology

Solar illumination changes with latitude and season. Thus, direct comparison of scenes taken at different locations and times is not possible without radiance corrections. Figures 12a and 12b show this variation.

When a SPOT scene has been obtained with off-nadir viewing, another correction must be introduced to take into account the modification of the sun-sensor angle. In addition, atmospheric effects cause the signal received by the sensor to deteriorate. Figure 12c summarizes these effects in the visible and near IR range. In the thermal infrared channels, water vapour absorption introduces an error on the temperature determination. Perfect correction of these errors is difficult, without in-situ measurements of reflectance on a reference panel, and many theoretical or experimental methods have been developed.

In agrometeorology, the main requirements can be summarized as in Table VII. Ideally, albedo should be known at $\pm$ 5% and surface temperature at $\pm$ 1°C.

3.4 Data Transmission, Reception and Preprocessing

3.4.1 Data transmission and reception. Data collected by the sensors are digitized on-board and transmitted in real time when the satellite is in view of a receiving station. Figure 13 shows the reception areas for LANDSAT. If a magnetic tape recorder is installed in the payload, data can be collected over remote areas, then transmitted to the ground when the satellite is in view of a receiving station.

The future* NASA plans imply the use of 2 TDRSS telecommunication satellites. The LANDSAT pipeline shall then be: LANDSAT-D → TDRSS → White Sands receiving station → DOMSAT → GSFC → EROS Data Center → NOAA → USER. The LANDSAT satellite is now operated by EOSAT on a commercial basis*.

---

*EOSAT intends to launch a modified version of LANDSAT on a platform OMNISTAR. LANDSAT-7 was scheduled for launch in December 1988. It will carry the Enhanced Thematic Mapper (ETM) with a 15 m panchromatic band and also potentially a low resolution (500 m) instrument. The LANDSAT-7 mission is set for 1991 and will carry the ETM with potentially a multi-band TIR capability in the 8 - 11.5 micron metre band.

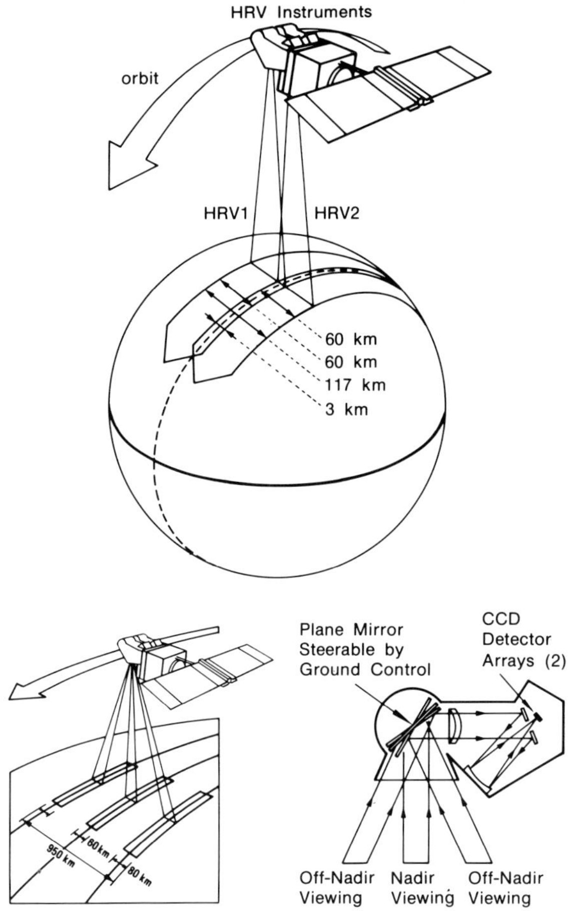

Figure 9. Nadir and off-nadir viewing. The two HRV can be pointed so as to cover two adjacent strips attaining a total swath width of 117 km (nadir) with an overlap of 3 km. The distance between adjacent ground tracks at the equator is about 108 km, thus complete Earth coverage can be obtained (SPOT documentation).

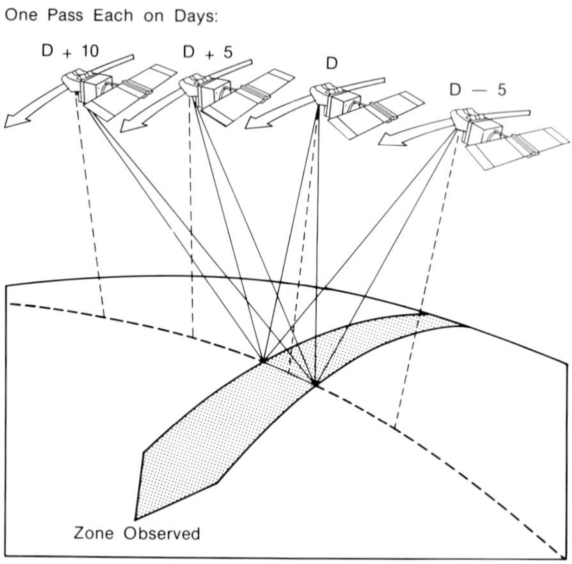

Figure 10. Revisit capabilities (SPOT documentation).

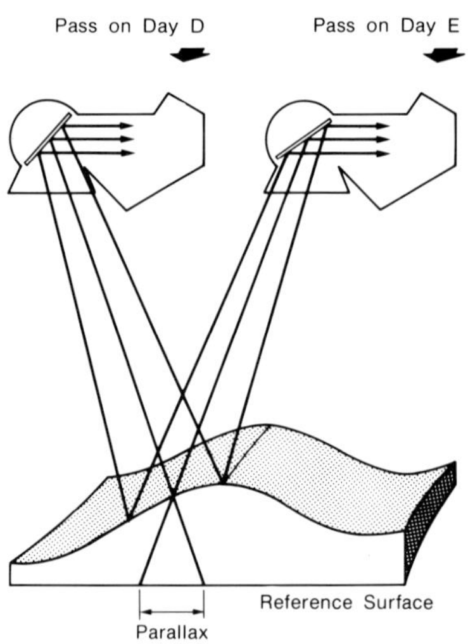

Figure 11. Stereoscopic viewing capabilities (SPOT documentation).

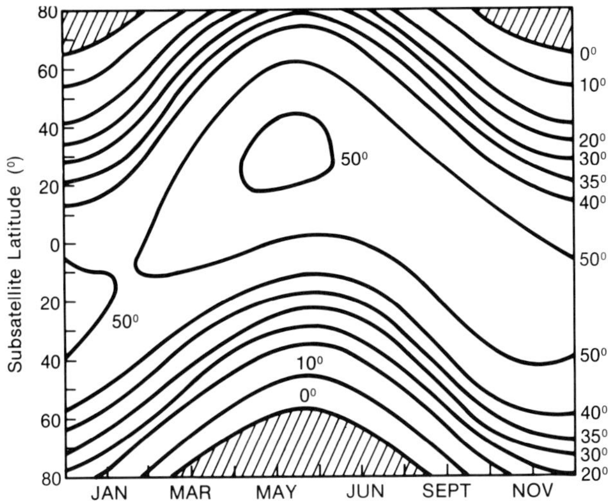

Figure 12a. Solar elevation angles at different latitudes and times of the year for the LANDSAT satellite (NASA LANDSAT Data Users Handbook).

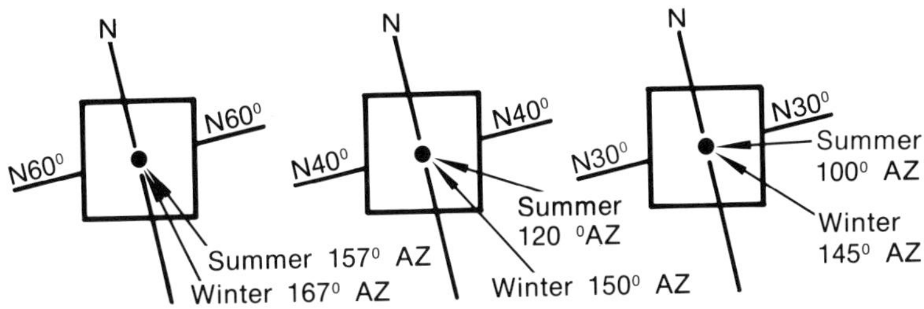

Figure 12b. Variations in azimuth of solar illumination with season and latitude. (Characteristics of the LANDSAT multispectral data system, U.S. Geological Survey, Open-file Report 78-187).

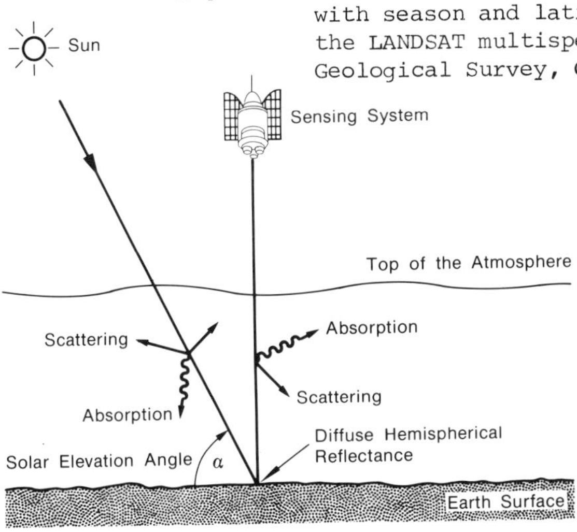

Figure 12c. Atmospheric effects on the energy path between the sun and a satellite.

TABLE VII
Satellite data requirements

	Requirement	Cases	Correction method or solution
Geographical accuracy	1 km for primary variables in homogeneous areas. .1 pixel for vegetation stress.	Day-night registration Multitemporal analysis	Orbit determination Preprocessing methods
Channel-to-channel registration	Better than 10% between VIS and TIR	Vegetation stress and crop yield estimation	Satellite sensor design
Radiometric accuracy .internal calibration .noise level of the sensor .atmospheric effects .illumination effects	Ideal = ± 5%		Satellite sensor design and associated electronics. Models of ground measurements. Tables, models and statistical methods

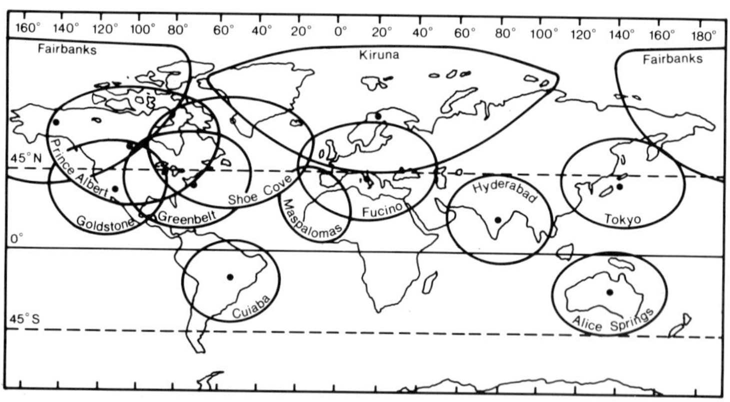

Figure 13. LANDSAT receiving stations.

3.4.2 Preprocessing. Before delivery, data must be corrected from primary effects (not including atmospheric illumination effects) and presented into a standard form. Preprocessing for SAR is a very complicated task, and is accomplished for EARTHNET-ESA by RAE-Farnborough and DFVLR-Oberpfaffenhofen (SEASAT and SAR-580). The actual processing capability for LANDSAT-D MSS is 133 scenes per day (SPD) with a 48 hour turn-around, and 50 SPD for the TM.

3.5 Future Platforms

While improved versions of geostationary and polar meteorological satellites and earth observation satellites (LANDSAT and SPOT) are being prepared, a new concept is under study: the space station, a large platform carrying a very diversified payload, refurnishable from the ground. In order to cover the whole world, a polar orbit is necessary and it appears that altitudes of the order of 700-800 km are desirable.

These constraints have led to the concept of a polar component of the space station. European countries (through the European Space Agency) have agreed to collaborate with NASA and the principle of a morning platform (ESA) and an afternoon platform (NASA) is under analysis.

Figures 14a and 14b show the envisaged solution.

4. REQUIREMENTS, CONSTRAINTS, PERSPECTIVES

Table VIII shows that the conditions of European agriculture require an improvement of the performance of sensors for the observation of individual fields. 10 m ground resolution in multispectral mode seems to be the optimum compromise (between fine detail discrimination and data flow). In this respect, the sensors of the 3rd generation are satisfactory = SPOT-4 HRV, polar platform sensors HRIS and SAR; on the other hand, global observations require the repetitivity offered by geostationary satellites with the performance of AVHRR designed with channels better adapted to vegetation observation. Is such an instrument technically feasible and can it be incorporated in an advanced METEOSAT? It is not possible to answer this question at present.

TABLE VIII. Estimate of field size distribution

	Kansas	France (Garonne)	Ratio
Cumulative % - No. of fields (50%) → field length	350 m	200 m	1.75
Cumulative % - No. of fields (50%) → field width	220 m	150 m	1.45
Cumulative % - No. of fields (50%) → field area	14 ha	6 ha	2.33

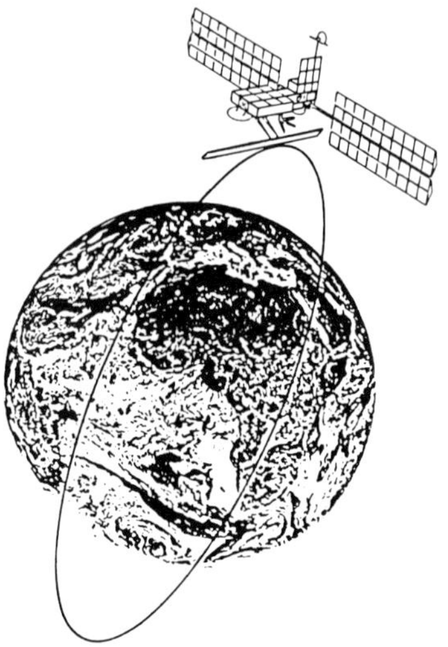

Figure 14a. Earth observation programmes; class of orbits. Various requirements (global coverage, constant viewing range and scene illumination conditions, standard synoptic observation times and avoidance of diurnal aliasing) lead, for an altitude of 500-1000 km, to circular or sun-synchronous orbits which are retrograde with inclination between 97 and 100 degrees (ESA documentation).

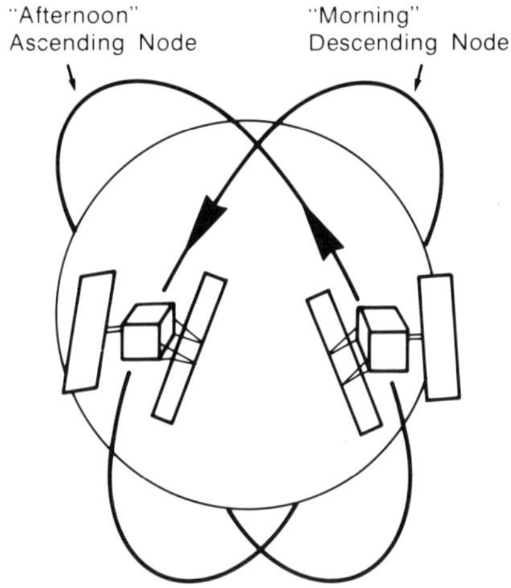

Figure 14b. Earth observation programmes; class of orbits. The need of data up to 4 times/day leads to two platforms in "crossed" orbits with identical instruments, with the benefit of additional redundancy in the observation system (ESA documentation).

Agrometeorology is a basic need for agriculture management and crop yield forecasting. Remote sensing methods offer solutions which can improve today's conventional measuring network. But in order to obtain the performances desired, technological development of the sensors and a parallel progress of the methods to derive the required data is needed. This is a real challenge for remote sensing.

ACRONYMS USED

ARGOS	The name of the DCLS developed by France for TIROS-N.
AVHRR	Advanced Very High Resolution Radiometer.
HRIS	High Resolution Imaging Spectrometer (ESA).
HRV	High Resolution Visible Imager.
IR	Infrared.
SAR	Synthetic Aperture Radar.
SPOT	Satellite Pour l'Observation de la Terre.
TDRSS	Tracking and Data Relay Satellite System.
ERS-1	ESA Remote Sensing Satellite - 1.

THE LUXURIANT IMAGE - AN INTRODUCTION TO REMOTE SENSING DATA PROCESSING

A. Hubaux
Former Scientist of the Joint Research Centre of the European Communities
Ispra (Va) - Italy

ABSTRACT. The text is an introduction to remote sensing data processing by computer, and is addressed to naturalists (geographers, geologists, pedologists, agronomers, oceanographers) with no particular computer science background. Emphasis lies on computer use to process and visualize remote sensing data, and on its inherent limits and possibilities.

The nature of the remote sensing images (actually a data matrix) is first described; these images may be acquired as photographical products (transparencies) or digital products (computer tapes). The chain of events which generate digital data, from the spaceborne sensor to the user magnetic tape, and data organization on these tapes, are briefly reviewed. Data visualization is then examined in detail. The relationships between digital data and the pictures which may be obtained from them on a screen or a hardcopy are complex and somewhat surprising. It is actually impossible to see the digital "image" in its integrality. But it is possible to derive an impressive number of quite different presentations of this image. For this, the role of transfer-functions, which are of fundamental importance, are introduced and analysed, and their relationships with channel histograms are illustrated through some examples. First, the visualization of a single channel image in black-and-white is studied. Then the manners in which the screen may be coloured, and how up to three channels may be seen together (but not more) are analysed.

The notion of channel space is then introduced and some treatments which can be directly derived from the study of this space, such as derived channels, are deduced. The derived systems: Intensity-Hue-Saturation, the Tasseled Cap, Thematic Channels, and Principal Components are briefly reviewed and the notion of robust algorithms is proposed.

Radiometric corrections and geometric corrections, which are necessary to obtain workable data, are briefly explained. The text ends with a very short introduction to context-free (i.e. pixel by pixel) classification procedures.

F. Toselli (ed.), Applications of Remote Sensing to Agrometeorology, 57–128.

FOREWORD

Since the early days of the computer era, hardware costs have steadily decreased by a factor of ten in some eight years. What seemed out of reach a few years ago is now quite common. And there is no sign that this trend will abate. Image processing equipment figures among the newest computer peripherals on the market, but it is presently developing at a spanking rate and the number of laboratories acquiring it is growing fast.

The aim of this paper is to bridge the gap between the languages spoken by naturalists on the one hand, and by computer specialists on the other, and thus to help the geographer, geologist, agronomer, oceanographer, etc. to see whether it is worthwhile for him to make an effort to obtain the collaboration of a computer specialist to solve his problem. The machines are here indeed, and improving every day, but learning what we can retrieve from them and learning the jargon necessary to understand the technical explanations, requires a far from negligible investment in time. And time, in this epoch of Greed and Haste, is becoming more valuable every day.

The present notes are the result of some six year's experience acquired by the author at the Laboratory for Image Processing of the Joint Research Centre, Ispra. They contain the first notions necessary to solve remote sensing problems with the computer. Some familiarity with computers will help in reading these notes, but is not a prerequisite. The intended readers are scientists with a background in some environmental science (geography, pedology, agronomy, geology, oceanography) who want to know more about digital processing possibilities for remote sensing data.

The text does not include any mathematical development, but explains a few mathematical methods in non-mathematical terms; no mathematical background is required. The emphasis is always on the possibilities offered by the computer and on the associated constraints.

The examples are geared toward the study of remote sensing imagery but researchers in other disciplines such as medicine or biology might find here some useful infications.

In the fledgling technique of digital image processing, there is no commonly accepted corpus of basic knowledge which any introductory course such as this one should necessarily include. The present notes, therefore, merely represent the author's views on what should be included in a preliminary text; they are biased toward the personal experience acquired by the author during his research activities. No apologises are offered for this bias, however. It merely reflects the state of the art in the pedagogical aspect of this new discipline.

These notes should be complemented by the examination of remote sensing pictures, which are available not only in the specialised literature, but also in popular magazines, posters, calendars, etc. After reading these simple notes, "non-linear stretching" and "false colour composite" will be taken for what they are, un-mysterious but somewhat vague descriptions of the methods used to obtain the final picture from the digital data stored on magnetic tape.

1. INTRODUCTION

1.1 Remote Sensing Is a Dialogue

Remote sensing has all the characteristics of the present time: it is oriented not toward cheaper energy, but toward better information; it involves enormous amounts of data; it is the fall-out (alas!) of military research, it is evolving fastly and it requires efficient and close collaboration between specialists of different disciplines. This last point is particularly important. From satellite and sensor design, satellite launch, data reception, data pre-processing and distribution, image final processing and interpretation, down to the final report arriving on the desk of the deciding authority, the chain of events is long indeed and requires the use of completely different techniques. Nobody on earth can claim to master them all. Among the dialogues necessary between technicians of various specialities, that which interests us most directly is the dialogue between the "informaticians" (computer specialists, mathematicians, programmers, experts in digital image processing) and the "thematicians" or "naturalists" (agronomists, geologists, pedologists, oceanographers). The naturalist comes to the informatician with a problem: "Could the soil humidity of a given region be mapped?", "Could satellite data help pinpoint where we should drill for oil in this area?". The question seems clear enough to the naturalist; as experience grows in the collaboration however, he realizes that the informatician has only partly grasped the naturalistic aspect of the problem; actually, what the informatician is looking for is an algorithm, i.e. a series of well-defined and unambiguous steps to process the data, so as to extract all the pertinent information, and only that information, from the huge amount of data available (defining algorithms is a major objective in many scientific and technical disciplines, to which considerable sums and manpower are dedicated).

The topic we will develop then is the dialogue between the informatician and the naturalist. Of all the links between specialists of different disciplines, this is probably the most difficult to achieve. We shall try to build a bridge between the thematic problem and the computer, in order to develop a common language to be shared by the man coming with his problem and the man who might possess the key to its solution.

1.2 Analog Versus Digital Processing

There are two extreme positions:

- The extreme naturalist will say: "Why should I use the computer at all? I can buy films (transparencies) and study them as I like; and I can do a much more refined job than the computer, at a fraction of the cost; all the more so since working with informaticians implies a long and painstaking preparatory work where I will have to explain exactly what I expect, but expressed in terms the other party can understand. I know from past experience that this is always more time-consuming and laborious than expected.".

- The extreme informatician has an opposite point of view: "To solve the problem formulated to me, I shall find and develop an algorithm which will be entirely free from any further human decision, and will therefore be completely objective, as opposed to human interpretation which is heavily dependent on the interpreter's knowledge. Also, it will then be possible to process future satellite data automatically to solve the same problem, and, once the algorithm is entirely defined and proved to be adequate, this automatic processing (the magnetic tape with the remote sensing data is read by the computer and, hey presto!, here comes the final map containing all the necessary information and nothing else, presented in a clear and unambiguous way which will be much faster and will yield more objective, more consistent and less costly results than those obtained by a team of human interpreters.".

Who is right? We shall try to answer that as we go along; in other words, we shall be present at the duel between analog (by human interpretation) and digital (by computer) remote sensing data treatment.

1.3 Overview

Throughout the present notes, the amphasis will be on computer use to process image data, its limits and possibilities.

We will first see what a remote sensing "image" is and answer the question of why these images are sold under two formats, as photographic products (transparencies) and digital products (computer tapes).

We will then look briefly at the chain of events which generates digital data, and we shall devote a word to data organisation on magnetic tapes (chapters 2 and 3).

In the remaining chapters we will turn our attention to pixel-by-pixel treatments, i.e. to what can be done on the image by processing each pixel (image point) individually, i.e. context-free processing. What can be done when each pixel is analysed within its surroundings, by taking its neighbours into account, will be examined in the next lecture, given by Dr. J. Dejace.

The complex and somewhat surprising relationships between the data as they are written on tape, and the visualization thereof on a screen or as a hardcopy, will then be examined. We shall see why it is actually impossible to see the digital "image" complete (chapters 4 and 5). But it is possible to derive an impressive number of quite different presentations of this image.

For this, the role of "transfer-functions", which are of fundamental importance, will be introduced and analysed in some detail. The relationships between these transfer-functions and histograms will be of assistance in understanding this visualization process (chapters 6, 7, 8 and 9).

We shall then introduce the notion of "channel space", and deduce some treatments which can be directly derived from the study of this space, such as derived channels (chapters 10, 11 and 12).

The corrections necessary to obtain workable data will be briefly reviewed; these corrections fall into two categories: "radiometric corrections" and "geometric corrections". We shall see, among other things, why these terms, in fact, cover two different kinds of "corrections"; we shall see that to speak of "corrections" to improve data geometry is a good example of those abuses of language which abound in remote sensing jargon (chapter 13).

The present course is not directly addressed to classification procedures and we shall, therefore, limit ourselves to a short introduction to these techniques (chapter 14), the emphasis being always on the possibilities, difficulties and limits of computer use in this field.

1.4 A Word on Programming

As a remote sensing study proceeds it is soon realized that more software is needed and that the on-site software, even if abundant, is not exactly adapted to the case. Then either commercial software must be bought, or some programs must be written in-house. Neither of these solutions is easy. Simply looking at what exists on the market is deceptively time-consuming. Reading the software commercial description often does not clearly show whether it solves our particular problem. Then implementing the program on our computer installation may well be either childishly easy or reveal some unexpected compatibility problem which requires fixing. Finally, some time must be allowed to learn the possibilities and limits of the program, and to understand its messages.

Writing the program in-house raises other kinds of problems. Experience shows that the time actually used to make the program operable is often much longer than was thought at the start. This is in part due to the wide-spread belief that programmers write programs. What an illusion! It is almost the same as saying that a professional photographer devotes his time to shutter pressing! In program realization, program writing proper is a nearly negligible part of the total time. The diagram proposed by Brooks (1975) is applicable to all programming efforts, with the exception of very small programs for personal use with a short life-time expectancy (see Figure 1).

Another aspect of the programming problem is that the time necessary to implement the various steps of the algorithm is often much less than the time dedicated to solving input/output problems: formats, compatibility, improving throughput, etc. These nagging I/O questions are especially time-consuming (and most often unrewarding) in such fields as remote sensing where the data flow is considerable.

2. WHAT IS AN IMAGE?

2.1 The Invisible Image

People who are accustomed to working in the computer realm know quite well that a number of current terms are used there with a special

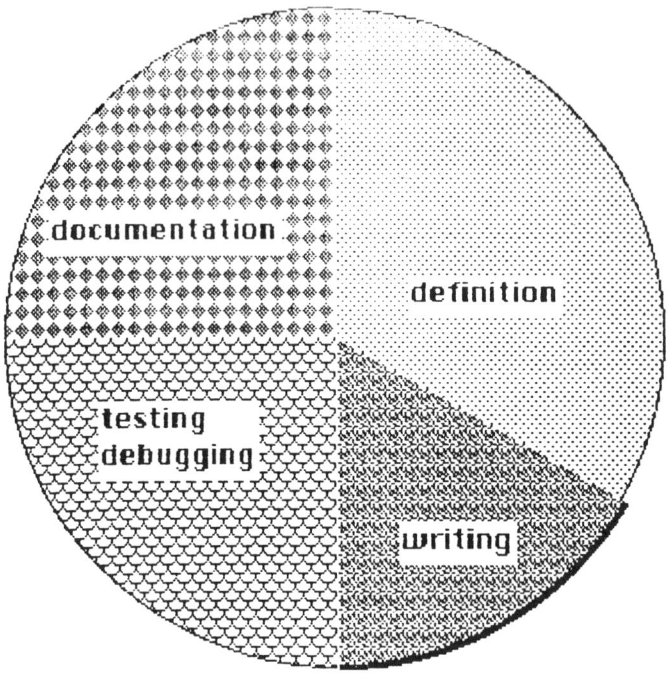

Figure 1. Brooks diagram; program writing represents only 1/6 of program development efforts.

meaning. It is only by considerable semantic stretching that we can say that a computer "reads" or "writes", that it has a "memory", etc. In programming language the equal sign "=" is often used not to specify equality but "move", "reals" are only approximations of mathematical reals, etc. In remote sensing, the same stretching occurs and probably in a more far-fetched way. One example is the word "image". We have been trained to believe that an image is something to be seen. Wrong! Here the word covers something which nobody has seen in its entirety, because it is simply impossible to see a whole image! Why this is so will become progressively clearer as we explore the concept. To this end, the chain of events from the sensor to the user must be briefly reviewed.

2.2 From Satellite to User

Earth-observing satellites have one or several sensors on board; in the present text we will take as example the Multi-Spectral Scanners which are today the most widely used type of sensors; several developments are also applicable to other types of sensors, however.

Multi-Spectral Scanners make measurements on the ground below, line after line, uninterruptedly. The long strip of land thus analysed by one sensor is divided into scenes. A scene is generally square in shape, with some obliqueness due to the earth's rotation during data capture; a Landsat MSS scene is 180x180 km and is made of some 2340 lines, each line comprising some 3240 measurements (pixels), in four wavelengths (channels); the image corresponding to a scene takes some 30 seconds to be acquired. The fields, forests, cities, lakes, shores, etc. are the objects of study. Hereunder the work "object" will be used with that meaning exclusively. Objects of particular importance are, of course, the cultivated fields; we will use the word "field" in this sense in the present text. It should be clear that the objects we want to study are almost always somewhat subjective; if we are making an inventory of land use, the forests will be one object for us; but a forestry specialist will need to differentiate as much as is feasible between different kinds of forests, and for him deciduous trees will have to be separated from conifers, whereas all arable fields and indeed everything which is not forest, might just as well be grouped into one category: "non-forest".

The image is made up of an ordered series of measurements. The piece of ground measured at one instant is called a pixel ("picture element"). In the multi-spectral scanners, the measurements are generally the intensity of electro-magnetic energy (for example, reflected sunlight) coming from each pixel, in various wavelengths, or, more correctly, in narrow windows of the electro-magnetic spectrum. SPOT, in its multi-spectral mode, measures light in three wavelengths: green, red and near-infrared. These spectral slots are called channels. Thus, SPOT has three channels, LANDSAT MSS has four and the Landsat Thematic Mapper has seven. By the same kind of semantic slipping by which we speak both of a china cup (the container) and a cup of tea (the contents), the word "channel" is extensively used not only to indicate a certain window in the electro-magnetic spectrum, but also to designate the set of all measurements obtained through that channel for a given image. In that last significance the word band is used as a synonym. Thus, a Thematic Mapper image is made up of seven channels or bands.

For LANDSAT the data are acquired line after line with a rotating mirror; these lines are perpendicular to the satellite trajectory. SPOT works slightly differently; the whole line is acquired at one time by a series of Charge Coupled Devices (CCD). From our present point of view the result is the same however, a huge amount of data (measurements) is acquired and transmitted, via radio signals, to a receiving station having a dedicated antenna (a 10 metre parabolic dish, oriented toward the passing satellite and receiving data as long as the satellite is above the horizon, i.e. some ten minutes). These raw data, which are stored on special, high density magnetic tape, are normally processed at the receiving station to obtain system correction data. We shall see briefly, furtheron, what these "corrections" are. The whole process is called "pre-processing", because the user is supposed to perform the processing proper.

2.3 Photographic and Digital Products

From these data, two series of products are then put on the market, photographic (transparencies and paper prints) and digital products (magnetic tapes). Why these two forms of presentation? What are the advantages of each? The aim of the present notes is precisely to answer these questions.

The distributing agencies offer a whole range of products; we list here the most commonly used.

2.3.1 Photographic products.

- Quick-looks: small paper prints obtained from raw data at an early stage of the pre-processing; they are used to check reception quality and, above all, to give an idea of the cloud cover, a most important hindrance to optical data use.
- Black-and-white transparencies (films), which are derived from system corrected data, each transparency representing the image obtained in one channel; they are used by laboratories without access to digital equipment. A number of treatments may be carried out on them, but we shall see furtheron that they contain only a fraction of the measured data. Nevertheless, they are the best format for archiving the data in a non-digital manner.
- Colour-composites (transparencies or paper prints), obtained by combining three channels, each with a colour (e.g. red, green, blue). Most are false colour composites, so called because they differ from what we see with our eyes; except for the Thematic Mapper, there is no blue channel on satellite sensors, but there are one or more near-infra-red channels; furthermore, the attribution of a colour to a channel is quite arbitrary; the most common false colour combination is: green channel in blue; red channel in green; one near-infra-red channel in red; vegetation then appears with a bright red colour.

2.3.2 Digital products.

- Computer compatible tapes (CCTs): magnetic tapes to be processed by image processing laboratories. These tapes are said to be "computer compatible" to distinguish them from high density tapes which are used at the receiving stations to store the data received from the satellite, but are unfit for computer processing. These CCTs are the standard remote sensing data digital support and are widely used. The possibility of replacing them with optical disks (compact disks) is being closely studied. These disks will offer a number of advantages over tapes: they are permanent (tapes may become unreadible after some years), more compact, more robust (they may be read any number of times) and are accessible non-sequentially, whereas tapes must be read in their entirety to obtain information written at the end of the tape.

As an illustration, the prices of some of these products sold by Eurimage are, in 1987 (1 Accounting Unit = 0.8 US dollar):
- one Landsat MSS scene on CCT: 900 AU;
- one Landsat Thematic Mapper full scene on CCT (all 7 bands): 5100 AU;
- idem, but a quarter scene: 1620 AU.

We will now have a closer look at the image in its digital form; in the next chapter we will see how it is stored on CCT, and in the other chapters, how it can be visualized and processed.

3. THE DIGITAL IMAGE

3.1 255, the Mystical Number

In every creed, religion or technique, a few numbers are recurrently used: it was 12 for the Ancient Romans, in the Bible we find 3, 7, 12 and 40, with logarithms it is 10 and e, and in remote sensing it is 255. This number 255 is one of the "magic" numbers of image processing, but it has no direct connection with remote sensing proper; it appears simply because the byte is the basic unit for most computers; on one byte, which is made up of 8 bits, an integer from 0 to 255 may be stored; $2^8 = 256$, but we start from zero, so the last number is 255. Hence, 255 does not derive from some remote sensing necessity, but simply from the fact that most computers and their peripherals are nowadays organized around the byte.

3.2 The Image on Tape

As we have seen in the preceding chapter, a digital "image" is actually a large quantity of measurements obtained by a satellite sensor. To store these huge amounts of data, the measurements are recorded on tape as integers, in most cases as one-byte integers. This is more than sufficient to record the range of possible values obtained in one channel because, in most cases, this range is between 60 and 100 values. This possibility of storing a range of 255 values is often mistakingly interpreted to mean that the range of actual values is 255, but this is incorrect. The range is less than half that number in most cases.

These one-integer data are often called grey levels or grey values (in French "niveaux de gris") because the most immediate way to visualize these values is to present them on screen as one-channel black-and-white picture as will be described in chapter 5. From now on we will use the word "value" to designate these grey values.

To enable the user to return to the original measurement values expressed in physical units, linear conversion factors are specified on the tape, among the ancillary data (i.e. all data which are not pixel measurements).

3.3 Tape Formats

A digital image, then, is best conceived as a series of large tables of one-byte figures, one table per channel, all tables being made up of an equal number of lines and columns (see Figure 2); in computer terms it is a three-dimensional matrix, with one dimension corresponding to lines (sensor swaths), the second dimension corresponding to the number of pixels in a line, and the third dimension being the number of channels (bands).

Normally the pixels have the same dimensions in all channels; in some cases however, one channel has a lower resolution; a good example is channel 6 of the Thematic Mapper, which has pixels of 120x120 metres; one pixel in that channel corresponds to 16 pixels of the other channels; in order to keep the same number of lines and columns in the image matrix, the values of channel 6 are repeated 16 times.

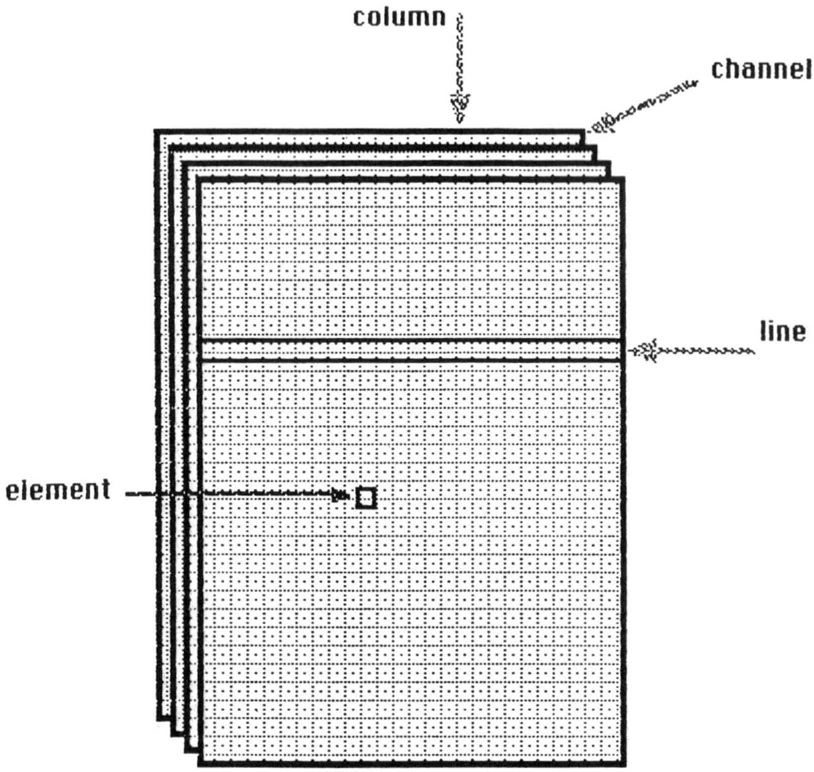

Figure 2. An image is a three-dimensional data matrix.

On the CCT, the bytes are written sequentially and some mode of organization of the data matrix must be defined. The detailed organization of a magnetic tape is called its format. The two most widely used types of formats are the Band Interleaved by Lines (BIL) and Band Sequential (BSQ).

In the BIL format, the lines are given in sequence, each line containing the data of all channels, whereas in the BSQ format, all data of channel 1 are written line after line, followed by all data of channel 2, etc. (see Figure 3).

The overall data organization being thus rather simple, it would seem that CCT formats used by distributing agencies would be fairly easy to read, but alas this is not the case!

A number of stumbling blocks may be encountered:

- the format descriptive brochure is made up of some 50 pages or more of highly technical indications and contains format specifications intermixed with parameter definitions and physical considerations;
- this brochure does not exactly correspond to the tape received; some minor format changes have been introduced; to track these changes may take several days;

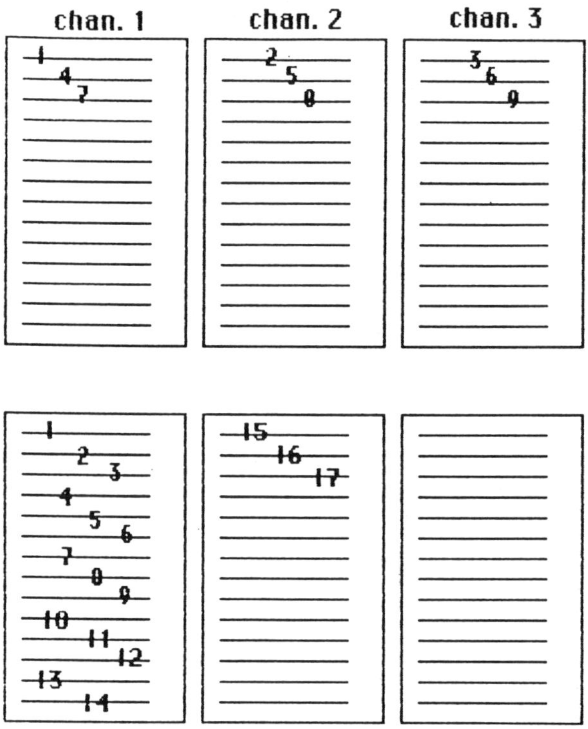

Figure 3. The two most common tape formats: top - band interleaved by lines (BIL); bottom - band sequential (BSQ).

- the ancillary data are written under a great variety of formats, often somewhat ambiguously explained;
- for some mysterious reason, some tapes contain files separated by double end-of-files, which are not easily passed over, etc.

For these reasons, if an experienced programmer takes one week to unravel the mysteries of a new tape format, he has done a quick job.

To alleviate these difficulties, the Landsat Ground Station Operators have developed a "Universal Format" which should be used for all Landsat formats, and perhaps for others too. In another approach, a symbolic language has been proposed to describe all CCT formats (Hubaux et al., 1984).

4. IMAGE VISUALIZATION

4.1 The Luxuriant Image

We have said in chapter 2 that the digital image cannot be seen in its entirety. But we must now qualify this frustrating statement. We shall examine what can be seen of that elusive image and we shall discover that a considerable quantity of quite different pictures may be obtained. Indeed, so numerous are the possible image presentations that it is practically impossible to see them all on an interactive console.

As a comparison, consider a cathedral: a very great number of pictures (photographs) may be taken of it, each one presenting a different aspect, but it is impossible to put them all on a single photograph. By buying a photographic product, such as a B/W transparency, we acquire only one picture, something like one photo of the cathedral; but by buying the CCT, we could say that we acquire the whole cathedral! Well, not exactly the monument itself, but something like a super-hologram, or a series of holograms, something which gives us the capacity to generate any photo of that cathedral from any view angle, with any focal length.

Before we can discuss these techniques of generating various presentations from the digital data, we must first look at the visualizing hardware. We shall then dedicate some time to the presentation of a mono-channel image in black-and-white, both because this is a good introduction to more complex cases, and also because this type of picture is very useful in image analysis.

4.2 Images and Pictures

The words "image" and "picture" are synonyms, or almost so, in current usage. But we will reserve the first here for "digital image", i.e. an organized collection of measurements stored on some computer medium, whereas the second word will be used for what we actually see with our eyes; a picture is then for us an image presentation. This distinction is not commonly accepted, but we shall introduce it here. We can then say that:

From one image, a large number of pictures may be derived.

Although the rest of these notes will be devoted to explaining what kinds of pictures may be created, only a fraction of the possibilities will be mentioned; others will simply be hinted at.

4.3 Why Do We Need to See the Image?

But some computer enthusiast might object: "Why do we need to see the image at all? After all, the image is just another data matrix, somewhat bigger than the usual matrices dealt with by the computer, but not that much bigger; and the data are all integers, which are treated by the computer most efficiently. Certainly, there must be ways to check the validity of the processing algorithms without necessarily presenting the results in a pictorial form; and I can always print part of these results on the line-printer, and even present them by that means as a picture if needed." There might be some semblance of truth in this, and indeed, in the very early days of digital image processing, people attempted to produce pictures with the line-printer, but nobody does that any more. The problem, as I see it, is that the eye, or rather the "eye-and-mind", is such a formidable tool that we cannot do without it.

For instance, we can immediately recognize our son or brother in a group photograph, even if the group is large. This ability is called Pattern Recognition, a technique which, when transposed to computers, is still in its infancy. The computer computes tremendously well, but the human eye-and-mind is a fantastic tool for pattern recognition! The ability with which we can use this tool is generally taken for granted, but actually it is incredibly fast and accurate. This is true also in the animal world, where pattern recognition - recognizing rapidly the predator or the prey - is a matter of life and death. This is no coincidence; Desmond Morris (The Naked Ape) is right: built in ourselves, we still have tools which were necessary to survive in a primitive environment. And among the skills our prehistoric ancestors developed, the only one we still use quite extensively is precisely pattern recognition. There are very few human activities where one form or other of pattern recognition is not involved, whether we drive a car, play tennis, meet a friend, read, or watch TV. It took something like one billion years of evolution to bring this skill to its present state of perfection; no wonder then that the computer cannot compete!

When we see a picture (actually, a quite sketchy drawing is sufficient), we immediately see what it represents, or even what it could represent. We even have difficulties in not seeing the represented object, but in seeing instead what the picture really is, lines, spots and blobs of various colours and shapes. The computer works in a vastly different manner; it needs to be "told" every step of the algorithm we want to be accomplished. But we, as human beings, do not work with algorithms; we can recognize minute topographical details on an aerial photograph, but we are unable to explain to somebody else how we did it, much less to instruct a computer to do it for us. We need, therefore, to visualize what we obtain with our image digital processing.

4.4 Visualization Devices

Around the computer, a number of peripheral devices are used to store and present the image:

- magnetic tapes, also called Computer Compatible Tapes (CCTs) are a permanent storage medium;
- magnetic disks serve as temporary mediums for computer processing; they hold both the original image and its transformations;
- visualizing consoles have:
  - a working memory (generally of limited capacity) containing that part of the image which is to be currently analysed,
  - a screen, often of 512x512 pixels, somewhat like a TV screen, but with a much better resolution,
  - a pointing device (mouse or track ball) to pinpoint a screen pixel,
  - a series of look-up tables, the fundamental role of which will be clarified in chapters 6 and 7;

- hard-copy devices are of two types:
  . devices producing photographic documents (B/W or colour films),
  . printers producing pictures directly on paper; they are handy and useful, but their outputs are not of the same quality as photographic products, especially because their number of grey levels is limited.

All visualizing consoles offer the possibility of enlarging part of the picture. This may be obtained with two quite different methods. In the simple method, each pixel of the local memory is doubled on the screen, more precisely, it is linked to a 2x2 square of screen pixels. The picture is then enlarged linearly by a factor of two. By the same procedure, the picture may be enlarged by four, eight, sixteen. The series does not normally go much farther because with big enlargements, the screen contains just a few big squares, and the eye no longer recognizes the scenery.

The more sophisticated method offers true zooming, i.e. enlargements by continuous factors. This means that some algorithm must be used to define the relationship between a memory pixel and its screen representation. The value of each screen pixel must be interpolated.

It should be added that the whole field of image processing is undergoing a tremendously rapid evolution; to be able to follow the market becomes almost a full-time job!

5. ONE CHANNEL IN BLACK AND WHITE

5.1 The Mono-Channel Image in Black-and-White

Most images are made up of more than one channel, but it is often necessary to visualize the channels separately, in black-and-white. Topological details, especially elongated patterns such as roads or railways, may be quite easy to detect on a B/W picture, even if they are narrower than one pixel, and may be blurred or completely invisible on a colour composite. We will, therefore, examine first the visualization of a single channel in B/W.

The image to study resides on disk, accessible by the computer to which the console is linked. It is generally too large to be copied entirely into the console memory, which often has the same capacity as the screen, for example three planes of 512x512 points or "pixels". An image subset has first to be chosen. Of course, this subset may consist of some regular sampling of the image, such as a pixel every two pixels, on alternate lines, allowing the vision of a 1024x 1024 pixels sub-scene.

From the original image then, we first extract a selection and copy it on the device local memory, using the standard software which goes with the device. We shall call it "the local image".

5.2 Grey Level Histograms

On the display unit, it is possible to obtain a graph of the subset histogram for a given channel. It consists in counting the number of pixels belonging to every grey value and in drawing the corresponding curve (see Figure 4). This histogram, as we will now see, is extremely useful.

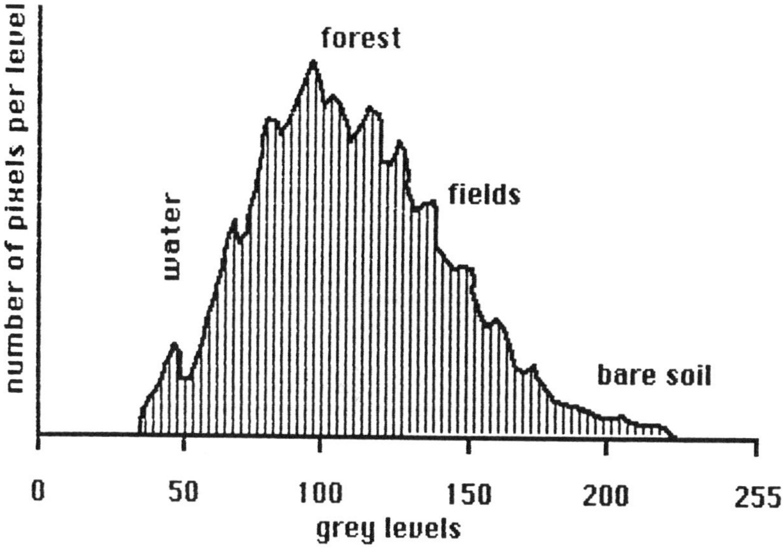

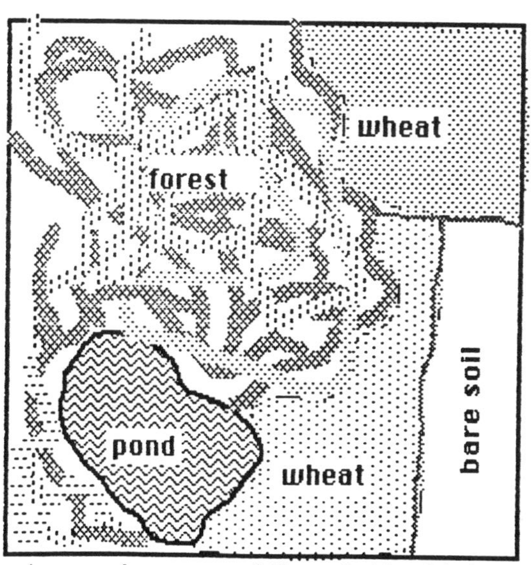

Figure 4. An imaginary sub-scene and the relative histogram in one channel.

A look at it indicates whether the value distribution is uni-modal, bi-modal or multi-modal, i.e. if it has one or more maxima. As an example, let us suppose that the local image contains three types of objects: a lake, a forest and a few cultivated fields, some with dense and green vegetation, others with bare soil (see Figure 4 bottom). The histogram of a given channel, a near-infra-red channel say, might look something like Figure 4 top. Let us note already here that the number of maxima on the resulting histogram will depend not only on the nature of the image objects, but also on their relative abundance; if, for example, the lake is very small, a simple pond, the water maximum will not be distinct on the histogram.

Simple as this example may seem, it already illustrates a number of interesting facts:

- There are three types of mixed pixels, i.e. pixels straddling two categories (say forests and fields) in the histogram:
  1) pixels which are on the border between a forest and a field, and which necessarily have intermediate characteristics;
  2) pixels coming from an isolated tree or hedge; should they be counted with the forest?;
  3) spurious pixels giving the same spectral response as a certain object, but belonging to another object; there are some algae in the ponds, say, giving the same response as the field vegetation.
- The categories we had distinguished at the start: water, forest, fields, are not quite so distinct on closer analysis, and some arbitrary decisions are necessary.
- The scene studied, on which we may define an algorithm of some sort, may prove to be not quite so representative as we would wish it to be; for example, the lake might be made of clear water and thus quite distinct, whereas in the same region, outside our scene, some ponds contain much algae and others contain quite muddy water, and they might be confused with vegetation or bare soil.

It thus appears that we are dealing with natural objects, not laboratory creations, and whatever approach we take, our results will necessarily be more approximate than in situations where we have full control. These inherent uncertainties make it difficult for physicists who come from more exact fields of research, to accept remote sensing.

5.3 Channel Dynamics

We have seen that we may regard an image in two quite different ways. We can consider that it is a series of physical measurements, or we may look at it as a potential source of pictures from which we will derive useful information. In both cases, the total number of grey values (for the scene under study and for the considered channel) is, of course, important; if there are more levels, the image is richer in potential information. This is sometimes called the channel dynamics.

In image processing it is often useful to reduce the number of levels, as we will see. Note, however, that the number of levels cannot be increased, except by quite artificial procedures, which would only add noise but no real information.

In an image considered as a picture source, the important factors are the relationships between the grey values of the various objects, not their absolute values. The images with the histograms of Figures 5, 6 and 7 will produce essentially the same pictures, despite the fact that the histograms do not coincide, because the differences in response of water, forest and fields remain constant. Therefore, any transform of the values may be applied to the image, without modifying its contents, as long as the number of levels is not diminished. This transform may be linear or non-linear.

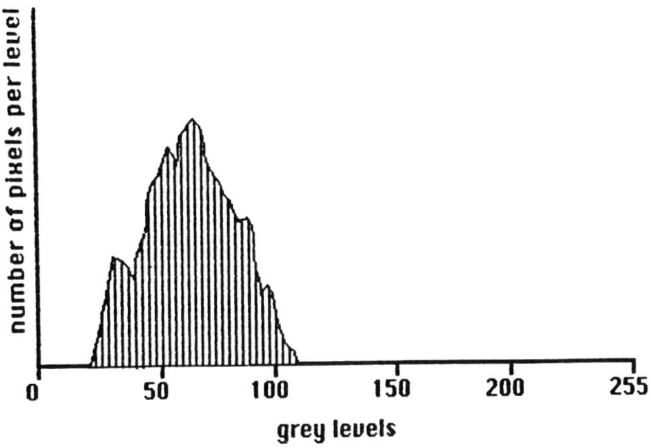

Figure 5. Images with histograms of Figures 5, 6 and 7 will give the same pictures.

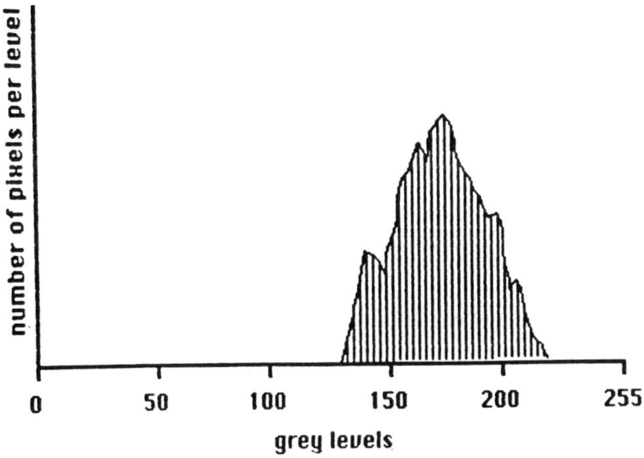

Figure 6. Same histogram as Figure 5, but with higher values.

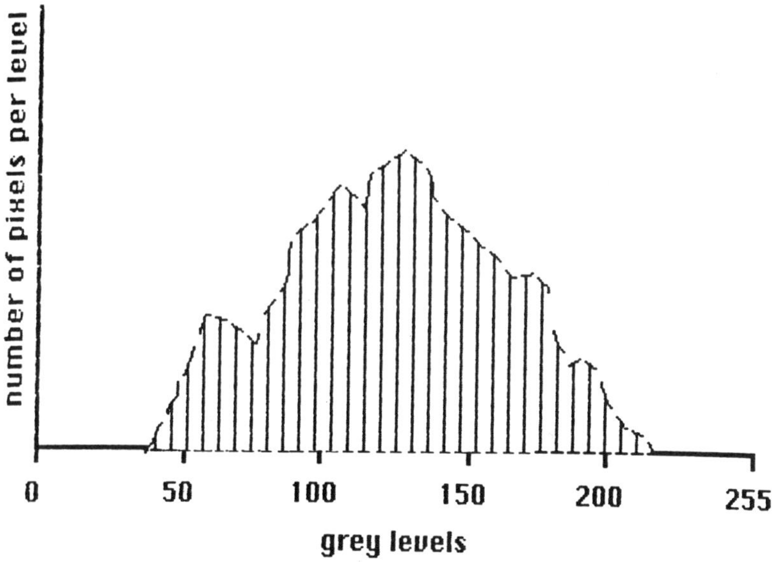

Figure 7. Same histogram as Figures 5 and 6, but spread on a range of level values twice as large. Here, every other level is empty, so that the total number of the image grey levels is the same.

6. TRANSFER FUNCTIONS

6.1 The Ever Present Transfer Function

To obtain the screen picture, every screen point is connected to an image pixel of the device local memory and receives a level of light intensity linked to the grey level value of that pixel (see Figure 8). This link between the pixel grey level of the digital image and its screen counter-part is commanded by the user who may modify it in various ways. For that purpose, the user uses transfer functions which are of fundamental importance; their use must be mastered by anyone wanting to use visualizing consoles. Fortunately, the concept is quite simple once it is fully understood.

The essence of the problem lies in the fact that, as we have seen,

> one pixel measurement in one channel (one value) is often comprised within a range of 60 to 100 values, whereas the eye cannot distinguish more than 16 or, at best, 20 grey levels.

It is, therefore, simply impossible to represent all grey level values as distinct intensities. The process has much in common with photographic work. Amateur photographers know that the B/W negative transparency contains more details than can be printed on the positive paper. Not because it is a negative, but because it is a transparency.

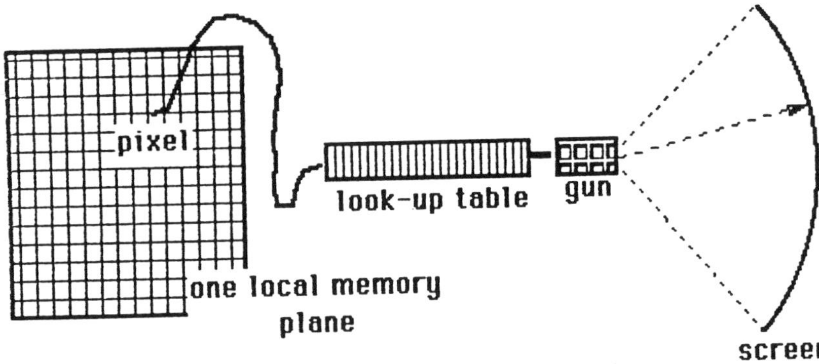

Figure 8. Display device. Simplified view of the local memory to screen connection. Every pixel of the local memory is linked to its screen counter-part through a look-up table (for a more accurate diagram, see Figure 1).

This is due to the fact that the number of possible grey levels which the eye may separate on a transparency is of the order of 20, whereas on the paper-print a smaller number of levels may be distinguished. Thus, it may be said that the film contains more information than the paper-print.

A choice must be made, and this choice is obtained via a transfer function. In a photographic film the choice is made by the technician preparing the film, whereas on a digital visualizing console it is obtained interactively by the user.

So the transfer function plays a fundamental role in all forms of image visualization.

What is it?

It is simply the association of every possible screen light intensity level to every possible grey value. It may be represented by a graph such as Figure 9, where the input level values - those forming the original digital image, normally from 0 to 255 - are in abscissae, and the screen intensities are in ordinates, from "completely black" to "full white". These intensities are often also numbered from 0 (black) to 255 (white).

Through this transfer function, one screen intensity is associated to an input value. The user may easily change the transfer function, in an interactive way, most often by using the track-ball or the mouse.

6.2 Linear Transfer Functions

The function used for the memory to screen transfer may be of almost any shape, provided that every input value has always one, and only one, output value (intensity).

The most common family of transfer functions is made up of three straight lines, two horizontal (one at the bottom, one at the top) with an oblique line joining them (see Figure 10). The grey level enhance-

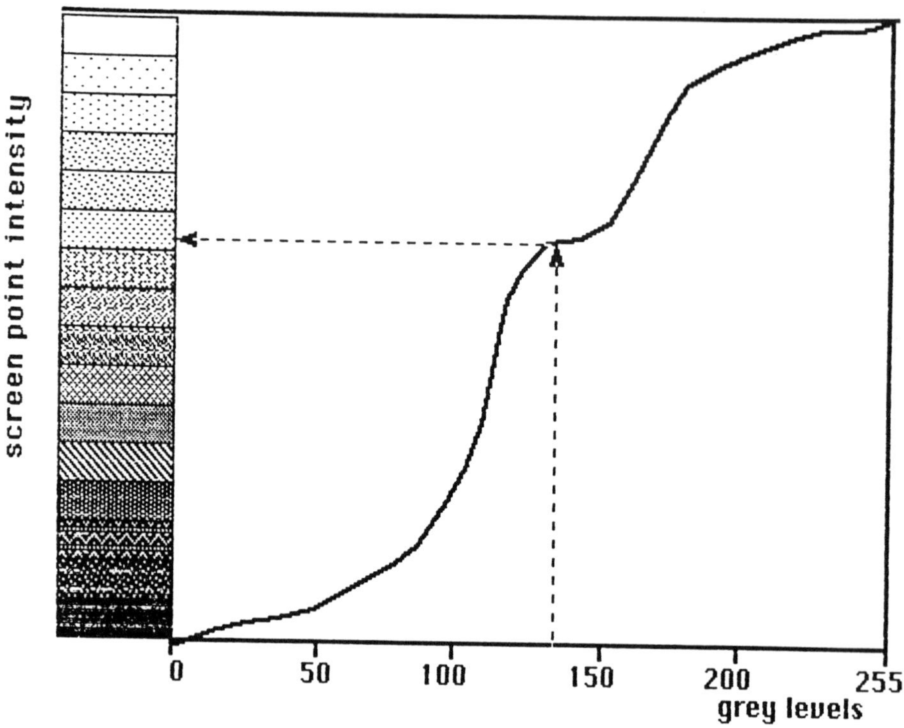

Figure 9. A transfer function. The link between every image grey level and its screen intensity may be represented graphically. Here, for example, level 140 will correspond to a medium grey intensity on the screen.

ment obtained through the use of such linear transfer functions is often called linear stretch.

The simplest transfer function is the "identity" (see Figure 11): 0 is represented as black, 255 as white, and the intermediary values are linearly interpoled. With the reverse, where 0 corresponds to full white and 255 to black, we obtain a "negative" picture. The notions of "positive" and "negative" are, however, less distinct than is the case in photography, precisely because the same local image may give a positive or a negative picture, at user's whim.

6.3 Transfer Functions and Histograms

There is a great advantage in considering the local histogram and the transfer function together; they may be superposed on the same graph. The range of grey levels occupied by the image values then appears clearly, and this range may be used to choose an appropriate transfer function.

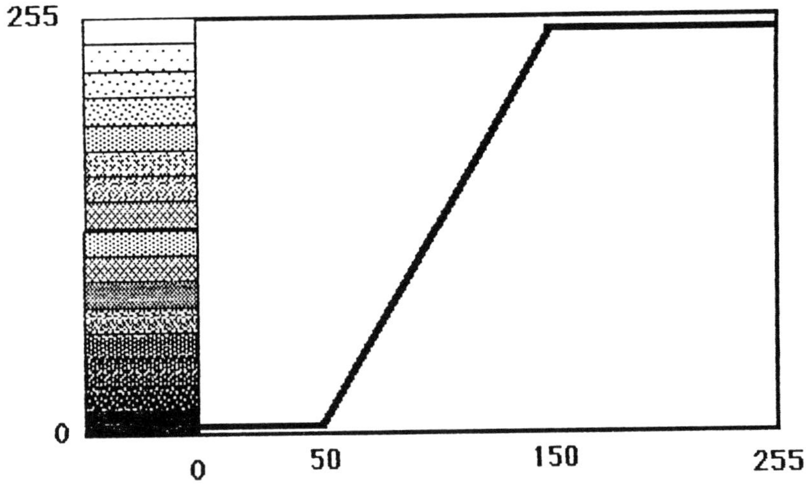

Figure 10. A simple transfer function. All pixels with level under 50 appear as completely dark on the screen, and all above 150 as full white. The whole range of screen intensities is used to differentiate levels comprised between 50 and 150. Such transfer functions are frequently used.

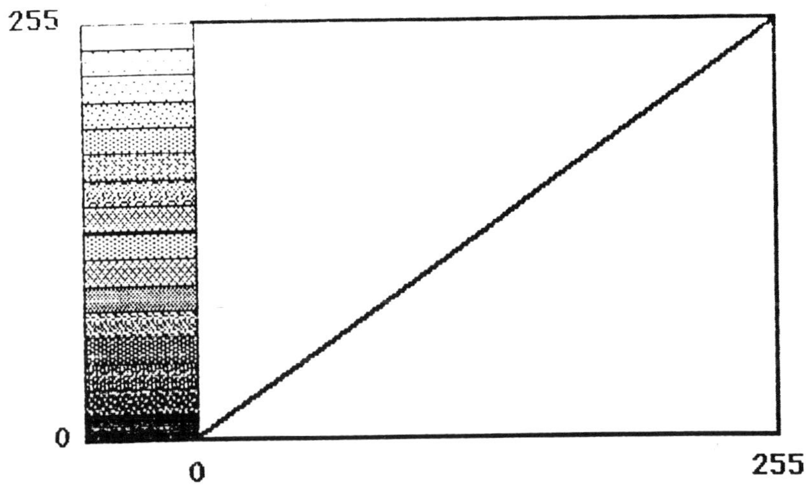

Figure 11. The identity transfer function. "Full black" corresponds to a level of 0, and "full white" corresponds to 255.

As a first example, consider the identity transfer function, together with the image histogram (Figure 12); the resulting picture will contain a few intensity levels only, and will be of poor quality. A much

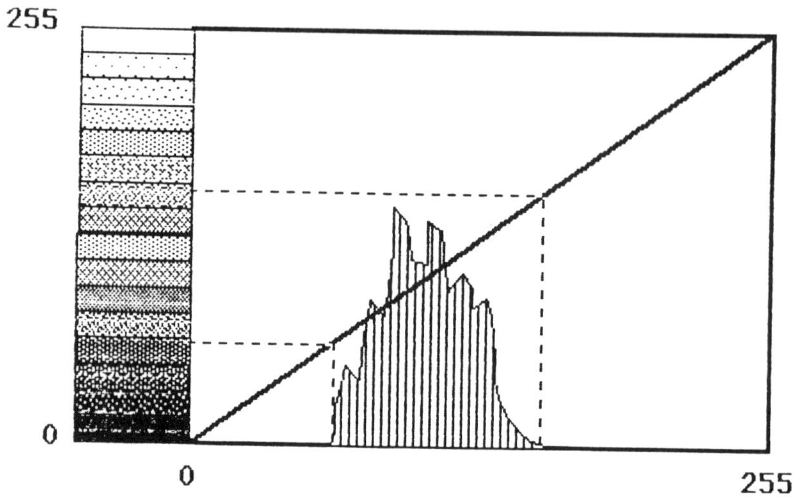

Figure 12. If the identity transfer function is applied to an image with a histogram such as this one, where many low and high levels are empty, a poor picture is obtained, with a restricted number of grey shades.

better result is obtained with a linear stretch such as that of Figure 13, where the whole range of intensities is used.

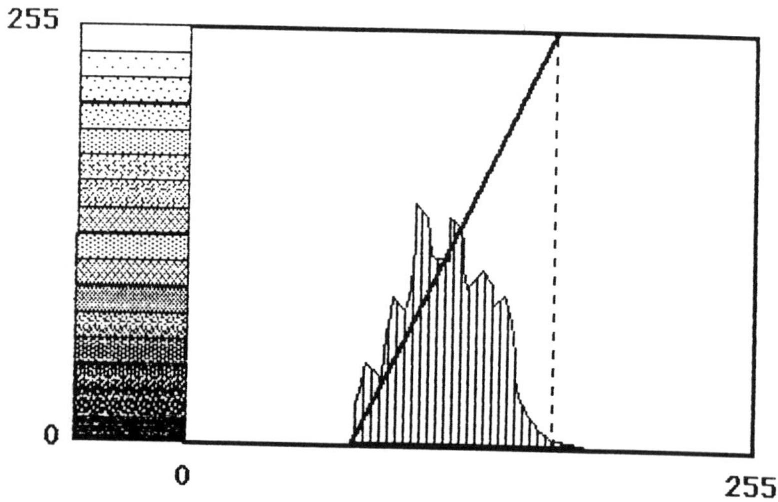

Figure 13. This transfer function is much more appropriate. The whole range of screen intensities is used.

Just as in photography, a transfer function may give an over-exposed or under-exposed picture. It may, for example, give detailed information in the low grey levels (the shadows of a normal photograph); the high levels will then be over-exposed (the sunny parts of the photograph will be "burnt") (see Figure 14).

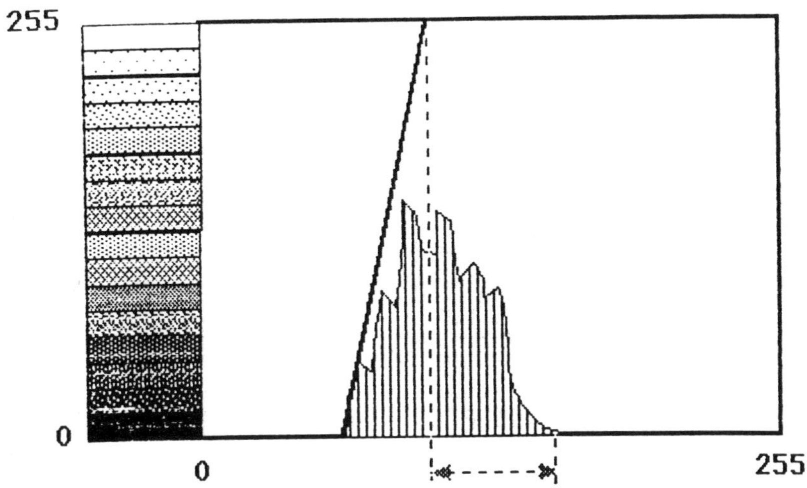

Figure 14. This transfer function will yield an over-exposed picture; high image values (indicated by the arrow) will all be full white.

Also, a "hard" picture may be obtained, corresponding to printing on hard photographic paper, by using a steep transfer function: most input values yield either very dark or very bright intensities; only a few values have intermediate, i.e. grey levels (see Figure 15). Vice-versa, a "soft" picture may be created with a "flat" transfer function: the contrast between dark and bright parts is attenuated, and the overall appearance is a "flat" picture, lacking in "relief" (Figure 16).

Scale-like functions may also be used, whereby monotone areas are created (see Figure 17); but this technique, called level-slicing (in French "seuillage") is better obtained with colours, and we shall see how this is possible in chapter 8.

6.4 How Transfer Functions Are Obtained

The transfer function is implemented in the device by the use of a look-up table. Look-up tables are frequently used in programming to implement a function. In a computer program, a function (logarithm, sine, etc.) is most often introduced as an algorithm including computations on the input value, sometimes with branchings done on intermediary results. If, however, the number of possible input values - and hence of output values - is restricted, it may be advantageous to use, instead

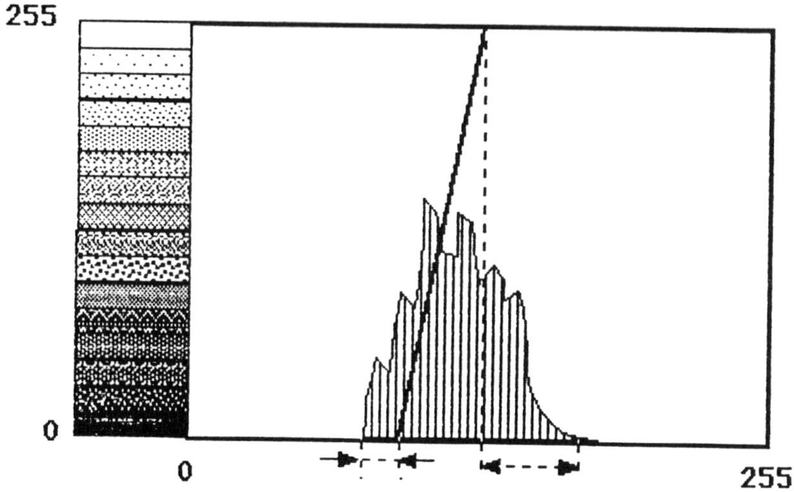

Figure 15. This transfer function will yield a "hard" picture, mostly made of black (low values) and white (high values).

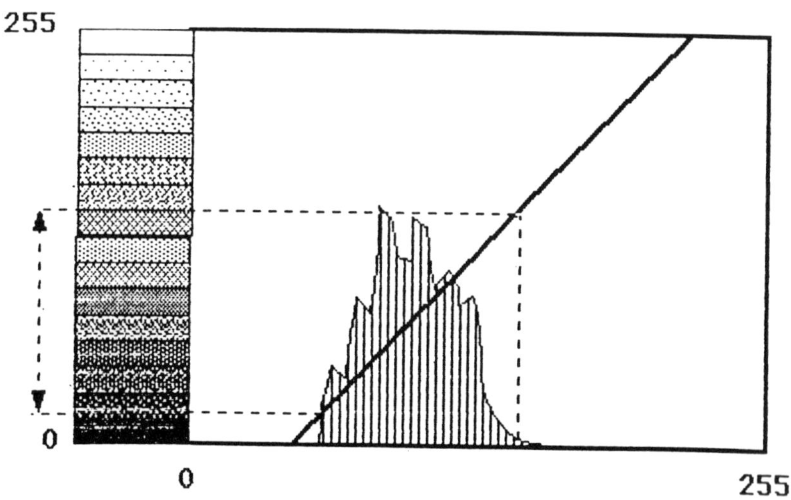

Figure 16. Here a dark, "flat" picture will be obtained, made of dark grey shades only, earmarked by the arrow.

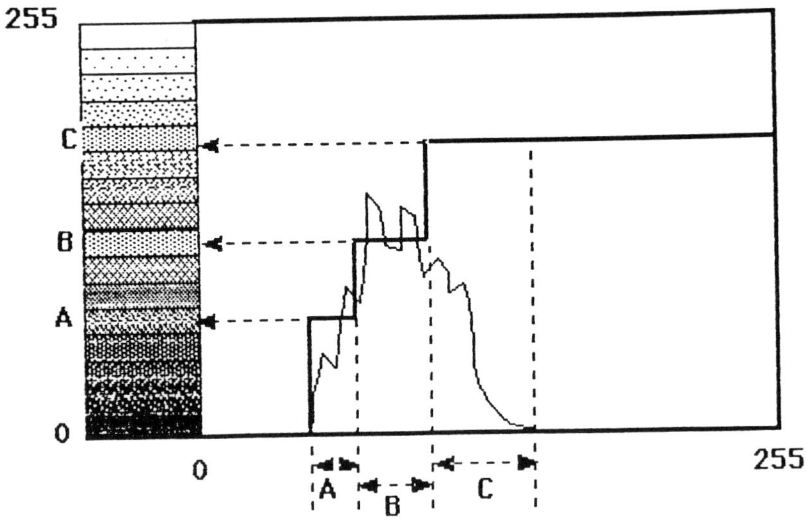

Figure 17. Density slicing: all input values within range A will have the same grey shade (i.e. the same intensity), A, on the screen.

of the algorithm, a look-up table. It consists in a table where the output value corresponding to every possible input value is stored in memory. This technique is well adapted to transfer functions because, as we have seen, the input values may vary from 0 to 255. It is, therefore, sufficient to have a vector (a row) of 256 bytes where each byte contains an output value. Each byte occupies a given position within the vector; in other words, each byte has an address; the input value will then indicate the address at which the corresponding output value is to be found. For example, suppose the beginning of the look-up table is something like:

address:	0	1	2	3	4	5	6	7	8	9	10	11	...
output :	0.20	0.23	0.27	0.35	0.48	0.39	0.55	0.67	0.87	0.88	0.93	110	...

In this example, an output value of 67 will correspond to an input value of 7.

The screen picture is refreshed every 1/25 of a second; during that interval of time, all local memory elements are read one after the other. The sequence of events is the following: the value of an element is read; that value is used as the address in the look-up table where the output intensity is defined; with that intensity, the gun "paints" the screen point corresponding to the memory element; then the next element is read, etc.

Normally, the same transfer function serves for all screen points.

6.5 Non-Linear Transfer Functions

Up to now we have seen transfer functions made of straight lines. Actually, there is no restriction on the transfer function shape, provided that every input value has one and only one output counterpart.

An interesting transfer function is the cumulative histogram function, i.e. the "integral" of the grey levels histogram (see Figure 18). Each function point is obtained by summing all histogram bars situated to its left, and adding this sum to the histogram bar at this point. The application of this transfer function to an image is called histogram equalisation.

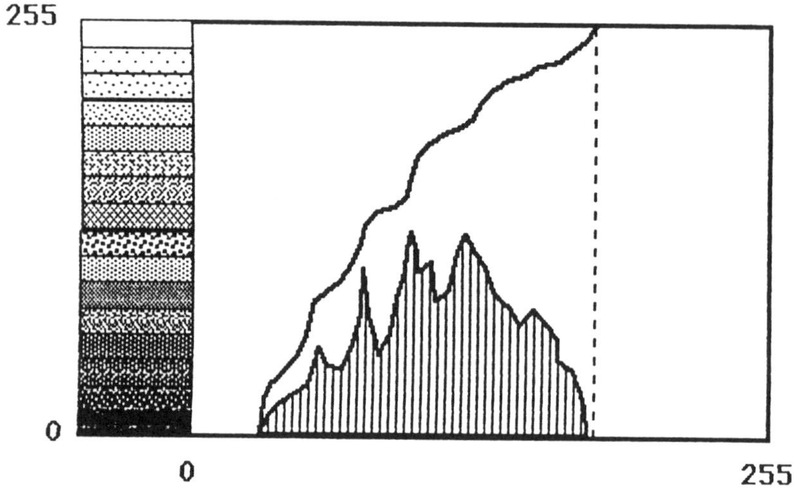

Figure 18. The histogram integral curve, scaled on 255, may be used as a transfer function.

This transfer function is at the right place and has the right spread, so it would seem that it is the ideal curve. It is indeed sometimes built into the system, so that, once a histogram has been computed, it is automatically implemented on the screen. Experience shows, however, that this transfer function is seldom the best one. For some reason, the picture is often "hard", with mostly very dark and very light areas.

The cumulative Gaussian curve, i.e. the integral of the Normal Distribution, computed on the histogram mean and standard deviation, has several advantages (see Figure 19); it has its steepest slope around the mean, in the central part of the histogram, i.e. precisely where detailed information is often needed; it is more regular than the cumulative histogram and, therefore, the resulting picture is more "intuitive", i.e. easier to interpret (an important quality, but not easy to define); at both ends, there are still some details in the very low and very high values.

More complicated transfer functions may be created, where one screen intensity corresponds to two or more input values, although this kind of

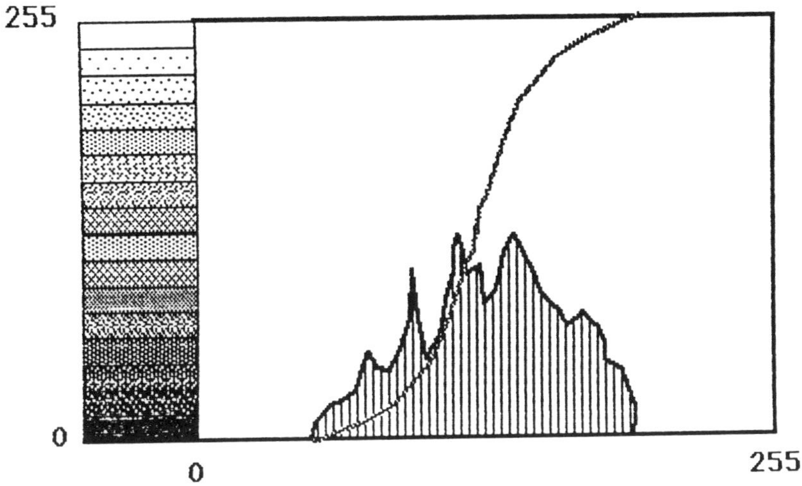

Figure 19. The Gaussian integral, computed on the histogram mean and standard deviation, gives a transfer function which offers several useful characteristics.

transfer function is not so frequently used; we will see an example in the next chapter.

6.6 Degrading Information

By now, the reason why we have made the distinction between "image" and "picture" should be clear. The image resides, at least partially, on the visualizing device memory. Transfer functions do not change it, but "extract" a part of its contents so that we can see it. By this transformation no information is added to the image, of course; on the contrary, some information is lost. But losing information is exactly what we are looking for, we would like to get rid of all the information which is not directly relevant to our problem, and to modify the pertinent information so as to highlight the aspects more closely connected to our study.

6.7 Returning From Picture to Image Values

We have seen that some transfer functions must always be used to create any picture, either on the screen or as a hardcopy (B/W transparency, colour-composite, poster, etc.). This is also true for colour composites, as we will see in chapter 9: one transfer function has been defined for every channel (i.e. for every colour, since there is one colour per channel). The transfer function is hardly ever indicated on the resulting document, however.

Therefore, the original values of a transparency cannot be traced back. In the rare cases where the transfer function is known, a microdensitometer may be used to find the grey levels present in the picture, but return to the original values is imprecise at best, because the number of grey level classes is less on the transparency than it is on the digital image. And this is the reason why transparencies contain but a fraction of the original information. The technician who created the transparency had to choose the "best" transfer function, only there is no such a thing as the "best" picture, and hence there is no "best" transfer function; actually he had to find a compromise between various transfer functions; should he enhance the differences between low levels, or between high levels, or find a middle road? His problem is that he does not know for what purpose the transparency will be most used. This explains then why some distributing agencies would like to get rid of film and colour composite production, so as to concentrate on CCTs, which are much more amenable to standardization, and which represent the future, as more and more laboratories will be equipped with digital image processing.

7. TWO EXAMPLES

7.1 Two Scenes

The concepts explained in the preceding chapter are not difficult, but they are of such basic importance that it is worthwhile reviewing them through the study of two pedagogic examples. We will consider one channel of two little imaginary scenes, acquired by a multi-spectral scanner. That channel could be, for instance, the SPOT infra-red channel, but it could also be another channel from another satellite. In all figures of the present chapter, the left half contains the histogram of the scene for that channel, a scale of screen intensities which is here, for clarity, reduced to eight levels, and a transfer function. To the right, the scene is presented as it would appear on a screen with that transfer function. The histogram, of course, is constant for a given scene, since we always consider the same channel.

The first scene is made of cultivated fields around a pond. Each field has a restricted range of values, indicated under the histogram of Figure 20. The scene is unrealistic from several points of view, the most important over-simplification being the lack of overlap between the field radiometric responses. With real fields, even if the plantations were different (wheat, barley, potatoes), the radiometries would not be disjoint.

In the second scene, we find the same pond, but surrounded by a forest; there are shadows in the forest canopy and, therefore, the radiometry varies from pixel to pixel. Here these variations are represented, somewhat artificially, as a few patches with different radiometries.

These two scenes are mere illustrative tools of the role of transfer functions.

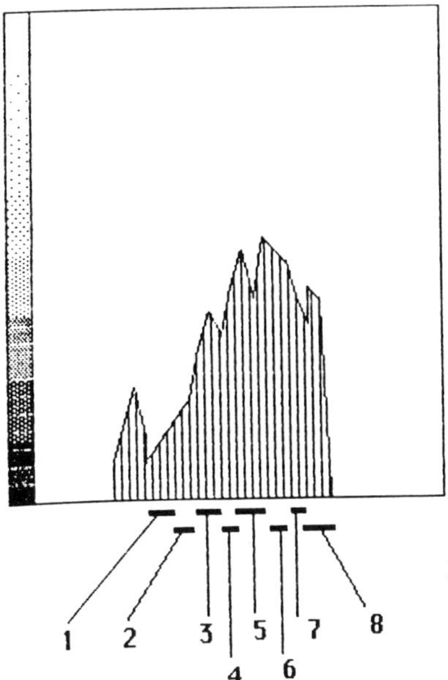

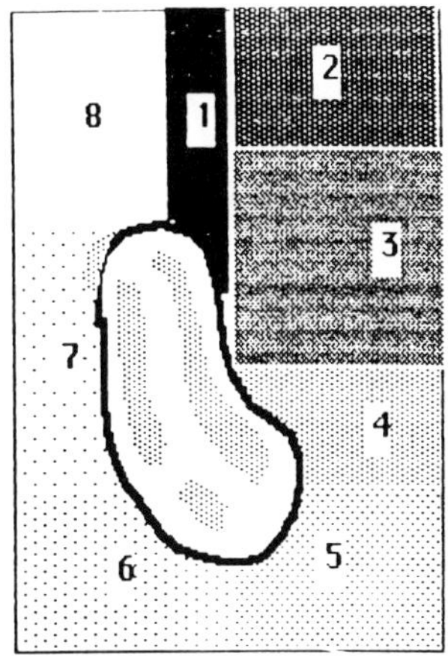

Figure 20. An over-simplified imaginary example; eight fields lie around a pond; each field has its own radiometric response in the channel for which the histogram is presented at the left. To simplify reading, only eight screen intensities are represented. The over-simplification lies in the fact that there is no overlap between the field responses.

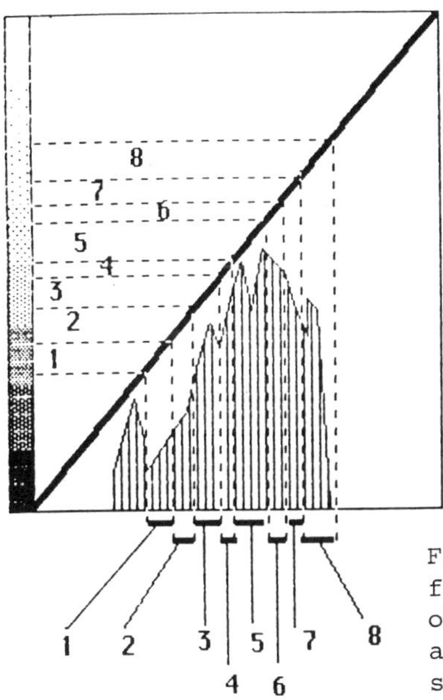

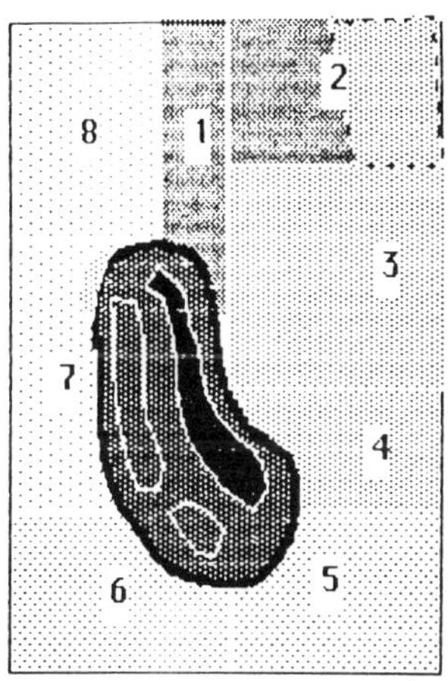

Figure 21. When the identity transfer function is applied, a "flat" picture is obtained, with only part of the available intensity range being used. Field 2 straddles two intensities.

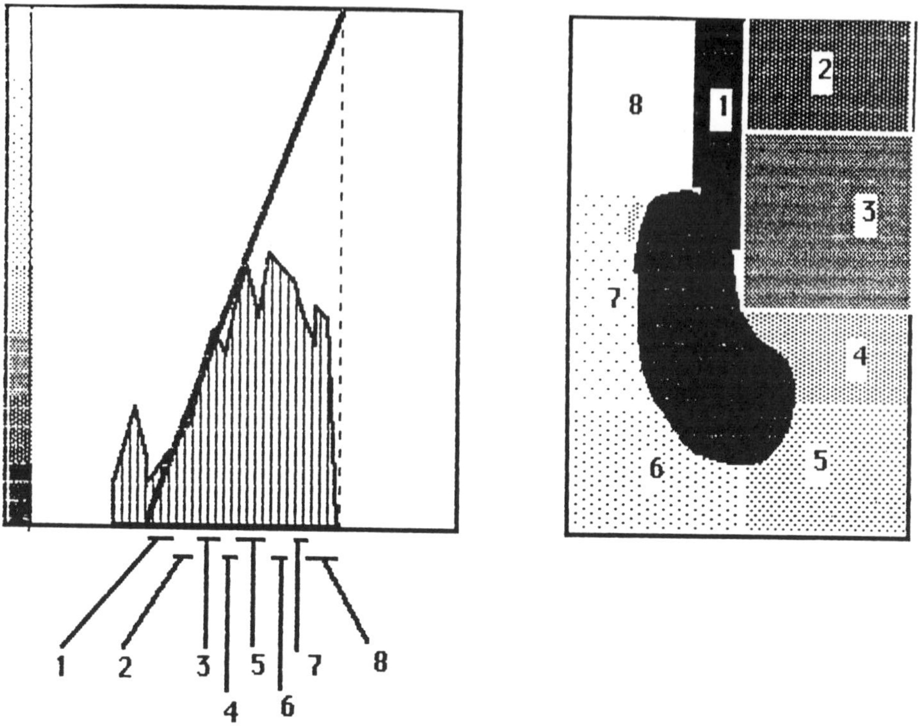

Figure 22. A linear stretch. Water and field 1 are at the lowest intensity (black). The other fields are distinct.

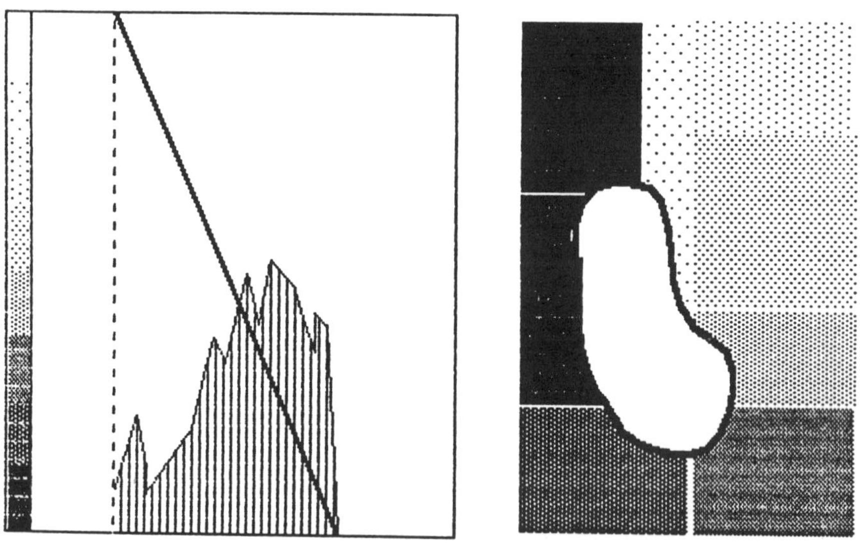

Figure 23. With this "negative" linear stretch, low responses (water, field 1) yield high screen intensities, and vice-versa.

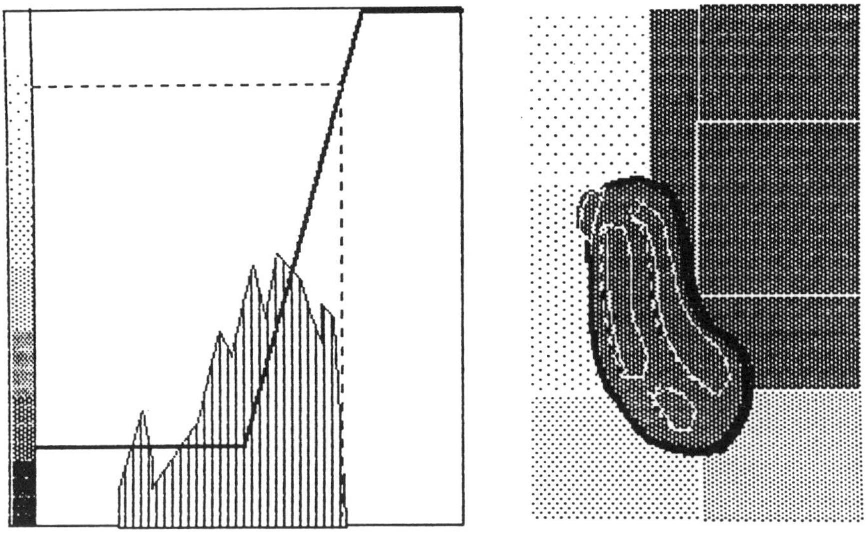

Figure 24. A really poor transfer functions; water and fields 1 to 4 are confused, and the other fields do not use the reamining intensity range completely.

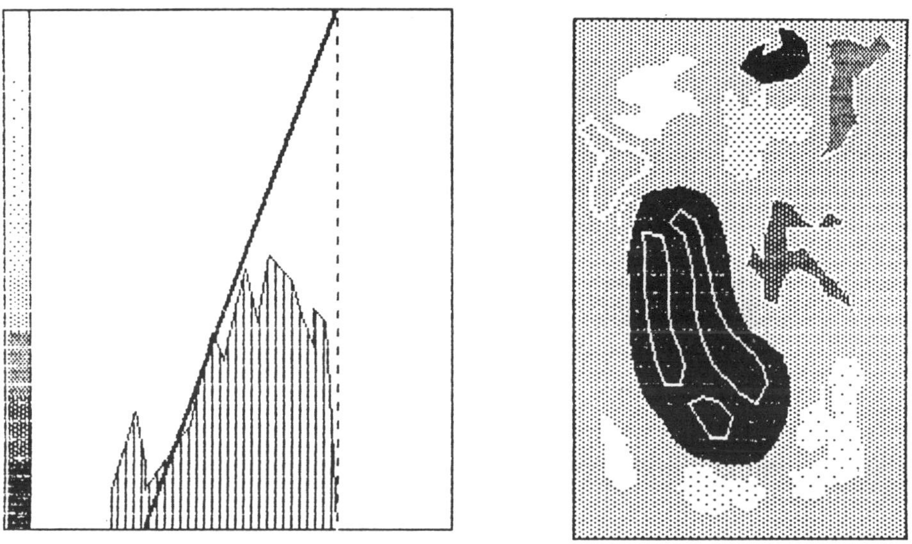

Figure 25. Another small imagenary scene; here the pond is surrounded by forest, with a canopy including varying degrees of shadow, and therefore with numerous and irregular spots, unrealistically represented as a few patches. A linear stretch underlines the radiometric differences between these patches.

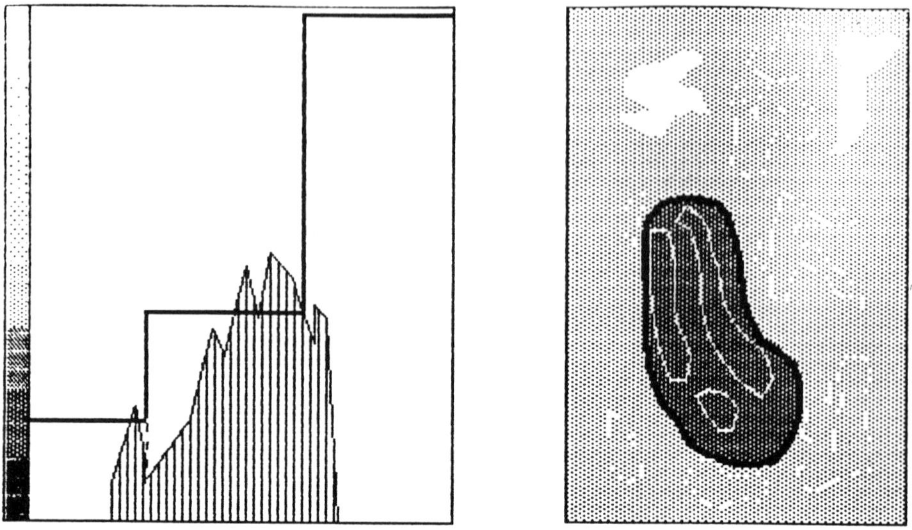

Figure 26. A level slicing (also called density slicing). One intensity is dedicated to water, and two (mid-grey and white) to forest.

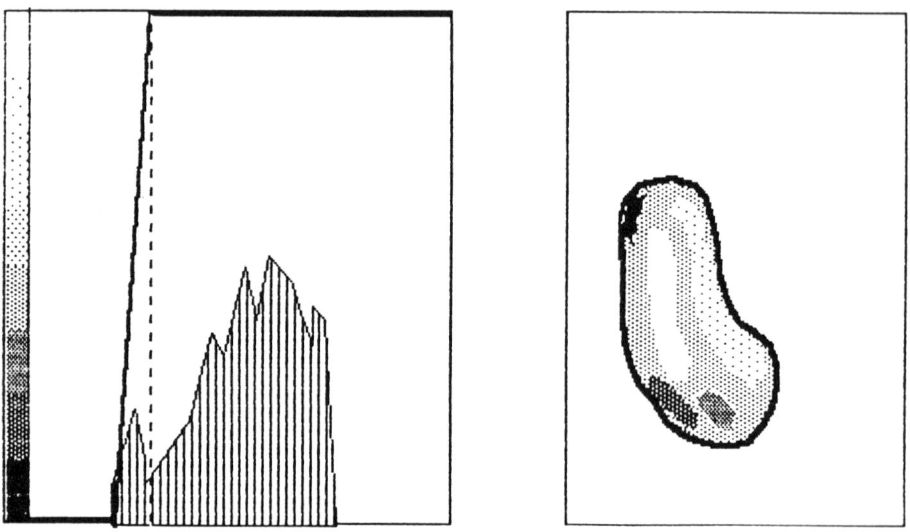

Figure 27. The whole intensity range is used for water. The forest is completely overexposed. This presentation reveals many details in the pond which remained hidden with the other transfer functions.

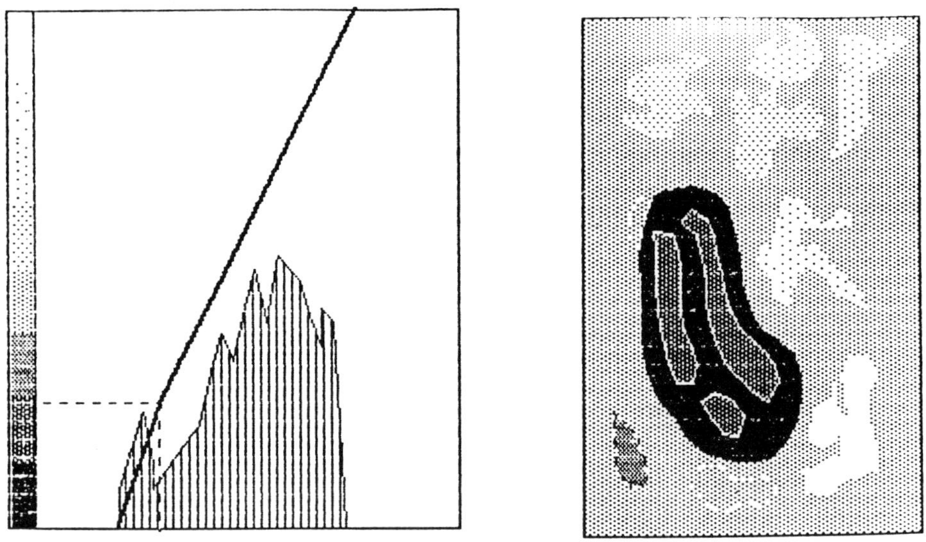

Figure 28. A compromise; one fourth of the intensity range serves for water, and the rest is used for the forest. Compare with preceding figures.

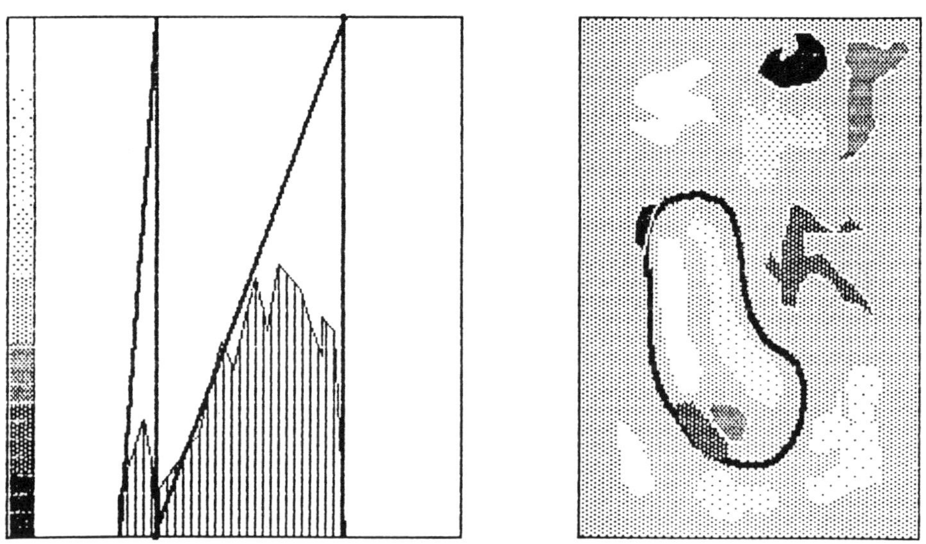

Figure 29. Another type of compromise; the whole intensity range serves both for water and forest. The picture is rich in details, but the same intensity (full white for example) may just as well come from the pond or from the canopy.

8. COLOURING THE SCREEN

8.1 It Is Much Nicer in Colour!

Until now, we have supposed that each grey value is transformed through a look-up table into one output intensity value, and that one-channel images thus give necessarily B/W pictures. Actually, the system is more complicated, but also more powerful.

Every screen point receives not one (grey) light intensity, but three, one in each of the three fundamental colours, red, green, blue. The input value read on the local memory then goes not through one but through three look-up tables and from there to a red, green or blue gun which refreshes the screen (see Figure 30). If these three fundamental

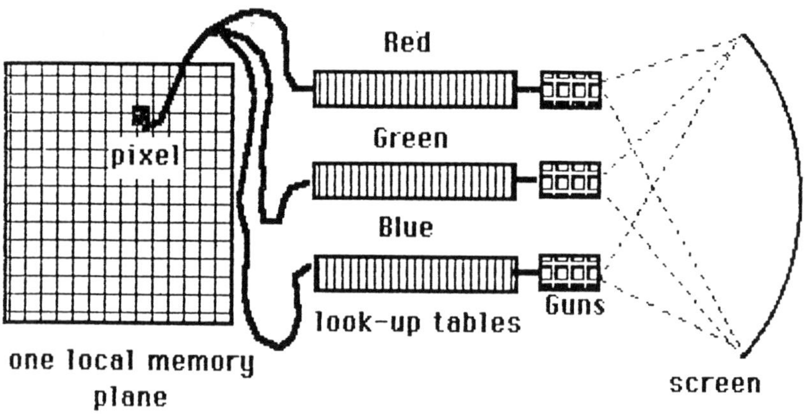

Figure 30. In the display device, all elements ("pixels") of the local memory plane are read sequentially. Each value is sent, not to one (cf. Figure 8) but to three look-up tables; the output of these look-up tables determine the intensities of the three funs which "paint" the screen.

colours are mixed with equal intensity (i.e. the three transfer functions are equal), a white or grey colour is obtained, and this is what we had implicitly supposed up till now.

(Let us note that actual display devices are still more complicated than the scheme indicated at Figure 30; there may be an input look-up table for each memory plane, and output look-up tables situated just before the guns. The device may also include computation capabilities between the memory planes, etc.)

But now you might say something like: "Well, this looks quite promising! If I want to scrutinize my mono-channel image where, as I have seen on the histogram, there are some 60 interesting grey levels, now, instead of having to be content with grey shades to try to differentiate all these levels on the screen, I may play with an enormous number of colours; indeed I may have 256 shades of red, and for each of these, 256 shades of green so that in all I may create, interactively, 256^3, i.e. over 16 millions colour shades, almost an infinity, and certainly more than enough for my 60 levels, right? Anyway, certainly the eye can distinguish more tints than grey shades, so let us put colours on the screen!". Indeed, the commercial prospectus going with the visualizing console does contain such affirmations, claiming that the apparatus has the potentiality to create an enormous number of colours.

But, alas, reality is different! It is true that I may create some 16 millions hues, digitally different. But what I want is to create shades which are distinguishable by the eye, on the screen, not by the computer. And this is a completely different matter. Suppose that

- colour A is made of red = 152; green = 202; blue = 253,
- colour B is made of red = 140; green = 218; blue = 240,

they will look exactly the same on the screen. No difference is visible.

How many visually different colours can be created? Really distinct colours, where none could be confused with any of the others? The answer will vary because it depends, among other things, on the areas we want to colour; the smaller the areas, the fewer colours are distinguishable. But, on average, it is difficult to create more than 15 distinct colours and that is a disappointment!

There is another point. The eye, when seeing a scale of grey shades, conceives immediately an order, from black to white. But a coloured scale is necessarily arbitrary, even if we go from cold colours (blue, green) to warm colours (orange, red).

Nevertheless, the colour level-slicing technique is used quite often to present final results and we shall see how this can be obtained, once we have recalled a few basic principles about colours.

8.2 The Fundamental Colours

Any colour may be obtained from a mixture of the three fundamental colours. In particular, mixing them two by two, with equal intensity, gives the following results:

red + green = yellow
red + blue = magenta (a kind of reddish violet)
green + blue = cyan (a light bluish tint)

This relationship may be illustrated with a star (see Figure 31). Also:

red + green + blue = white,

again if these colours are mixed in equal quantity.

There is yet another useful way to present these same relations:

yellow = white - blue
magenta = white - green
cyan = white - red;

it is, therefore, said that yellow and blue are complementary.

The trio yellow, magenta, cyan may also serve as fundamental colours from which any colour may be obtained. In particular, most if not all commercial slides are made of three layers, one in yellow, one in magenta and one in cyan; also colour printing is most often obtained with matrices of these three colours, with black as a fourth colour.

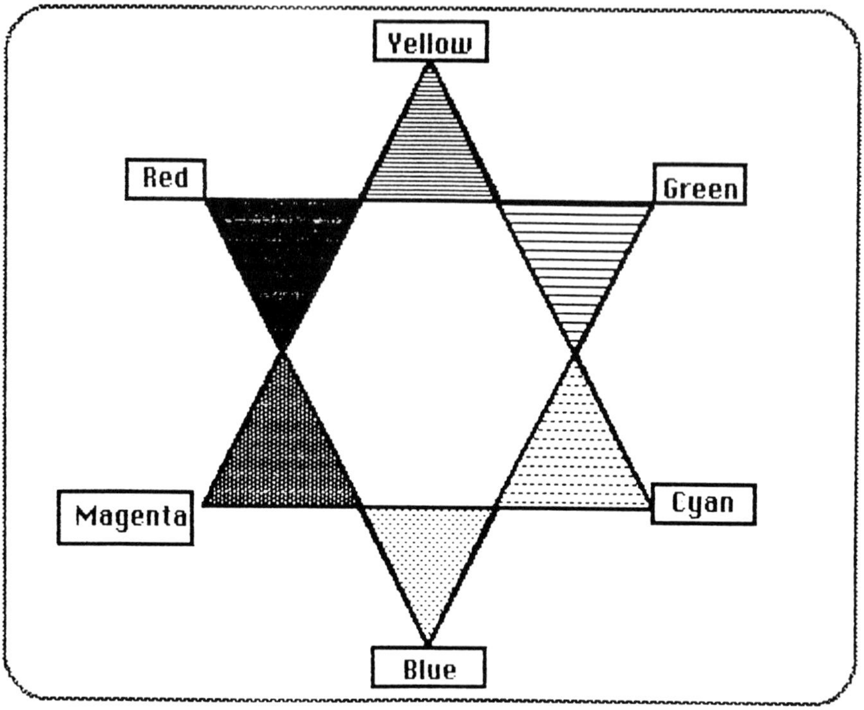

Figure 31. A mnemonic graph to illustrate the relationships between fundamental colours. Red and blue mixed in equal quantity give magenta. Also, blue = white - yellow.

Other colours may be derived such as:

much red and some green yield orange
much green and some red yields a yellowish green
much blue and some red yield violet, etc.

more generally, any tint may be obtained by mixing red, green and blue (or cyan, magenta, yellow) in appropriate proportions (see Figure 32).

8.3 More Colours Than the Rainbow

The technique used to create different colours through the use of transfer functions is readily explained. The software that the user has at his disposal to modify the three transfer functions linked to the three colour guns will vary considerably from one installation to the next, but the principle remains the same.

It may be obtained by specifying the three values which will create the colour, such as:

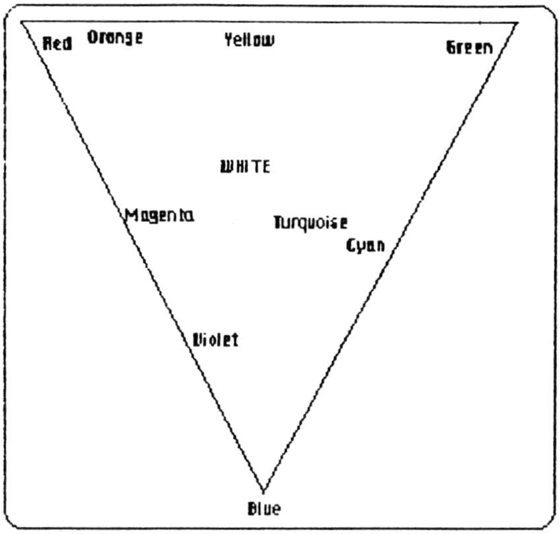

Figure 32. Any colour or hue may be obtained by mixing red, blue and green in appropriate proportions.

Red	Green	Blue	Resulting colour
180	0	180	magenta
120	0	180	violet
180	0	230	violet
120	120	180	light violet
200	200	0	yellow
220	170	0	orange
130	70	60	brown
100	120	140	greyish blue

These examples serve as illustrations only; one must actually play with the colour transfer functions to zero in on the exact tint wanted. Often, the transfer functions are accessible through the track-ball or mouse, and the process is interactive.

It is thus possible, by adjusting properly the transfer functions, to associate a given colour to a certain input value, or range of input values (see Figure 33).

8.4 Level Slicing

From this colouring method to the picture presentation technique called "level slicing" or "density colour slicing" (in French "seuillage"), the step is easy. The technique is often used to present final results on a mono-channel image. An example is the temperature distribution map obtained through the study and proper modifications of an infra-

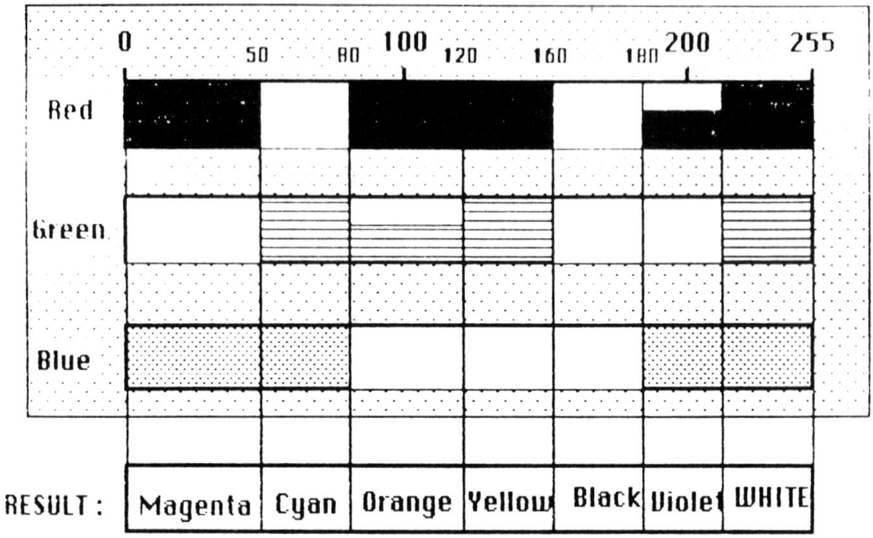

Figure 33. An example of a colour scale. From value 120 to value 160, full red and full green, with no blue, give full yellow. From 180 to 220, violet is obtained by mixing some red with full blue.

red channel, such as channel 6 in the Thematic Mapper. Suppose that we have obtained such a map; a black and white presentation such as those explained in chapters 6 and 7 will show the "relief" of the temperature distribution, its "hills" and "valleys", but the eye will not easily recognize the zones of equal temperatures. This is obtained by a presentation in colours; for that, we look at the picture and at its histogram (see Figure 34), and we cut the 0-255 level scale in the series of intervals which will best illustrate the spread of temperatures; the intervals may be regular, or contain different numbers of levels, the choice is ours. We associate a distinctive colour to every interval, and we try to find a set of colours which are both quite distinct and convey the idea "from cold to warm" (this may be far from easy and may require quite a number of trials).

We then implement these colours by modifying the colour transfer functions, interval after interval. To check our results and also to help the reader of the map, we must add a temperature scale to the picture. For this, we create a set of small rectangles, each having the medium value of an interval. Or, alternatively, we create a "grey wedge", an elongated rectangle where the pixels have values going linearly from 0, at one end, to 255, at the other end (see Figure 34). Then our temperature scale will automatically appear on this wedge.

A coloured picture is thus obtained from a mono-channel image. It highlights effectively the temperature or radiometric variations within one image; this is obtained at the expense of topographical details, however.

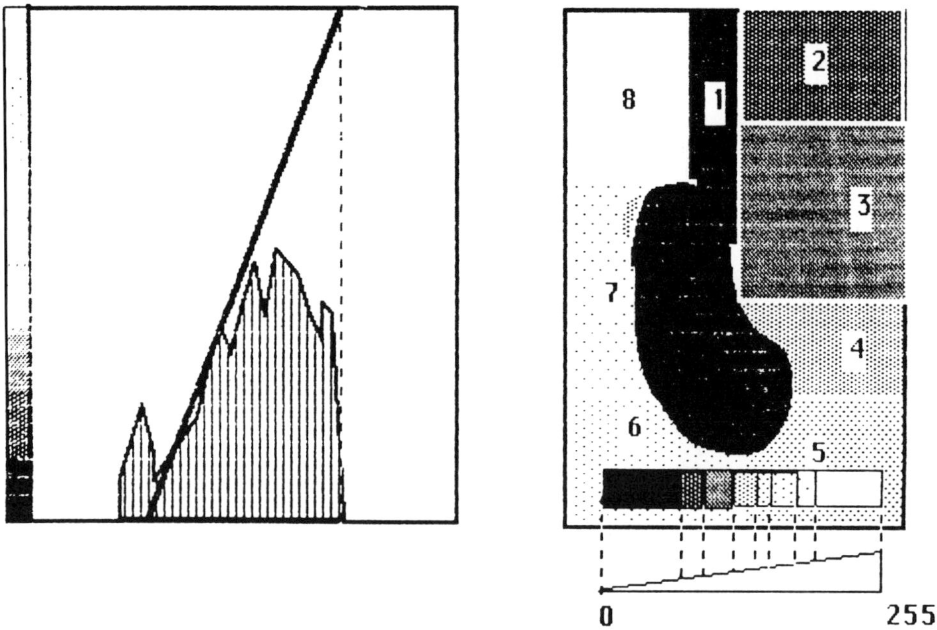

Figure 34. A "grey wedge" is added at the bottom of the picture. It consists in a small elongated rectangle with values going linearly from 0 to 255. It will reflect the transfer function effect and, together with a legend, it may be used as a scale of intensities, or colours.

Level slicing is used quite often to present final results. Publications in remote sensing contain numerous examples. They also illustrate the difficulty of finding a colour scale where the colours are really distinct, not only on the scale (where it is generally the case), but also in the small areas of the picture (and this is rarely achieved). The publication printer has always had a problem. The number of distinguishable tints is conspicuously narrower on paper than it is on a screen or a transparency. Even if he works as carefully as possible, there is some further quality loss when he goes from the photographic document to the printing matrix. Even now, good quality colour printing remains a delicate and expensive art.

8.5 An Example

Level slicing is a handy technique which is used quite often not only to present final results but also to analyse an image. Consider, for example, our little scene of Figure 25, together with its histogram. We said in paragraph 7.1 that the low values correspond to water, and the high values to forest. Suppose now that we would like to check this and also to study up to which point we can truly separate these two objects on the basis of their radiometric response in the visualized channel. We could do that with a B/W transfer function, but using colour will greatly expedite the research and will yield more reliable results.

There are a number of ways to approach this little problem. Let us look at one possibility. From the histogram (Figure 35), we see which level corresponds to the minimum; we tentatively take that value as the highest "water" value and we use a set of transfer functions such as that indicated at Figure 35; we then see a picture where:

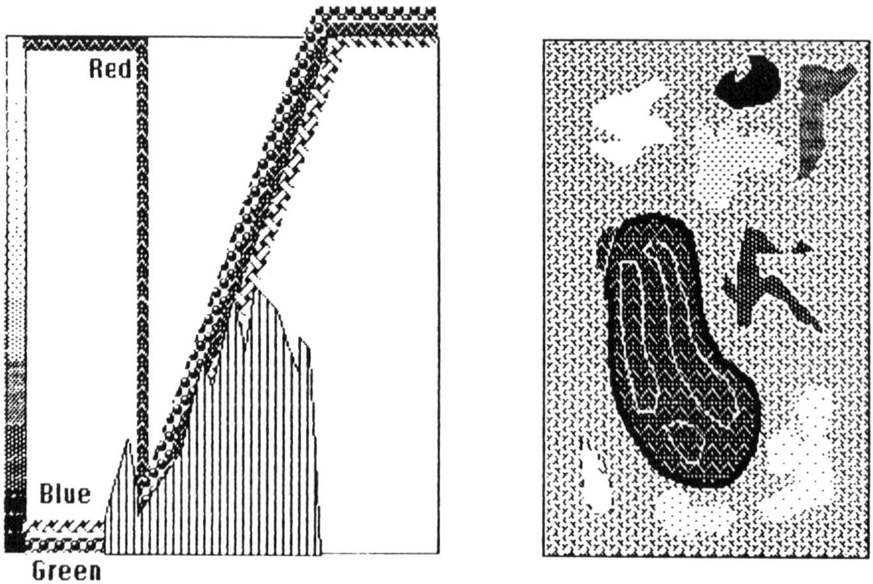

Figure 35. With this set of transfer functions, water will be in red, while the forest will be in B/W just as in Figure 25.

- high levels ("the forest") appear in grey shades, just as in Figure 25;
- everything under the "water" maximum appears in bright red.

We can now closely scrutinize the water-forest boundary and see whether the level we have chosen is the best separation we can find.

Let us suppose that we also see a few isolated red pixels in the middle of the forest. What the picture will not tell us is the reason for their presence. Are they due to very small ponds? Or are they perhaps the result of a shadow effect in the canopy? Of course, a first thing to do would be to study the other channels in the same way; the ambiguity might be solved by a confrontation of the resulting pictures, but it is not certain. Quite often, answering such questions which are quite frequent in image analysis ("I see here an anomaly, what is it?") would necessarily mean knowing the scene quite well, or having at hand an aerial photograph, or ground observations.

So it is quite feasible and easy to obtain mixed pictures, in part B/W and in part coloured, and it is often useful to do so. There is one circumstance, however, when this should not be done, one occasion where a fully coloured picture, possibly with bright colours, is an absolute must: when a Very Important Person comes to visit the image processing laboratory!

9. THE THREE-CHANNEL IMAGE

9.1 Three Channels Only

From a mono-channel picture, one goes easily to a multi-channel picture. There is an unexpected limit, however; indeed, no more than three channels may be visualized together. This is because there are but three fundamental colours. We may then create a picture with channel 1 in red, channel 2 in green and channel 4 in blue, say, but channel 3 must be excluded. Hence, with a Thematic Mapper image, which contains six channels (without considering the thermal infra-red channel which is better studied by itself), one may compose (6x5x4)/(2x3) = 20 different presentations. Actually, much more than that can be created because, first, any channel may be associated to any of the three colours, and second, for every channel the transfer function may be modified at will; and this is not the end of the story because "derived" channels can also be created, as we will see at chapters 11 and 12.

9.2 Colour Composites

In a three-channel representation, the three planes of the display device local memory are used; each plane contains one channel of the selected subscene and each plane is connected, through a transfer function, to one of the colour guns (see Figure 36). The user is free to change the colour of each channel interactively and to modify the transfer functions. It is generally advisable to look first at each of the three channels separately, in black and white, and to choose the transfer function giving the "best" aspect for each channel. That "best" aspect will depend upon what kind of object we would like to highlight; it will also partly be a matter of personal taste. Normally, a simple transfer function, giving a rich range of grey shades in the objects of study, should be tried first.

When the three channels are visualized together, it may happen that the picture has a dominant colour; there is a general greenish tint, say, because the colours are unbalanced, and the transfer function for the channel we have decided to present in green yields a general intensity which is stronger than the other two channels, or, on the contrary, the green may be too low and the picture will have a dominant magenta. This lack of equilibrium between colours is easily corrected by modifying the transfer functions.

An interesting colour combination is the one which comes close to the colours as we see them with our naked eye:

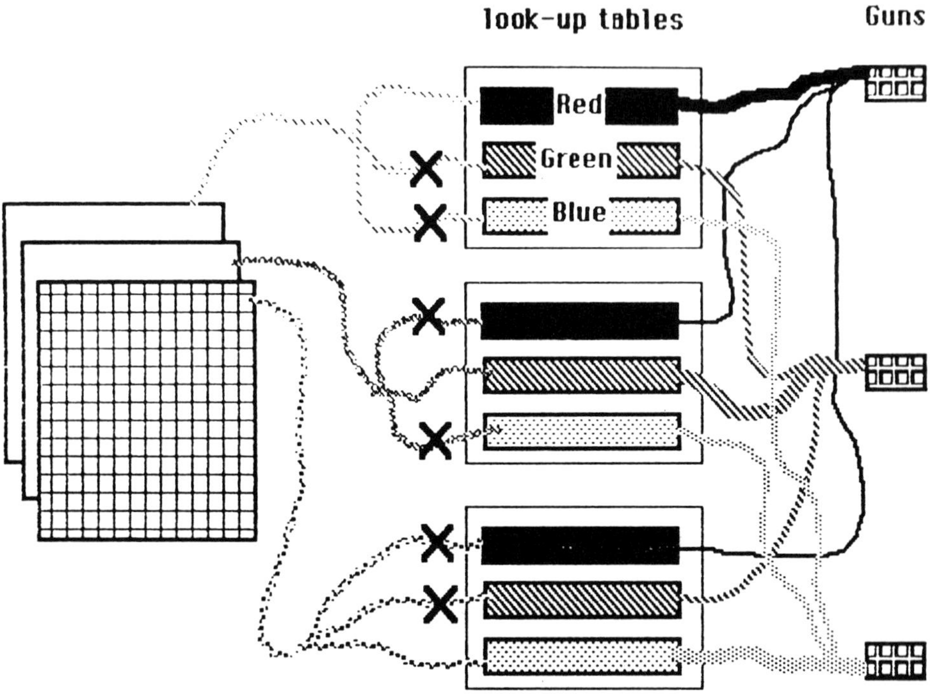

Figure 36. Display device with the connections for a colour composite (simplified view). Each of the three local memory planes is connected to a gun through a look-up table.

the blue channel in blue
the green channel in green
the red channel in red;

pictures obtained through this combination are called True Colour Composites. At present, only one sensor has the blue channel necessary for this, the Landsat Thematic Mapper. In all other sensors there is no blue channel because blue is strongly absorbed by the atmosphere. All other combinations are called False Colour Composites (although there is nothing really "false" in them).

An often used False Colour Composite, created by analogy with infra-red aerial photographs, is:

the green channel in blue
the red channel in green
a near-infra-red channel in red;

vegetation then appears in bright red because it absorbs incoming red sun light and reflects the incoming near-infra-red intensively; surprisingly, the fact that it also reflects green light (hence its colour to our eyes) is of secondary importance only; we shall return to this vegetation behaviour in chapter 11.

9.3 Channel Correlation

Most channels are positively correlated two by two; a measure of this correlation is the correlation coefficient used in statistics. As is well known, these coefficients vary between -1 and +1; a value of zero indicates a lack of correlation, i.e. the channels are independent or, in other words, they contain different types of information, whereas a value of +1 would indicate a complete redundancy: the information would be identical and one of the two channels would be superfluous; This extreme case never happens in practice, but high positive correlations are common.

As an example, the correlation coefficients of LANDSAT MSS channels over French scenes are:

Channel	1	2	3	4
1	1.0			
2	.95	1.0		
3	.43	.35	1.0	
4	.12	.02	.91	1.0

The three SPOT MSS channels have a correlation of the same order, except for channels 1 and 2 which are extremely positively correlated; some searchers even consider that SPOT data are actually two-dimensional and that it is, therefore, sufficient to consider either channel 1 or 2, but the difference between these channels, although limited, may be significant.

This correlation is in the nature of the objects studied and hence unavoidable. Of course, we would prefer to have uncorrelated channels, because each would then express a different aspect of the scene, but this is not the case with the original channels. We will see, however, that uncorrelated channels may be derived from them, see chapter 12 "Principal Components".

Although statistical correlation coefficients are commonly used to express the links between channels, some caution should be used in deriving conclusions from them. Indeed, the basic statistical assumptions which should be satisfied to compare them and study their significance are far from being true in the present case: the two marginal distributions and the joint distribution should all be Gaussian, and the measurements should be randomly sampled from the whole population. Not only are we very far from that, but the very notion of pixel population is actually much more elusive than is generally admitted. The whole subject certainly requires more theoretical study (there is nothing so practical as a good theory!); while we are waiting for that, correlation coefficients and the whole gamut of classical statistics should be used as handy and practical tools, but statistical tests should be used cautiously, if at all.

9.4 Two-Dimensional Histograms

The concept of mono-channel histograms may be extended to two dimensions; in two-dimensional histograms the joint distribution of pixel values in two channels is illustrated (see Figure 37. Unfortunately, it is not possible to visualize three-dimensional histograms, at least not easily).

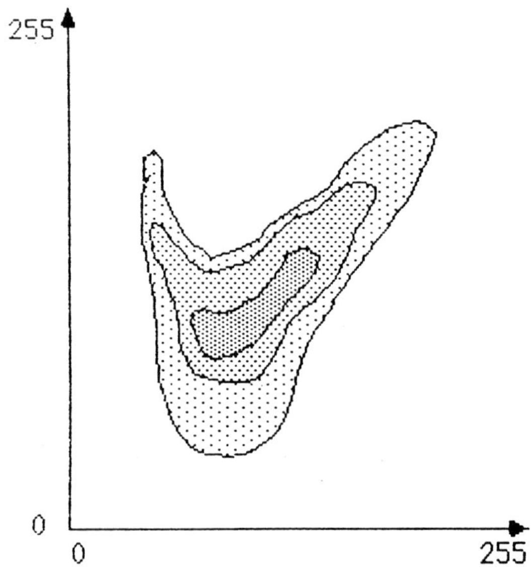

Figure 37. A two-dimensional histogram. One channel is in abscissae and another in ordinates. Each image pixel is represented as a point. The set of these points form a cloud from which many relationships between the two channels may be studied.

These histograms show clouds of points (pixels) of varying density and several conclusions may be derived from their study:

- what is the shape of the overall joint distribution.
- are there more or less separate clouds which would not be so readily apparent on mono-channel histograms (see Figure 38)?
- how far are the two channels correlated?

Careful study of these histograms and of their relationships with the various objects yields a rich harvest of information.

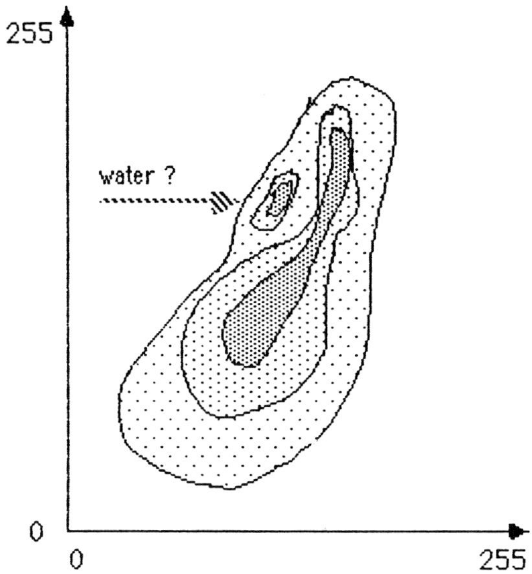

Figure 38. In this two-dimensional histogram, a small cloud appears which would not be visible on the histograms of each of the two channels. It most probably indicates that an object (perhaps water?) has a radiometric response which is different from the rest, and may therefore be isolated.

10. THE CHANNEL SPACE

10.1 A Multi-Dimensional Space

Two-dimensional histograms have one channel as abscissae and the other as ordinates. From there, conceptually, it is possible to imagine diagrams with as many axes as there are channels. These axes span what is called the channel space, also called the feature space. This notion of channel space is extremely useful and serves in a number of developments which we will see only in part.

It should first be noted that, useful as it is, it would be an error to attach too much physical meaning to this space; it is closely linked to the characteristics of the sensors and it is by convention only and for the sake of simplicity that the channels are represented as orthogonal axes. But reasoning in this space clarifies many relationships between the radiometric characteristics of objects under study, as we shall now see.

10.2 The Pixel Cloud

In the channel space, every image pixel is represented as a point the coordinates of which are the values of that pixel in all considered channels. The whole image thus becomes a cloud of points which is conceptually easy to imagine in a three-dimensional space. The good thing is that reasoning in this three-dimensional space is useful and valid when we have to deal with "hyper-spaces", that is with more than three channels, where we cannot visualize (nobody can visualize a four-dimensional space), but where computations are just an extension of the three-dimensional case.

Each image object contributes to the cloud, having points more or less densily concentrated in some region of the cloud. How happy we would be if there were several clouds, one for each object, neatly separated from the others and gallantly floating in the channel space, waiting for us to discover and define their contours! It would then be possible, and relatively easy, to obtain a map with all pixels belonging to a given object - and those pixels only - receiving a distinct colour. What a help that would be for all thematical research! Alas, this is not the case with present spaceborne or airborne sensors. We must, in most cases, conceive one big cloud, with points of a given object concentrated mostly in one region, but overlapping the regions of the other objects. And for each object it is better to speak of a sub-cloud.

The points of a radiometrical category, meaning by that the set of all pixels where the spectral response is essentially the same (except for random variations) will often be included in an elongated ellipsoid with one axis markedly longer than any of the others; it will be a kind of cigar (yes, imagine, if you can, a four- or six-dimensional cigar); when studying an image it is often fruitful to study how this cigar is placed, relative to the other cigars. This can only be done progressively, by careful comparison of all the two-dimensional histograms. Some cigars may be somewhat flat (somebody trod on them); in other words, the second axis is conspicuously larger than the third. However, there the comparison stops because unlike cigars, our ellipsoids intersect and partly occupy the same region in space. But they do not, fortunately, entirely coincide, or remote sensing would probably be useless.

It is not even always true that one object will be represented in only one region of the cloud. It may happen that what we consider as an object, wheat fields, say, is actually composed of two or more radiometrically distinct objects; some wheat fields may be still green, while in others the wheat is already wilting and yellow. Or some ponds may be made of clear water and others may contain weeds, or turbid water. Let us note also that what we will consider as one class of object in one study, will have to be divided in several classes in another study. There is always an arbitrary component in the categories we create for our researches. With remote sensing data we only have radiometrical categories and they generally do not exactly coincide with the categories we are looking for.

10.3 Mixed and Spurious Pixels

The major hindrance to the distinction between categories is the existence of "impure" pixels. We have seen that pixels belonging to more

than one radiometric category may be divided into three groups:
- border pixels, straddling the contact between a field and a forest, say;
- spurious pixels, belonging to one category but having a radiometric response which cannot - with the data in hand - be distinguished from the pixels of another category;
- mixed pixels where two objects contribute to the response, such as a wheat field in spring time, where the young plants do not yet cover the ground entirely, and the resulting pixel has intermediate characteristics between bare soil and vegetation.

The first kind will always be present because they reflect a reality. But their proportion may be diminished by increasing the sensor spatial resolution. This has been done for the LANDSAT family by going from the MSS pixel with its 60 to 80 metres resolution to the Thematic Mapper pixel with a 30 metres resolution; SPOT has gone farther, with its 20 metres resolution in multi-spectral mode, and 10 metres in panchromatic mode.

The second kind could, and probably will, progressively disappear, or at least diminish markedly, by improving the radiometric resolution, notably by increasing the number of channels; again this has been done in the LANDSAT family by going from four (MSS) to seven channels (TM).

The third kind is in the natural order of things, and we should accept it; the only reason why it seems embarrassing is because, in every human activity, we are accustomed to think in distinct categories and our maps (just as our laws and everything we produce) deal with separate objects; in mathematical terms, we work and we think in a discrete manner; but Nature is not always discrete and often offers us a continuum. We should be glad to have a tool allowing us to describe continua, but we are not, because we just love discreteness, and abhor continuity.

10.4 The Main Diagonal

A line of particular interest in the channel space is the main diagonal; this line starts at the origin; each of its points is equi-distant from all the axes (see Figure 39). This is easy to visualise in three dimensions, as the main diagonal of a cube with three edges forming the three axes spanning the space. To understand its role, imagine two pixels of the same object. An example could be a wheat field, where all plants are at the same phenological stage, say in full bloom. But the wheat field is large and on a hill slope; the surface of the first pixel is perpendicular to the sun and intensely illuminated. The surface of the second receives sunlight more obliquely and its reflected light has a weaker intensity. How will the two points representative of these two pixels be placed relative to one another in the channel space? For both of them the contribution of each channel remains the same and, therefore, they will lie on a line passing through the origin (see Figure 40).

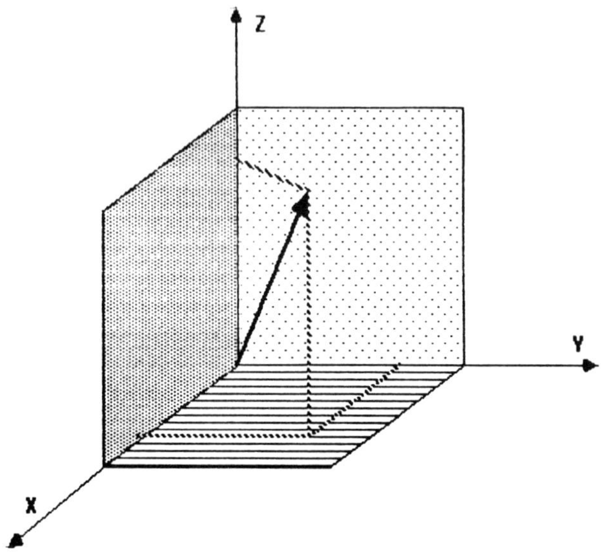

Figure 39. The main diagonal. Its points are equi-distant from the three axes. It may not be drawn for more than three dimensions, but it may easily be conceived.

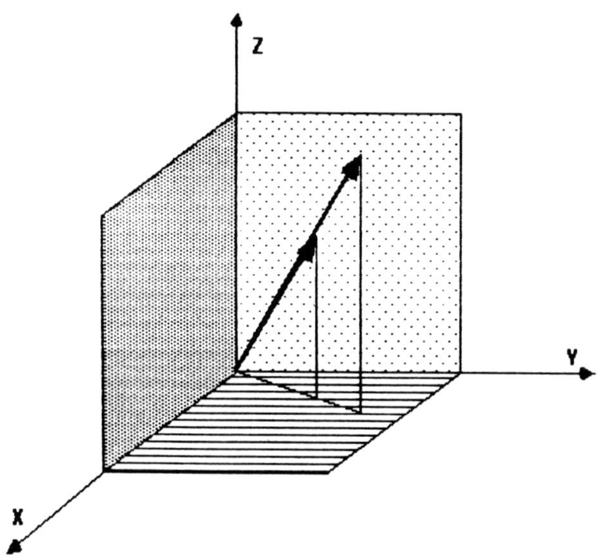

Figure 40. Two pixels of the same object, but reflecting light with different intensities. The representing points will be on a line going through the origin.

The distance of the point to the origin is thus representative of the total intensity reflected by the pixel. It is not an exact measure of that total intensity, because we have measurements in a few wavelengths only, but it is the best approximation we have.

In two dimensions (two channels), points of equal intensity will be on a circle centered on the origin; in three dimensions, they will lie on a sphere; in more than three dimensions, it will be a hypersphere.

Let us consider further the two-dimensional case. Since all grey values are positive, the points will, of course, be in the first quadrant. Actually, because extremely low or extremely high values are extremely rare, the points will be mostly grouped not too far from the main diagonal. It is possible then to consider that points of equal intensity are situated, not on an arc, but on the tangent to the arc which is perpendicular to the main diagonal (see Figure 41). In other words, we approximate an equal intensity circle by its tangent drawn at the main diagonal. This, admittedly, is an approximation, but it greatly simplifies computations because we may compute x+y instead of SQRT(x^2+y^2). Also, the main diagonal then may be considered as a new axis which represents the total intensity because all points with equal x+y (and hence on the same tangent) may be considered as having the same intensity.

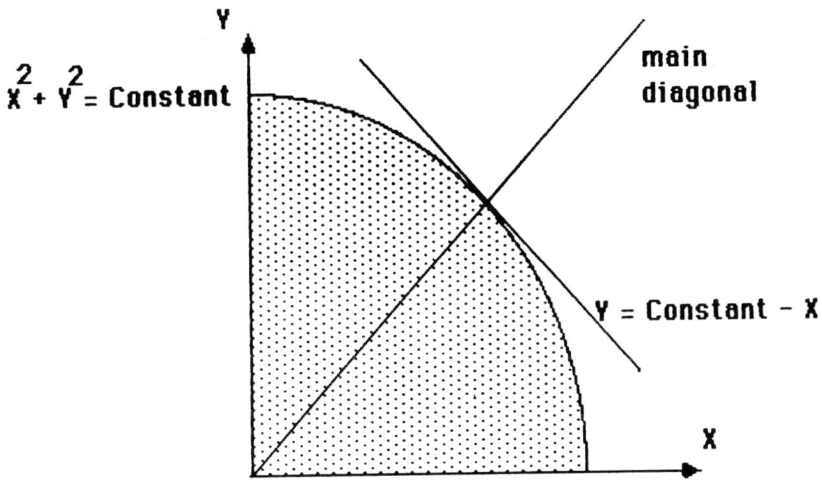

Figure 41. In two dimensions (two channels), points of equal intensity will lie on an arc centered on the origin. This arc may be approximated by its tangent perpendicular to the main diagonal.

The extension to more than two dimensions is immediate. In three dimensions we shall consider the planes perpendicular to the main diagonal and tangent to the spheres centered on the origin.

One important group of objects is particulary close to the main diagonal: bare soils; their representative points are mostly aligned in an ellipse (or ellipsoid, or hyper-ellipsoid) with the main axis almost parallel to this main diagonal.

Intensity is closely linked to relief, as the above example has illustrated. And so if we create a new channel as the sum of all channels, i.e. for every pixel we compute x+y+z ... and store the result as a new band, the resulting image, when presented on the screen, will give a good idea of the scene relief. This is a first example of deriving new channels by computation done on the original image; we will see more of them in the next chapter.

10.5 Let the Axes Rotate

The remark presented about mono-channel images, that only the relationships between grey values are important, is also true here. What counts is the disposition of the object sub-clouds, relative to one another. We may, therefore, rotate the channel space axes, stretch one axis and shrink another, without modifying the information content of the channel space.

However, just as looking at a statue from different points of view will give us new aspects of its form and proportions, and is indeed a necessity if we want to study the statue closely, rotating the axes and moving the origin, thereby creating new, derived channels, is a valid approach and has led to interesting developments, as we will now see.

11. DERIVED CHANNELS

11.1 New Channels

It is often advantageous to create new or derived channels, computed from the original bands. These new channels present the information in a quite different and often useful manner, as some examples will illustrate. They may be appended to the original image on disk, thus increasing the number of channels; or they may be stored as a new image.

These channels may be combined in various ways. The computations are done on values, say x,y,z..., of all image pixels, on pixel at a time. Two or more of these values are combined to create a derived channel, such as $u = x/z$, $v = x+y+z$ ... One or more new values are thus added to the original values of each pixel. The derived channels thus obtained have, of course, the same number of lines and columns as the original channels.

Other kinds of derived channels may also be created, where the pixels are not considered one at a time, independent of the other pixels, but where neighbouring pixels enter into the computations; these kinds of derivations are, however, beyond the scope of this introductory course.

11.2 The One-Byte Constraint

Before looking at some examples of derived channels, a computer constraint must be mentioned. We have seen that, to be visualized, an image must be made of one byte values. This obligation to store all image data in a single byte, as integers comprised in the interval 0-255, may be called the one-byte constraint; it is met in most kinds of image computations and notably for the computation of derived channels; it may create some programming problems. Consider as an example, $v = x+y+z$; the sum of three integer values each of which may be as high as 255, may, of course, be greater than 255, but the result must be held on one byte and hence may not be greater than 255. Something must be done in the program, such as dividing the sum by the number of channels, or subtracting a well-chosen constant, or bringing all values greater than 255 to an arbitrary value, for instance 255.

Because of this constraint, some information may be lost. Consider for instance the derived channel $u = 1/x$. Let us first note that, fortunately, x will never be equal to zero on actual grey values, so we do not have to consider this special case, although, just to be sure, the program should check this unlikely event; u will then be positive and smaller than one.

To bring it into the range 0-255, we must transform it into $u' = u*C1+C2$, where C1 and C2 are well-chosen constants; and then change u' into an integer $u'' = \text{INTeger}[u']$; these u", computed on the image, are stored as the new channel values. If C1 and C2 are adequate, all u" will be spread into the 0-255 interval and there will be as many different values of u" as there are different values of x. The program must, of course, check that all u" are indeed smaller than 255, and bring all values greater than 255 to a fixed arbitrary value, for example 255; but by so doing something is lost: the channel dynamics diminish (see chapter 5) or, in other words, the number of grey values is decreased; if we return from the new channel to the old, by the inverse transfer function $u = (u''-C2)/C1$ and $x = 1/u$, we will not find all the original grey values (let us note that the loss of precision obtained by taking u" instead of u' is practically irrelevant if there is a sufficient number of different u" values).

Whatever solution is taken to scale results into the 0-255 interval, it introduces some degree of arbitrariness. This is another example of the nature of remote sensing data processing; it is a fine blending of sophisticated mathematical and computer procedures with raw and even rough approximations, and it therefore acquires a smack of the kitchen recipe which is quite surprising and often frustrating!

11.3 Single Channel Transforms

If I create a new channel by adding a constant to all values of a channel, the two channels will not be really different (except, of course, for the possible loss due to the one-byte constraint). The same is true if I multiply by a constant, see chapter 5.

But by creating a new channel which is a non-linear function of a single original channel, could I create a really new derived channel?

In other words, is it worthwhile to consider derived channels such as $u = x^2$ or $u = 1/x$ or any other function of x? Let us suppose that we choose the constants C1 and C2 of the preceding paragraph such that no u" is higher than 255 and no trimming is necessary. If I visualize the resulting image on the screen, I can do it with a transfer function which will be the inverse of the first function, x = SQRT(u) or $x = 1/u$. The picture will be very similar to that of the original image seen through the identity transfer function. To the derived channel, I may therefore apply the transfer function which will transform the derived channel back into the original channel. Hence, there is nothing to be gained, really, in deriving channels from a single channel, although it may sometimes be useful to do it for presentation purposes, or when two images are combined.

But combining two or more channels yields much more interesting results, and we will briefly consider a few examples, beginning with two-channel combinations.

11.4 Channel Ratios and the Normalised Vegetation Index

Channel ratios of the form $u = x/y$ are often used. Let us first note that $u = x/y$ and $v = y/x$ are equivalent, because one is a function of the other: $v = 1/u$, and we have just seen that the resulting images are the same in that case.

Of course, here also what we have called the "one-byte constraint" must be taken into account, but this does not present any major difficulty.

The number of possible channel ratios grows rapidly with the number of channels; for LANDSAT MSS it is $(4x3)/2 = 6$, whereas for the Thematic Mapper some $(7x6)/2 = 21$ channel pairs may be conceived. It is better, therefore, to look for possible physical or phenomenological reasons for choosing some of the possible ratios rather than others. This is the case of the Normalised Vegetation Index.

We have seen at chapter 9 that healthy vegetation absorbs red radiation and rejects (reflects) near-infra-red radiation quite strongly, whereas in sick or wilting vegetation the difference in reflection is much narrower; vegetation is the only object which behaves in such a manner. This phenomenon is of fundamental importance in many remote sensing applications. The physical and agricultural aspects will be examined elsewhere, but we will briefly consider here some implications derived from the computation of an index based on this phenomenon.

There are various ways to express this vegetation behaviour; for example, the difference, for every pixel, between one near-infra-red channel (NIR) and the red channel (R): NIR-R may be computed. The greater the difference, the healthier and denser the vegetation; if the index is below a certain threshold, the pixel probably contains no vegetation. Experience shows that this is a vlid approach; it is used, among other things, with AVHRR data (TIROS satellites). This index is, however, sensitive to variations in total reflected intensity which, in turn, is influenced by the ground slope and its orientation and, therefore, the Normalised Difference Index:

$$NDI = \frac{NIR-R}{NIR+R}$$

is often preferred.

The denominator is a measure of total reflected energy, a somewhat coarse measure admittedly, but experience shows that it is satisfactory. The resulting pictures have proven to be of extreme importance and there is considerable literature on this index, so it is useful to consider its computing aspects.

(The index is also used to process ground measurements obtained with a hand-held spectrometer, but we will limit ourselves here to the creation of channels derived from digital images.)

First, NIR and R, as written on the CCT, are comprised between 0 and 255, hence the index varies between +1 and -1, so we also meet the one-byte constraint here; we will multiply the index by a constant factor and add a constant; this will introduce some degree of arbitrariness but by now we are familiarized with that problem, and these constants may be furnished with the resulting picture.

Second, there is a one to one relationship between (NIR-R)/(NIR+R) and NIR/R; given the value of one, the other is easily computed; indeed, if we write z = NIR/R,

$$NDI = (z-1)/(z+1).$$

This relationship is not linear but this, actually, is not important, and the pictures obtained are identical; nevertheless, the form (NIR-R)/(NIR+R) is always the one which is considered in the literature.

Third, since we start from integers and want the result expressed as integers, it is possible and advisable if computer time is to be spared, to program the index computation so as to deal with integers only, instead of doing the computations on intermediary reals; but this requires some care in programming because integer division is not the same as real division.

12. SOME DERIVED CHANNEL SYSTEMS

12.1 Derived Systems

Various systems based on the concept of derived channels have been proposed and it is worthwhile reviewing some of them, because they illustrate a number of points we have encountered so far. They have in common a rotation of coordinate axes in the channel space; each presents a consistent system of derived channels, obtained for the most part by linear combinations of original channels (the sole exception is the intensity-hue-saturation system, which includes polar coordinates). Each has its advocates and opponents, and a choice between them will be partly a matter of personal taste. But together they illustrate the fact that an impressive number of different presentations may indeed be obtained, many of them containing useful and quite original information.

12.2 Intensity-Heu-Saturation

A system which is often used, especially by German researchers, is the intensity-hue-saturation system; it is better presented in three dimensions. We will suppose that the three original channels are in the green, the red and the infra-red, respectively. Strictly speaking, this channel combination does not specify the "true" colour, i.e. the colour as the eye sees it; for that, we should consider the three channels blue, green, red, but, as has already been mentioned, blue is rarely used in satellite sensors, whereas one or two channels in the near-infra-red are quite frequent. We shall, therefore, use quotation marks and speak of the "colour" of a pixel to specify the relative contribution of each channel.

The intensity-hue-saturation system has the main diagonal as its first axis and this axis, as you will recall, represents total reflected intensity (see chapter 10). To visualize the two other axes, we must first imagine that we fix one point, arbitrarily, on the main diagonal; we will call it T (see Figure 42); we draw the plane perpendicular to that diagonal and passing through T. This plane intersects

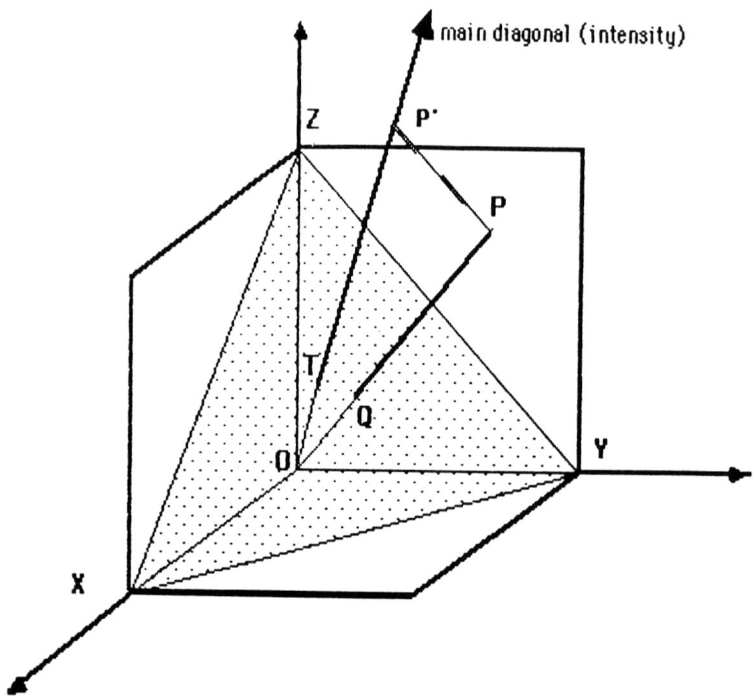

Figure 42. Intensity-hue-saturation (1). Point P represents any pixel in a three-dimensional channel space. Any plane (X,Y,Z) perpendicular to the main diagonal, cuts this diagonal at a point (T). OP is defined by its projection on the main diagonal (OP', pixel intensity), and its intersection, Q, with plane XYZ.

each of the three axes at a certain point which we shall call X, Y and Z, respectively. These points X, Y and Z form a triangle which has T at its centre. Let us recall that all pixels which have their representative point on this X,Y,Z plane have the same intensity. It should be clear also that all pixel points lying on this plane are inside this triangle and not very far from T.

Let us consider any pixel; we shall call P its representative point in the channel space. Let us join P to the axes origin, O. In the intensity-hue-saturation system we consider that P may be defined by:

- its projection on the main diagonal, parallel to plane X,Y,Z: P';
- its intersection with plane X,Y,Z; let us call it Q.

OP' represents the reflected intensity. Q characterizes the "colour" of the pixel or, in other words, the contribution of each channel X, Y or Z; the nearer Q is to one of these points, the greater the role played by that channel, i.e. the more important is that particular wavelength in the energy reflected by pixel P. In the X,Y,Z plane we will now draw polar coordinates, with T as their origin. If Q is very near to T, the contributions of X, Y and Z are almost equal and the resulting "colour" is almost "white". A whitish colour (pastel colour) is said to be saturated, so TQ length represents the degree of saturation; TQ is the radius of the polar coordinates. We can take TX as the axis of origin; the angle between TX and TQ represents the "colour", or more appropriately, the hue of the pixel.

The pictures obtained from these new channels are conspicuously different from the original pictures and each yields a different aspect of the image.

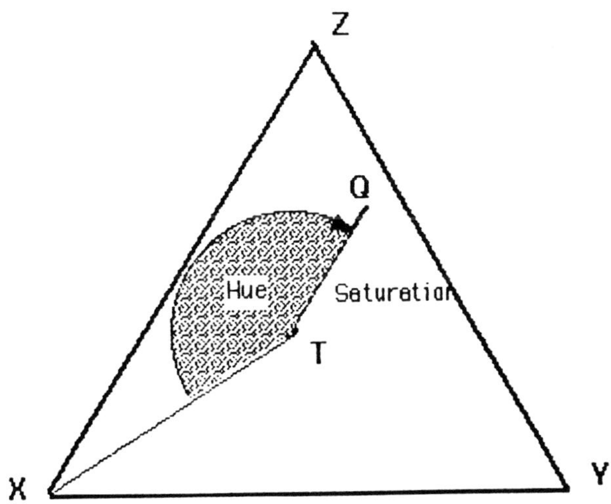

Figure 43. Intensity-hue-saturation (2). In plane X,Y,Z, the length of vector TQ represents the saturation, while angle HTQ characterizes the hue.

12.3 The Tasseled Cap

Other systems attempt to link the radiometric responses to natural objects. One such a system is the tasseled cap which was proposed by Kauth and Thomas (1976). The system was specially designed to study agricultural scenes with LANDSAT MSS data. It is an orthogonal system, defined in the four channel space of the LANDSAT MSS. It is rather too complicated to be fully explained in all its details here, but, in a first approximation, its main lines may be described as follows.

The system has four new axes. The first axis goes from the origin (which remains unchanged) to a point representative of the soil. The authors propose a standard American soil, but results would not be very different with another soil. This axis does not exactly coincide with the main diagonal, but is not very far from it (see Figure 44). The second axis, perpendicular to the first and called "greenness", points toward the vegetation maximum; you will recall that vegetation reflects intensively in the near-infra-red and, therefore, this axis points towards the NIR region. The third axis, "yellowness", is perpendicular to the first two and indicates, as the name implies, the direction which will be followed by vegetation when it is ripening and wilting.

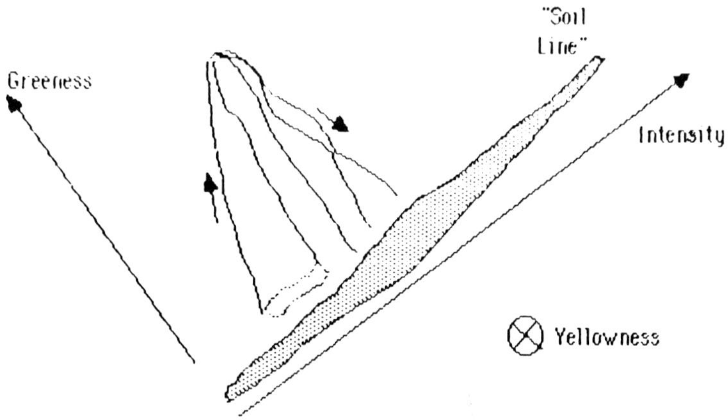

Figure 44. The tasseled cap system. In the LANDSAT MSS four-dimensional space, four new axes are defined. The first is parallel to the "soil line", and is near to the main diagonal. The second points in the direction of pixel radiances with increasing biomass. The third (yellowness) is perpendicular to the first two and points in the direction of wilting.

These three axes contain practically all the useful information; the fourth axis, perpendicular to the first three, contains mainly residuals, noise such as striping, etc. For that reason it has been named, somewhat poetically, "nonesuch".

The name "tasseled cap" serves to indicate the path followed by vegetation in the channel space, from emergence, where the young plants response is insufficient to mask the soil response, to full development, where the biomass is important and the soil completely covered, to final wilting and decay, where the signal is again that of bare soil or very near to it. One vegetation pixel thus starts from some point on the bare soil "cigar", goes toward the maximum vegetation point, and from there goes down back to the bare soil. As the points start at different places of the bare soil cigar, they go up in a sort of flattened cone with the maximum vegetation point as its summit (the cap) and from there they go back to the soil in various directions (the tassel). Unfortunately, with the limited experience I have on this subject, it seems to me that the tassel is not as separate from the cap as we would wish it to be, but is contained inside the cap.

The system has been widely accepted for the study of LANDSAT MSS data, especially in agriculture, and has proved to be of great help when reasoning in the MSS channel space. It has recently be revised to fit the enlarged opportunities offered by the Thematic Mapper channel space, where soil humidity may be considered.

12.4 Thematic Channels

There is yet another way of presenting the cloud. Let us first remark that our channel space has - by definition - as many dimensions as there are channels. We may, therefore, design new axes which, if they are as numerous as the original channels and if they are independent (not lying in the same plane), may be used to describe and study the cloud just as well as the original axes. If we can give a meaning to each new axis, in particular if we can relate it to a given class of radiometric objects, the interpretation of the resulting pictures will be facilitated. In the systems described above, the new axes form an orthogonal system; i.e. these axes are perpendicular two by two in the original channel space. This may be convenient, but is certainly not a necessity. The original axes are always presented as an orthogonal system, but there is no particular reason to do so, except that it facilitates drawing. We are allowed to devise oblique new axes if there is some benefit in doing so, because they will be oblique in the original channel space, but they may just as well be represented as orthogonal, if we so wish.

We have proposed (Berg and Hubaux, 1984) a system where each axis is chosen so as to represent a given class of objects, and goes from the origin to the centre of gravity of that class. As an example, let us take a SPOT scene: three channels, hence a three-dimensional space, therefore, we may create three new axes (see Figure 45); suppose that the scene is composed of essentially three kinds of objects: water, vegetation and bare soil (this is quite a common case). On the scene we want to study, we will search for pixels representative of "pure" water, as far as can be achieved, and we will compute the mean of these pixel values for the three original channels; this will define point W, the centre of gravity for water in the channel space. In the same manner, we define V for vegetation and S for soil. Of course, we try to find pixels belonging to only one category so that, in our vegetation pixels for instance, the soil is completely covered by the plants and thus does not contribute at all to the pixel radiance. We have now

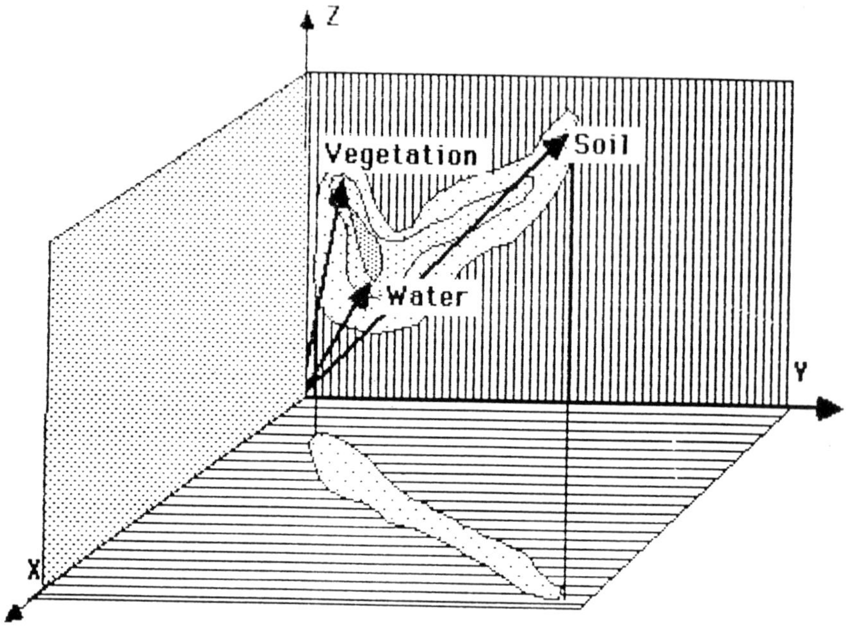

Figure 45. Thematic channels. The new axes go from the origin to a point in the channel space representative of a single object, for example water. The number of axes remains unchanged.

three new axes, OW, OV and OS. We create - rather we let the computer compute for us, using an appropriate program - the new coordinates of every pixel. Mathematically speaking we have rotated the axes. We obtain - on the magnetic disk linked to the computer - a new image representing the same scene as the original, with exactly the same number of lines and elements, and the same number of channels.

Each of these new bands has something to tell us. In the water channel, water-rich pixels have high values and all the other pixels have low values. If the scene includes swamps, they will appear quite clearly and the degree of "swampiness" or wetness will be evident.

The vegetation channel will yield a picture quite similar to that obtained with the Normalised Vegetation Index; zones with dense and healthy vegetation will coincide. Some difference will be noted only for low values, but this is normal; below a certain threshold, the NVI, although it may be computed, has no physical meaning; and this is true in the vegetation channel: low values are not particularly significant. This coincidence for high values may surprise the mathematically minded because the vegetation channel was obtained through a linear transform, whereas NVI is a quotient and hence not linear, but a close reasoning on the cloud of vegetation pixels would show that the result is to be expected; we will not go farther here, because the developments would take too much space.

In the soil channel, bare soils are highlighted and it may happen that some topological detail on this soil, such as a path or a road, which was completely invisible on the original channels, is now clearly delineated. This illustrates the fact that it is possible, by simply changing the axes, to reveal hidden image features.

The interest of the method lies in its simplicity, both conceptually and computationally. It is, in fact, fairly easy to program (it requires a simple matrix inversion and, as always, the rescaling into the 0-255 interval), its execution requires a minimum of computer time (the new image may be computed entirely with integer arithmetic), and it yields single channel images which are easy to interpret in terms of the scene objects and may reveal unexpected details. It is, therefore, a highly intuitive method. It should be clear, however, that some caution must be exercised in interpreting these images; if, for example, part of the vegetation is wilting, it may have a considerable response in the "soil" channel, not because some soil is visible under the vegetation, but because completely wilted vegetation has a response similar to soil, so wilting vegetation will have an intermediate response between vegetation and soil.

12.5 Principal Components

Another kind of axis change, often used to obtain particular image presentations, is the "principal components" system, where axes are simply rotated. In mathematical terms, the new axes (or channels) are the eigen-vectors of the correlation matrix computed on all channels. While this may be clear to some, it will leave others completely at sea. Fortunately, the concept is easily explained in two dimensions.

The pixel cloud on a two-dimensional histogram often looks something like Figure 46. On this diagram we might draw several contour-

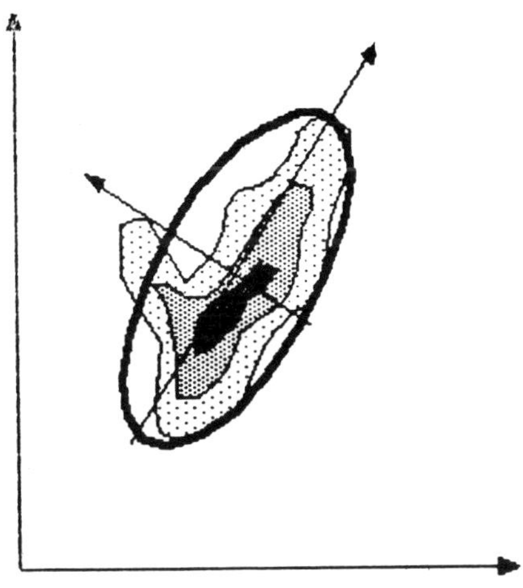

Figure 46. Principal components. The smallest possible ellipse enclosing 95% of the pixel cloud is drawn. Its two axes are the principal components.

ellipses which would enclose most of the pixels, say 95% of them; we will draw the smallest possible ellipse which fits this condition (note that we may always draw, or compute, this ellipse, even if the cloud has a shape departing conspicuously from an ellipse). This ellipse has two perpendicular axes which cross each other at the cloud centre of gravity. These two axes are the principal components in two dimensions; they are linear functions of the two original axes. It may be demonstrated that they are statistically independent, i.e. that they have a correlation coefficient equal to zero. In more than two dimensions, the cloud and contour ellipses (or more correctly ellipsoids) may not be represented on paper, but the principle remains the same; the axes of the contour ellipsoid are the principal components and are statistically independent. The resulting mono-channel pictures are interesting: the first one, with the longest axis, contains most of the total information. The last pictures, those corresponding to the shortest axes, contain almost nothing but noise (striping for instance), and may, therefore, be discarded from further study. The method is thus useful for reducing the number of channels and hence speeding up further computations.

But let us first remark that we have again applied a classical - and powerful - statistical tool without respecting the underlying theory; the points making the cloud should be randomly chosen from the global population, whereas we take the points belonging to a given scene. We should not complain if our results contain some inconsistency!

So, attractive as the method may seem, it has inherent limitations; notably, the axes are scene-dependent. This dependency will serve to illustrate, in a negative manner, the concept of robustness.

12.6 Robust Algorithms

The object of most research in remote sensing is the development of algorithms, as we have already seen. Among the qualities of these algorithms, one of particular importance is robustness. I take the term from statistics, where a "robust test" is one which may be applied even in cases which depart from the basic hypotheses upon which the test theory lies; for example, the population is supposed to be Gaussian, but a robust test will give reliable results even for non-Gaussian populations. Transposing this concept to remote sensing we may say that a method, or algorithm, is robust if it may be applied to other regions, on other dates, than those on which the method was developed. Admittedly, the applicability domain of a new method will be difficult to delimit exactly and we should be satisfied with unprecise descriptions of this domain. But, even if difficult to assess, the concept is important.

In most cases we want our method to be applied in future years, where conditions will certainly not be equal to those of the past or present year. The principal components method illustrates the idea quite well, in a negative manner.

To clarify the point we will take a simplified scene as an example; we will suppose that this scene is made of two objects only, wheat fields in full bloom and bare soil. We will study this scene in two channels only, red and near-infrared, simply because it is much easier to draw in two dimensions (see Figure 47). We now cut the scene into two sub-scenes, quite arbitrarily, a Northern and a Southern part, say, and we will suppose that the proportion of bare soil is greater in the North. If we now draw the two-dimensional histograms of the two sub-scenes and the two ellipses encompassing 95% of the pixels, what do we see (Figure 48)? The ellipses occupy quite different positions; in other words, when we go from North to South, there is a far from negligible rotation of the axes of the ellipse! So, even at the same date but on different parts of the same scene, the principal components are not steady but rotate according to the proportions of the objects which we encompass in our computations; we are not masters of this rotation because we cannot fix the proportions of every object.

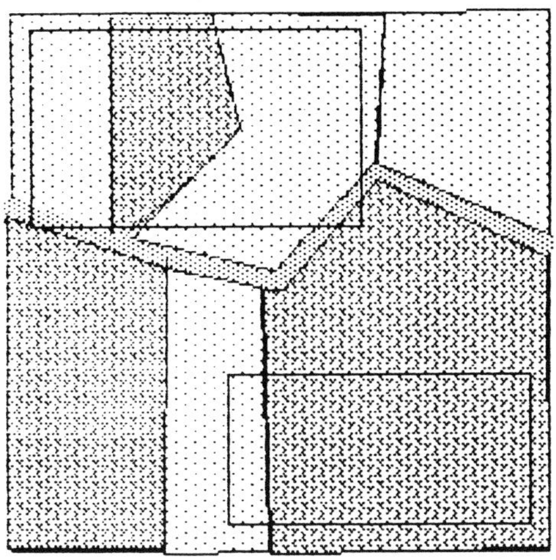

Figure 47. A scene made of wheat fields (darker grey) and bare soil. The two rectangles indicate two subscenes, the histograms of which are presented in Figure 48.

So the classification algorithm which included principal components would not be robust. And indeed I do not think it has been proposed. But the PC method is nevertheless interesting because it yields image presentations which are quite different from those obtained with the original channels.

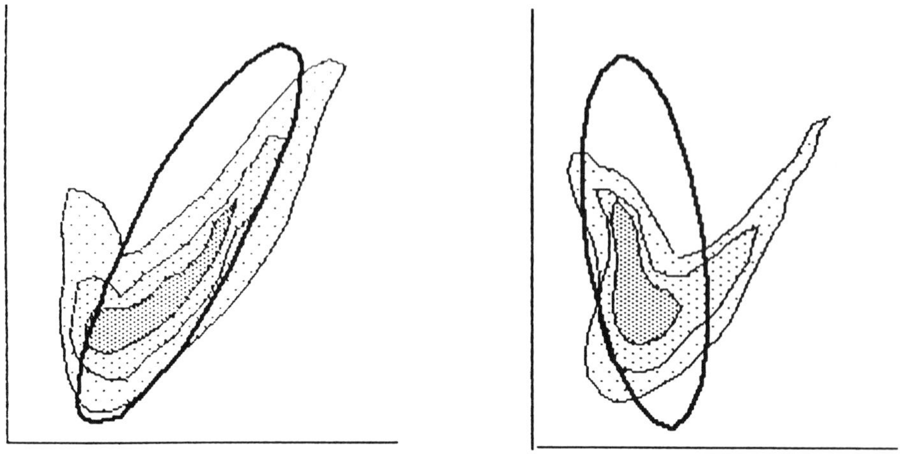

Figure 48. The histograms of the two subscenes on Figure 47, with the Northern scene on the left, which contains mostly bare soil. A shift in proportions of bare soil and vegetation results in quite different principal components.

13. RADIOMETRIC AND GEOMETRIC CORRECTIONS

13.1 Corrections or "Corrections"?

Now that we know how the image is visualized, we can better understand how this data matrix can be processed on the computer.

A first kind of treatment, often applied, consists in some improvements of the original data. These improvements are, somewhat improperly, called "corrections". We will see that the expression is unfortunate. The word "correction" conveys the idea that the product was somehow "bad" in its raw form and that its overall quality has been improved by properly "correcting" its defects. But this simplistic vision may be misleading.

An image, as we have seen, has two aspects. It may be considered as a source of different pictures and visualized as we have seen in the preceding chapters, or it may be regarded as a data matrix of electromagnetic energy measurements. Unfortunately, in most cases, by improving the pictorial quality of the image, its physical measurement quality is degraded, as we will now see. Actually, there is a trade-off between the two aspects. And, therefore, instead of speaking of "corrections", it would be more appropriate to speak of modifications; but the expression "radiometric and geometric corrections" is now part of the remote sensing vocabulary and we will use it here, even if we are unhappy with it.

13.2 System Corrected Data

The distributing agencies (ESA, NASA, NOAA, SPOT-Image) often sell the data in two forms: raw data or system corrected data. The first are the data as they were received by the station, practically without modifications; they are normally sold as CCTs, because they are not fit for picture generation. For the second, the distributing agency applies corrections using the parameters that the receiving station has at its disposal (satellite ephemerides, pre-flight and in-flight sensor calibration), but not ground observations. These first data modifications are called pre-processing because it is presumed that the final processing will be done by the user.

13.3 Radiometric Corrections

When the raw image is visualized, various imperfections mar the aesthetic quality of the resulting picture. One such defect, quite frequent, is striping, where the intensity of the lines composing the picture (horizontal in the case of LANDSAT MSS images) varies considerably, somewhat like a poor TV picture.

This is due to the fact that the image was acquired line by line, by several sensors, each of these sensors having its own characteristics. In LANDSAT MSS for example, there are six sensors for every channel.

There are two ways of correcting this striping, one "external" and one "internal". The first consists in using the calibration of the sensors, so as to homogenize their responses. For that, use may be made of the pre-flight calibrations, or better, of the in-flight calibrations, obtained because the sensors regularly measure on-board calibrated lamps or equivalent calibrated sources for thermal infra-red data.

It may happen, however, and this is not unusual in practical cases, that these calibration references are missing or inadequate. The data may then be "corrected" without using any external information. In the case of striping of LANDSAT MSS images, the following algorithm could be applied for each band:

- the mean and standard deviation is computed for every line;
- the sum of every sixth line mean is computed; six sums are thus obtained, one for every sensor. The same is done for the standard deviations;
- if there were no difference between sensor responses, these sums should be practically the same because they are computed on the whole image and the random variations between lines should largely compensate each other. It is then assumed that the differences between these sums represent the variations between sensors; from the mean differences the offset characteristic of each sensor is computed, and from the standard deviation differences the factor characteristic of each sensor is computed;
- the corrected values of all pixels are computed, applying these offsets and factors to their respective lines.

The resulting picture represents a striking improvement over the original. The eye now sees a "pleasant" picture, looking very much like an excellent aerial photograph.

This second technique illustrates the fact that it is often possible to improve the picture quality by computations done on the image data themselves, without any outside information. By doing so, however, the values are modified and they lose some of their physical significance.

To improve or modigy some aspect of the image, more than one type of algorithm may be conceived. Two remarks must then be made: first, it is not easy to devise a criterion to judge what is the "best" algorithm; second, the results obtained by mathematically quite different algorithms are often almost the same, so a quality criterion is not so important after all!

13.4 Geometric Corrections

When we process or visualize the data, one of the implicit assumptions we make is that the pixels are perfectly equidistant and that each represents an area equal to the nominal pixel area. This is certainly not true for the original pixels. Let us first consider a single swath line. Pixels near the centre of the line necessarily have a smaller surface than pixels near the extremities. The fact that the Earth is not flat but round and curving away from the satellite sensor, accentuates the difference. If we now consider two adjacent lines and remember that they have not been acquired simultaneously, but one after the other, we will realize that their pixels are not perfectly aligned because the Earth, between the two acquisitions, has rotated by a fraction of a pixel; since it is a fraction, this slight difference may not be compensated for, because we deal with integers only. After a few lines, however, the accumulated differences amount to one pixel and the line must be shifted by one pixel.

There are many geometrical modifications to be introduced; they will be the object of the next lecture. One point, however, should already be remembered: most of these modifications are obtained by interpolation and this operation - which amounts to the creation of new pixels between the original pixels - implies that some kind of algorithm must be devised to define these new pixels' radiometry, i.e. to give them a value in each channel. As will be evident in the next lecture, there is something inherently arbitrary in this interpolation and part of the original pixel radiometry is lost in the operation.

14. FROM VISION TO CLASSIFICATION

14.1 Why Classify?

In many types of research some kind of classification is required. In its simplest form it is often necessary to isolate some object from the rest: wheat fields from "non-wheat-fields" (i.e. everything else), lake pixels from land pixels, or bare soil pixels from vegetation, etc.

To classify then is to group the pixels in a restricted number of categories. This, in turn, implies that some criteria be defined based on the properties of the objects to be classified as obtained from the digital image; these criteria must be included in a classifying algorithm.

An important "fallout" of the classification process is that the total area occupied by every class is then simply and quite effectively obtained by summing the number of pixels belonging to that class. This planimetric process is automatic and far cheaper than estimating the areas on a map with a planimeter.

Contrarily to research in geology where, since the rocks do not change with the seasons, the analysis of one scene may well give definitive results, for research in agriculture or forestry, what we are looking for is a classification algorithm; the scene chosen for study is only an example and the method, once defined, should be applicable to future scenes in regions with like characteristics.

A typical example where some classification scheme is useful is land-use inventory where the desired result is a map giving the distribution of land occupation, at a certain date, July 1986 say. It would be more satisfactory if this map could be obtained automatically, i.e. without human interpretation; the CCTs containing the images would be read by the computer and a carefully devised program, enclosing well chosen algorithms, would directly produce the classification maps on some output device. But that stage has not been reached yet; we will see now some of the major stumbling blocks which still litter the road toward that goal.

In its simplest form then, to classify is to divide the image pixels into two categories, A and B say. In this case, when we compare the results with ground observations (the ground truth) we obtain four groups:

1) pixels rightly put into A;
2) pixels rightly put into B;
3) pixels wrongly put into A;
4) pixels wrongly put into B.

These results are often presented in tables called confusion matrices, see Figure 49.

Before reviewing the problems inherent to classification procedures in general, we will illustrate the procedure by a simple example.

14.2 The Box Classifier

You will remember that, in our study of the mono-channel image in chapter 8, we saw that some pixel separation is obtainable, water pixels

		classified in	
		A	B	
ground truth	A	300	50	350
	B	25	400	425
		325	450	

Figure 49. A confusion matrix. A ground truth study has established that, in a certain area, 350 pixels belong to category A. On the same area, a classification done on radiometric data has correctly recognized 300 of these pixels, but has wrongly put 25 of them in B.

being generally lower than forest pixels. If now we reason on two channels instead of one, it will be clear that we can improve our "sieving" capacity considerably, as Figure 50 shows. We can cut the channel space (it is a plane in this case) into a series of rectangles, each of these rectangles corresponding, at least partly, to one of the classes we want to create. It is now easy to process all image pixels one by one, and to decide to which rectangle each pixel belongs. Instead of two, the procedure may be applied to three channels; instead of rectangles, we cut the channel space into parallelipeds or, to use a simpler word, into boxes. And, of course, there is no reason to stop at three channels if we have more channels at our disposal. The channel space will be cut into "hyper-boxes" (boxes with more than three dimensions), but the principle remains the same.

Let us review what we have done:

1) We have closely analysed the image, first channel by channel, then in three-channel colour composites.
2) We have compared the pictures obtained with what we know from external sources, about the scene.
3) We have created two-dimensional histograms.
4) For each channel we have defined certain values as limits between boxes, corresponding to the classes we want to create; in doing so, we have realized that it is not possible to define these limits in "empty positions", that is at places where pixel density is nil or quite low; we had to compromise and we realized that the first choice may not be the best one, but we made it tentatively.
5) With these limits we have classified the image pixels through a box

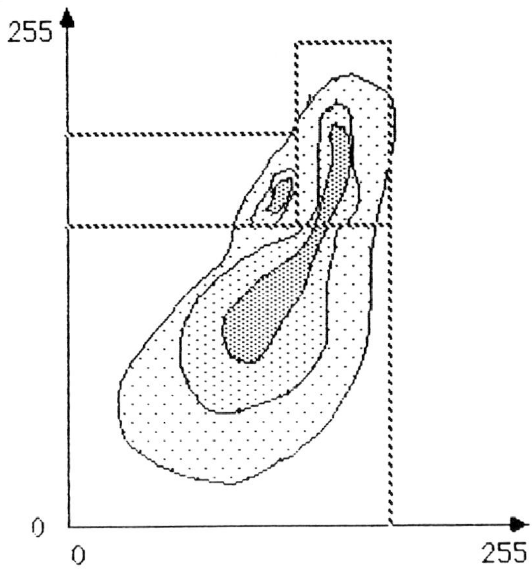

Figure 50. Box classifier. In a two-dimensional channel space, the cloud of points is cut into a series of rectangles ("boxes"), each enclosing, as far as possible, the range of radiometric responses which are characteristic of a given object.

classifier program.

6) We have looked at the results and seen that some limits should be modified; we have changed them accordingly and returned to point 4.

A point in favour of the box classifier is that, as the computations are reduced to a minimum, it is possible to repeat the whole procedure without increasing the computer time too much. It is even possible to devise a system by which the iterations are done locally, via appropriate transfer functions, the control being done directly on the screen, and the method is then really interactive. We have also seen in chapter 11 that it is possible to obtain derived channels which are parallel to the main axis of one of the object clouds; when this is the case and if the objects are those we are looking for, then the box classifier may be a valid method, particularly because it may be made interactive.

By looking at the box classification method we have met a series of problems which are to be solved for any type of classification:

- We have had to create a greater number of classes, based on radiometry, than our original number of thematic classes because, much as we would have liked one, there is not always a one to one correspondence between the thematic classes (water, forest, wheat fields, potato fields, bare soil) and their radiometric responses.
- We have had to iterate, that is to repeat, the procedure with modified parameters.

- We have seen that, whatever we do, mixed pixels constitute a major problem and that we have to take some arbitrary decision by putting our limits in their midst.
- Finally, it appears quite clearly that the results must be thoroughly checked and for that we need what is called ground truth.

14.3 Ground Truth

The results of any classification scheme must be validated, i.e. these results must be confronted with some external information. In most cases this means that some ground observations must be obtained, concomitant to remote sensing data acquisition. These observations are called "ground (or sea) truth" (in French: "réalité terrain"). Sometimes it is necessary to supplement these ground observations by aerial photographs. Whatever the quantity of "ground truth" data, they are never, of course, "the truth" (whatever that means), but an approximation to it; they cover in most cases only a small fraction of the whole scene; they are quite costly to acquire; and, for various reasons, they generally are not exactly contemporary with remote sensing data.

14.4 Unsupervised versus Supervised Classifications

Perhaps the most natural first step in attempting to classify an image is to start with the image itself, without any external information, notably without ground truth. This is called unsupervised classification. The computer program is made of two parts: first, the pixel distribution is studied, normally in the channel space and a first tentative partition of this space is obtained; this first part may either be done by the user or embodied in the computer program. In other words, the channel space is cut into zones and the centre of gravity of each zone is computed; second, all image pixels are analysed one by one and their distance to the centres of gravity are computed; there are various ways of defining these "distances", but it would take too long to review them here. The nearest centre of gravity defines the zone to which each pixel belongs. When all relevant pixels have been classified in this way, the whole process is repeated: now that we have a great number of pixels for each zone, the centre of gravity is recomputed and the whole classification process is repeated a certain number of times.

The procedure may seem quite attractive; we "let the computer find by itself the best possible classification, in a purely automatic and hence objective manner".

But many objections may be raised: first, of course, it is not the computer that completely stupid machine which takes the classifying decisions but the program; and the programmer who has written that program has had to take several a priori decisions; the "objectivity" is thus largely illusory. But it could be argued that the procedure is more consistent than the classifications obtained by various interpreters working analogically. I am not too sure that this is a strong argument either. If the results obtained by unsupervised classification are consistent but poorer than the interpreter's results, where is the advantage?

The procedure would be more satisfactory if two conditions were fulfilled: 1) the pixels should aggregate, in the channel space, into distinct, separate clouds, with only a few pixels midway between two clouds; 2) by luck, these clouds should correspond to the thematic classes we are interested in.

Actually, we have seen that the pixel distribution in channel space is a far cry from that and that these conditions are practically never attained. Furthermore, even if, by chance, it gives satisfactory results, could we be sure that, with data of another year, at a slightly different period of the year, on another scene similar to the original, with plants at slightly different phenological stages, we would obtain results of the same quality? I am afraid we must answer no; since we give full control to the computer program and do not know exactly what happens at each step, we become the slave and not the master; it is just like driving a car with our eyes closed, to use a comparison suggested by Mulder. For these reasons the procedure is rarely used and most classification schemes are supervised.

In a supervised classification, part of the external information - the ground truth in most cases - is employed to define the radiometric characteristics of each class, for instance: mean, standard deviation within each channel, and correlation matrix between channels. The image pixels are then classified on the basis of these characteristics. Here also, of course, the whole procedure may be repeated a certain number of times, with the characteristics being somewhat redefined (and, we hope, improved) at each iteration.

And here also, of course, as in any classification method, results must be compared with the ground truth. There is then a temptation to utilize the same ground truth data both to define the classes and to check the results, a procedure which was indeed used, alas, at least in the early days; of course, then the results are good if not excellent, but what does it prove? There is one word to describe this: "cheating".

14.5 Adding Internal Information: the Neighbouring Pixels

The only classification criterion we have considered up to now is the pixel position in the channel space; this is called context-free processing. This is perhaps the most natural approach and indeed historically it was the first to be tested. But now, on the basis of past experience, we know that, except for some favourable cases, to attempt to classify a mono-temporal scene purely on channel space criteria yields unsatisfactory results, say rarely better than 70% (i.e. of all classified pixels, only 70% are correctly classed).

A pixel should be seen, however, not only as a point in the channel space, but also, in its context, as a bit of ground most often surrounded by other pixels of similar nature, with similar radiometry; in other words, we should take the neighbouring pixels into account.

This is exactly what the interpreter does: he does not analyse pixels one by one, but he considers bigger units such as fields, or parts of a forest, etc. This is indeed the present trend in classification research. A number of pieces of research are analysing concepts such as texture, i.e. the particular pattern of the object surface, such as smoothness or roughness, which is one of the most characteristic features of an object, and one which is commonly used by the interpreter. Others are attempting to devise algorithms to recognize field boundaries automatically. The usefulness of this approach was demonstrated by Mégier and Mehl (1984) who created the "field by field" method; on scenes for which they had the cadastral field limits, they introduced these limits in the digital image (the procedure for doing this will be explained in the next lecture). They then considered these fields as the units to be classified, computing the mean and standard deviation field by field, and excluding pixels too close to the field boundaries. The improvement over pixel by pixel classifications is decisive.

It becomes now clear that often a mono-temporal image contains insufficient information to discriminate between objects of like nature and that, by considering two or three images, acquired at dates defined by the phenological stages of the plants under study, a considerable improvement can be obtained. More generally, to a single image, external information must quite often be added, be it to obtain a classification, or to study temporal variations, etc. This external information may be:

- an image of the same scene, acquired by the same satellite at another date;
- an image from another satellite;
- topographical data taken from a topographical or cadastral map;
- ground survey observations, etc.

But this means that the images of various dates must be registered, i.e. exactly superimposed.

To obtain these registrations it is necessary to effect a change of coordinates; one image (that of the first date, for instance) will be considered as the reference, and the other image(s) or map will be brought into those reference coordinates. The problems encountered for this operation will be examined in the next lecture.

To conclude this extremely short introduction to classification problems, we could state some classification "laws" (like "Murphy's law"):

"Whatever classification algorithm is used:

1) the results will always be partly correct;
2) it will not be easy to check what part is correct and what part is incorrect;
3) the method's potential adequacy vis-a-vis other scenes, at other dates, will be difficult to assess."

15. POSTFACE

We have seen that digital image processing is a young, immature and rough-hewn technique, with a surprising blending of mathematical theory and cooking recipes. Several statistical tools are frequently used, without respect for the basic assumptions on which statistical theory had been soundly grounded. Image processing is still waiting for its Gauss or its Hilbert. This may yet take some time, since the technique is closely connected to the hardware available, which is in full evolution, one could even say in full explosion.

At present it is also heavily influenced by the prodigious possibilities of our eye.

During our first steps in digital image processing, we have met a few ideas:

- It is not possible to see a digital image in its entirety, but an enormous quantity of pictures can be derived from it; so great is this quantity that probably not a single image has yet been studied in all its potential presentations.
- Whether on a screen or as a hardcopy (film or paper print), some transfer function has to be defined between the digital image and the resulting picture. Playing with transfer functions may yield quite different pictures out of the same source image. There is no such a thing as "the best" picture of a given image; the choice of the most adequate presentation must be guided by the theme of study.
- By reasoning in the channel space and by creating derived channels, one may obtain useful pictures, quite different from the original, and sometimes bringing out completely new details.
- But classification schemes based on the study of this channel space rarely yield satisfactory results, because the quantity of mixed pixels is significant.
- All geometric and most radiometric "corrections" while improving some aspect of the image, and notably its pictorial aspect, degrade some other part of the original information.
- Nevertheless, the registration of images and maps of various sources is often a necessity.

Remote sensing data processing is always a fight against an excess of information; in the jungle of data through which we try to find our way, we have to use an intellectual machette to chop away most information, so as to keep only that part (most often a tiny part) which is really relevant to our problem, and find the most convincing manner to present it.

The present trend consists in considering remote sensing data as a part only of the total information needed to solve most problems, and to incorporate them in Geographical Integrated Systems (GIS). This must, of course, be done with data in their digital format, and the use of the computer will certainly grow in the coming years. But whether the role of the human interpreter is declining, and whether completely automatic procedures will be found to solve most problems, remains to be seen.

REFERENCES

F.P. Brooks, 'The mythical man-month: essays on software engineering', Addison-Wesley, 1975.

A. Hubaux, W. Mehl, L. Fusco, J. Loodts, 'Format description for computer compatible tapes', Int. J. of Remote Sensing, **5**, 5, 819-838.

R.J. Kauth, G.S. Thomas, 'The tasseled cap - a graphic description of the spectral-temporal development of agricultural crops as seen by Landsat', Proc. LARS Symp., Purdue Univ. West Lafayette, Indiana, 1976.

A. Berg, A. Hubaux, 'Rotation of spectral channel axes; application to a rice growing area of West Africa', 18th Int. Symp. on Remote Sensing of Environment, Paris (France), October 1984.

J. Mégier, W. Mehl, 'Per-field classification and application to SPOT-simulated, SAR and combined SAR-MSS data', 18th Int. Symp. on Remote Sensing of Environment, Paris (France), October 1984.

IMAGE PROCESSING TECHNIQUES: FILTERING, EXOGENOUS DATA, GEOMETRICAL PROCESSING

J. Dejace
Former Scientist of the Joint Research Centre of the European Communities
21020 Ispra (Va) - Italy

ABSTRACT. These three lectures cover rather independent subjects. Elementary methods of filtering of digital images are described in a simple way with particular reference to smoothing and sharpening of images. Some simple geometric transformations are described. The use of control points for actual calculation of the transformation functions is explained. Some methods of resampling digital images are described and discussed. The importance of exogenous (i.e. not remotely sensed data) is inhanced. It is also discussed how to store and process these data in a remote sensing environment, to make them compatible with digital images.

F I L T E R I N G

The aim of filtering is to extract from the digital image a picture which, by enhancing some information contained in the image, will allow us to see more in detail some particular features. For example, an edge extracting filter will extract the lattice of the field limits whilst the homogeneous surfaces of the fields themselves are kept in the background. These filtered images will be used later on either to facilitate visual interpretation or for further computer-aided analysis of the image.

The processing methods described up to now by A. Hubaux in his lectures are pixel-by-pixel treatments. Filtering belongs to the class of local processing methods, i.e. the grey level value of an output pixel is now a function not only of one input pixel (same line and element) but of several pixels (of the same channel) in its neighbourhood.

1. INTRODUCTION

1.1 Convolution Product of Two Functions

Let f and g be two functions mapping R on R; the integral

F. Toselli (ed.), Applications of Remote Sensing to Agrometeorology, 129–155

$$F(x) = \int_{-\infty}^{\infty} g(u)f(x-u)du$$

is called the convolution product of f and g. A short notation f^*g is sometimes used. If g(u) is zero on all R except at a short interval in the vicinity of u = 0, we see that F is a kind of weighted average of the f value in the neighbourhood of x. The concept of convolution product is used, often implicitly, in many physical measurement processes. In this case, f is the "pure" signal to be measured, g is the transfer function of the measuring instrument and F is the result of the measurement (see Figure 1).

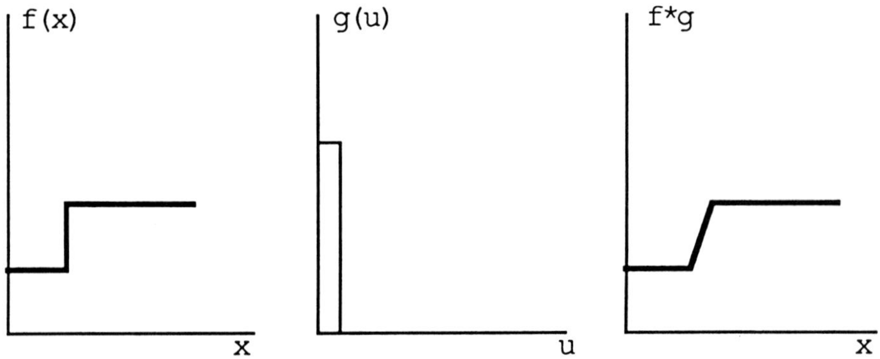

Figure 1. Example of a convolution product of a step function f by a rectangular slit g.

1.2 Convolution Product of Discrete Functions

A digital image (single channel for simplicity) may be considered as a mapping of N*N on R or N (i.e. at each ordered pair of integers, line and column numbers, it associates a real number or, in most practical cases, an integer coded on 8 bits). In this case, the notion of convolution product still exists but is expressed in a slightly different way by changing the integral into a sum

$$F(i) = \sum_k g_k f(i+k)$$

The set of coefficients g_k plays here the role of the g function in the continuous case. When the sum is limited to a few terms and the k index values are small, we see that the output value F(i) at point i depends on the input values of f at the surrounding points:

$$i-k_1, \ldots, i-1, i, i+1, \ldots, i+k_2$$

This operation, generalized to two dimensions, is called a convolution filter

$$F(i,j) = \sum_{k,l} g_{kl} f(i+k, j+l)$$

The set of g coefficients is called the kernel of the filter. It is often represented by a square table of numbers (3*3 or 5*5). Examples of such tables are given hereunder.

1.3 Fourier Space

We want to mention briefly, although this point will not be detailed in this course, an important result of convolution products. The Fourier transform of a convolution product is the <u>simple</u> product of the two Fourier transforms, so that the frequency spectrum of the Fourier transform of the kernel will describe the properties of the filter. A high frequency-rich spectrum will correspond to a high pass filter with edge enhancing properties. A low frequency-rich spectrum will give a low pass filter with smoothing properties.

2. EXAMPLES OF SIMPLE DIGITAL FILTERS

2.1 Derivative of an Image

If we consider the image as a function F(x,y), where x is the line and y is the element, the partial derivative

$$\partial F(x,y)/\partial x$$

is also a function of x,y and may be represented as an image. When the grey level F is constant, the derivative $\partial F/\partial x$ is 0. Where the grey level changes rapidly, the absolute value of the derivative is high; so the edges perpendicular to the x axis (i.e. parallel to the lines) will be put into evidence. As the function F is known only for integer pairs x,y, the derivative must be approximated on the digital image by a finite difference, in fact,

$$\partial F(x,y)/\partial x = \lim_{\Delta x=0} \frac{F(x+\Delta x,y) - F(x,y)}{\Delta x}$$

is replaced by

$$\frac{F(1+1,e) - f(1,e)}{1}$$

In terms of the convolution filter, the kernel is:

$$\begin{matrix} -1 \\ 1 \end{matrix}$$

or, still more simply, each pixel is replaced by the difference between itself and the pixel just above it (preceding line, same element). As the result of the difference may be negative, every possible precaution should be taken in the program to keep the result in the range (0,255). A similar operation in the perpendicular direction will emphasize the edges which are parallel to the columns.

2.2 Gradient

The gradient of a scalar function F(x,y) is a vector which has as components:

$$\partial F/\partial x, \quad \partial F/\partial y$$

The length of this vector is proportional to the "slope" of the function, in this case the rate of change of the grey level. The direction of the vector gives the direction of the maximum slope (direction perpendicular to the isolines of the grey level). On images, only the magnitude of the gradient is represented.

If, in each point of the image, we calculate the absolute value of the gradient, we will again find edges present in the image but now in any direction.

A popular algorithm for edge extraction based on gradients is the Robert cross filter

$$g(1,e) = \{(F(1,e)-F(1+1,e+1))^2 + F(1,e+1)-F(1+1,e))^2\}^{\frac{1}{2}}$$

which uses derivatives in directions with angles of 45° on the line and column directions. Due to the square root involved in each pixel, this algorithm is computer time consuming. A way to avoid the calculation of the square root is to use:

$$g'(1,e) = \max\{|F(1,e)-F(1+1,e+1)|, |F(1,e+1)-F(1+1,e)|\}$$

The results may be even better with the second formula. These filters may not be represented as convolution filters.

2.3 Laplacian

The Laplacian is a mathematical operator involving second derivatives defined by:

$$\Delta^2 F = \partial^2 F/\partial x^2 + \partial F^2/\partial y^2$$

In the case of discrete functions, such as digital images, this operator is generally approximated by a convolution filter with kernel:

$$\begin{matrix} 0 & 1 & 0 \\ 1 & -4 & 1 \\ 0 & 1 & 0 \end{matrix}$$

Involving second derivatives, the Laplacian is sensitive only to changes in the gradient of the image. A simple example in one dimension is (the kernel being then 1 -2 1):

$f(i)$	0	0	0	1	2	3	4	2	0	0	0	4	4	4
$\Delta^2 f(i)$		0	1	0	0	0	-3	0	1	0	4	-4	0	

The Laplacian alone is a good edge extracting filter. Generally, it is used for edge enhancement (sharpening) of images; the best results are obtained by subtracting from the original image the Laplacian divided by a constant greater than one (best values are between 2 and 5):

$$F - \Delta^2 F/k$$

The result is directly obtained by a convolution filter with kernel:

$$1/k \begin{matrix} 0 & -1 & 0 \\ -1 & 4+k & -1 \\ 0 & -1 & 0 \end{matrix}$$

As the sum of all the coefficients of the kernel is equal to unity, the average grey level of the image is not changed, but locally some values may result to be out of range, either negative or greater than 255.

This type of edge-enhancing filter is also called high-pass filter (see paragraph 1.3 on Fourier space).

2.4 Smoothing

When an image contains "noise", i.e. when a perturbation grey level is superposed in some way to the signal grey level, some features of the image may be obscured and become difficult to be seen. In this case, good knowledge of the physical causes of the perturbing phenomena will allow appropriate treatment. The correction of striping (the well-known six-line effect of LANDSAT MSS) by statistical adjustment, as is shown in another lecture of this course, is an example of what can and should be done. This way of tackling the problem is to be preferred whenever possible.

Only when noise is random and not correlated from pixel to pixel, filtering with averaging methods should be used. In this case, the grey level of the pixel (1,e) is replaced by a weighted sum of the surrounding pixels with, for instance, kernels of the type:

$$1/(4+k) \begin{matrix} 0 & 1 & 0 \\ 1 & k & 1 \\ 0 & 1 & 0 \end{matrix}$$

which adds to the value of the original pixel 1/k times the sum of the four nearest neighbours. More complicated schemes may, of course, involve the eight nearest neighbours or an even larger neighbourhood.

3. FILTERING EFFECT OF RESAMPLING

In the lecture on geometric processing of images, several schemes of resampling are proposed; the effect on radiometry and visible aspects of the image may be better understood looking at these schemes as if they were sort of filtering kernels.

The bilinear interpolation calculates the output grey level using the four nearest neighbours with positive weights; this explains why it has a strong smoothing effect. The effect of the cubic convolution scheme is more sophisticated; the four nearest neighbours have positive weight, the eight pixels along the edges of the square contribute with negative weight, and the four pixels in the corners have positive weight, but generally this weight is small

```
+  -  -  +
-  +  +  -
-  +  +  -
+  -  -  +
```

This explains the sharpening properties of thes scheme of resampling.

4. A LITTLE MORE ABOUT FILTERING

There exist, of course, more sophisticated methods of filtering which will only be mentioned here.

In adaptive filters, the kernel is not constant anymore but changes from point to point adapting itself to the local conditions of the image to process. Some filters (edge preserving smoothing), for instance, have a smoothing effect in the homogeneous regions of the image but not there where sharp variations of the grey level are encountered, and remove noise without blurring significant details.

Another possibility is to work directly on the Fourier transform of the image, modifying or cancelling frequencies due to undesirable effects and then to go back to the image by a second Fourier transform.

REFERENCES

Rosenfeld, C. and Kak, A.C.: 'Digital picture processing', Academic Press, New York.

E X O G E N O U S   D A T A

1. INTRODUCTION

At the beginning of the remote sensing history, in the early seventies, "promising results" were quickly obtained. Optimists hoped that some, if not most, of the problems of environment and land-use could be studied and solved by purely remote sensing techniques.

Ten years later, most of the results were still "promising results" and it became obvious that remote sensing could only add limited information to the knowledge which was already acquired patiently in-field during centuries.

We shall call these data which are not remotely sensed by satellites or other platforms "exogenous" data. This chapter aims at showing some aspects of how to handle such data in an image processing environment.

2. TYPE OF EXOGENOUS DATA

In general, what is registered on a geographical map may be, from the present point of view, considered as exogenous data.

2.1 Relief

The relief, i.e. the altitude, the magnitude and the orientation of the slope, are very important parameters for environmental studies, in particular in mountain agriculture and forestry problems. In many countries, large-scale topographical maps exist which are generally of good quality. Relief is represented by means of the well-known system of level curves or isolines.

During the last two decades, most of the cartographical institutes in Western Europe and Northern America made an enormous effort to computerize the map-making process.

In paragraph 7 some of the problems occurring in the transformation of these data into a digital terrain model (DTM), that may be integrated with satellite imagery, will be outlined.

2.2 Soils

The nature of the soil is another important parameter in agricultural problems and it is difficult to obtain detailed soil information from remotely sensed data. Where a map of soils exists, that information should be used if required by the problem, together with the satellite data.

On the soil map, the types of soil are generally represented by codes, coloured or shaded areas, each code corresponding to a soil type.

2.3 Other Thematic Maps

From a data processing point of view, most of the thematic data may be reduced to the preceding examples. If the feature represented is quantitative, such as average amount of rain, density of population, etc., the level curve technique will be used as in 2.1. If the feature is qualitative, such as type of forest, agricultural "ground-truth", the coded area representation may be used. However, the distinction between these two types of representation is fuzzy: when the level slicing of a quantitative character is coarse, specially when the isolines are interpolated between point measurements, the region between two level curves may be considered as a coded area.

3. AIM OF EXOGENOUS DATA PROCESSING

As has been said in other lectures of this course, remote sensing data are presented as digital images composed of pixels, and are generally multi-channel images; the channels may be the original spectral channels (also called spectral bands) or "derived channels" calculated from the original ones, such as vegetation index, or another combination of channels.

In an image processing environment, the exogenous data must be transformed into digital images. More precisely, one or more new "pseudo channels" have to be created that give for each pixel the numerical or coded value of the features of interest. It must be enhanced here that further computer-aided analysis of the data requires perfect alignment (or registration) of all channels; a pixel with the same line and element number in all channels must correspond to the same point on the ground.

4. DIGITAL REPRESENTATION OF SPATIAL GEOGRAPHICAL DATA

There are two distinct ways of storing spatial information for computer use, namely: "raster" format and "vector" format (more sophisticated techniques such as quad-trees or other types of tesselization are beyond the scope of this course).

4.1 Raster Format

The raster format is completely analogous to the way in which digital images are stored. The coordinates of points are not written explicitly but may easily be retrieved for each pixel from the order in which the data are stored. This format is only possible when the data are organized in a regular system such as, for example, the nodes of a regular grid. The way in which the data are ordered in storage has to be defined and described explicitly. One may think to pass from a map to a raster format by superposing a grid to the map, reading the feature value at each node, and giving the corresponding numerical or coded value to the appropriate pixel. The pixel size has to be defined taking into account the precision needed in the study on the one hand, and the availability of computer time and storage, on the other.

4.2 Vector Format

The vector format for storing data has been developed and is still used mainly in computer-aided cartography. In this format, a point is represented by its spatial coordinates which are explicitly written. A line is represented by an ordered sequence of pairs of coordinates which define points; straight segments linking these points approximate the line; a code is added to indicate what the line represents (road, river, etc.). A surface is represented by its contour line and a code.

4.3 Comparison of the Two Formats

A complete discussion of the advantages and drawbacks of raster and vector formats may be found in a paper by M. Jackson (1984). The principal arguments are summarized here.

4.3.1 The vector format is much less memory-consuming than the raster format. For instance, a quadrangular agricultural field is completely defined by a code and four pairs of coordinates (8 numbers which are stored with high precision in a few bytes). In the raster format, some tens, or hundred or even thousands of identical pixels are needed depending on the dimensions of the field and the pixel size required by the problem.

4.3.2 The maps plotted by specialized hardware using the vector format are of high cartographic quality and more similar to conventional hand-drawn maps.

4.3.3 The algorithms for processing vector structured data are more complicated.

4.3.4 Pixel-by-pixel comparison of several thematic maps or images is more straightforward in the raster format than in the vector format.

4.3.5 Software and specialized hardware, developed recently for image processing purposes, are immediately usable with the raster format.

However, today both formats are used at least as intermediate data storage formats and the dilemma "vector versus raster" is not yet solved.

4.4 Geographical Information Systems (GIS)

During the last ten years, much effort has been invested in formalizing and optimizing the use of computers in handling geographical data involving spatial and non-spatial information. These specialized software packages are called Geographical Information Systems. They are under development in several countries.

5. DIGITIZATION OF PAPER (ANALOGUE) DOCUMENTS

If the exogenous data which one wants to introduce in one's study are already in digital format, it will be possible (which does not always mean easy) to construct the pseudo-channel matching the satellite image using more or less sophisticated computer software but avoiding tedious manual work. Unfortunately, such digital geographical data are nowadays still rare and, probably, one will have to tackle the problem of digitizing a paper map.

5.1 Manual Digitizing

This is generally done using a digitizing table, most of the time an electro-magnetic model. The table contains a two-dimensional lattice of electrical wires. The paper sheet is fixed on the table, the point to be digitized is located by a cross-hair, in some models with the help of a magnifying eye-piece, and an electro-magnetic device coupled to the cross-hair locates the pair of wires crossing under the point, giving thus the coordinates in table-based coordinates. Simple programs calculate then, on-line, new coordinates in a map-based system.

The data resulting from such manual digitization are, of course, in vector format.

Manual digitizing is a tedious work, requiring high concentration of the operator in order to avoid (or, more realistically, to limit) all sorts of human error. It is considered a bottleneck in geographical data processing.

5.2 Raster Scanners

Presently, instruments exist which are able to scan an image, i.e. a photograph or paper map, either transparent or opaque, which gives a digital image as output. Unfortunately, most geographical documents printed on paper are not suited for raster scanning because they contain too much information, i.e. all the text, so that they need pre-scanning editing such as the masking of some features with coloured ink, and post-scanning computer editing. This explains why manual digitizing, where the human eye, brain and hand choose directly what to digitize, remains in many cases the only way to digitize.

6. GEOMETRICAL TRANSFORMATION AND RASTERIZATION

6.1 Geometrical Transformation

Once the data are available in the computer in digital form, they are probably not yet aligned among themselves and with the satellite digital image. The exogenous data generally originate from maps with different scales and different projections so that some geometrical processing is necessary.

It should be noted that, whenever possible, geometrical processing should precede rasterization:

- the amount of data to be processed is much smaller,
- no resampling is needed,
- the original precision of the vector data is maintained (see paragraph 4.3.1).

6.2 Rasterization

For coded area data, rasterization is a straightforward process using vector format data and a sophisticated computer program which is able to recognize included areas (islands) or other complicated situations. The final result is a digital image.

7. DIGITAL TERRAIN MODEL (DTM)

The digital terrain model, also called digital elevation model (DEM),is a good example of exogenous data.

7.1 Definition

A DTM may be defined as a data table giving the altitude above see level of the nodes of a two-dimensional grid (generally with a square cell). One immediately sees that a DTM is very similar to a digital image and may be organized in computer memory with the same type of storage (raster format). However, the difference is that the grey level, in the case of a satellite image, is the reflectance averaged on the whole surface of the pixel while the altitude in a DTM is a single point value.

7.2 Construction of a DTM

Building a DTM from stereo-pairs of aerial photographs is a specialized photogrammetric operation. Even the preparation of a DTM starting from a topographical map is not a simple mathematical problem. The input data table (the result of digitization of the level curves) has a fixed altitude as entry and variable coordinates as arguments. The output (the DTM) needs fixed equally spaced coordinates (in two dimensions) as entry and variable altitude as arguments. All this means a huge problem of two-dimensional inverse interpolation.

The choice of the grid cell size is an important point. The main criterion is the spatial frequency of the altitude variations in the region under study; the cell should be small enough to follow "reasonbly well" the relief as it is represented on the map without losing further information. The pixel size of the satellite digital image must also be taken into account, but a scarse quality DTM, with too large a cell, should not be built just to fit a low resolution remote sensing image. A final step of direct interpolation to fit the two grids together will increase the cost of the DTM only by a small amount.

7.3 Use of DTM

Once the basic DTM exists, many interesting models may be easily calculated with straightforward computer programs, i.e. the slope and azimuth of the maximum slope direction may be calculated at every point of the grid. Once these two new pseudo-channels are established, it is possible to calculate for each pixel the incidence angle of the sun on the element of plane tangent to the ground, and this at any moment, e.g. at the moment of image acquisition by the satellite. It is also possible to determine at that time all the regions in the shadow, which is of importance in mountain areas.

If one disposes of a large amount of computer time, more complicated ideas might be thought of, for instance the calculation of the total number of sunshine hours received at a given pixel location, eventually corrected for incidence angle (but unfortunately not for cloud cover!).

It will be shown in the lecture on geometrical processing that high precision image rectification for the high resolution land-use satellites, Thematic Mapper and SPOT, is possible only if a DTM is used as a master image.

8. CONCLUSIONS

Already existing exogenous data can, when they are integrated with remotely sensed data, help to solve problems of land-use and environment. Depending on the application aimed at, the exogenous data may even be the fundamental data and remote sensing may add only complementary information.

One may hope that, in the near future, the availability of digital geographical data directly delivered by National Geographical Institutes, and the development of GIS (Geographic Information System) to process them efficiently, will generalize that type of study which is now limited to pilot projects due to the tedious phase of data digitization.

REFERENCES

Jackson, M.J., 1984: 'Remote sensing and digital cartography', Lecture Notes, Ispra Course on Introduction to Image Processing for Remote Sensing Applications.

Marble, D.F. and Peuquet, D.J. (Eds.), 1983: Geographic Information Systems and Remote Sensing, in: Manual of Remote Sensing, 2, 2nd ed., pp.923-957, Am. Soc. of Photogrammetry.

GEOMETRICAL PROCESSING OF IMAGES

1. INTRODUCTION

As is explained in other lectures of this course, any serious study in the field of land-use, agriculture and forestry has to use multi-temporal data, either from the same satellite at different seasons or, more probably, data from different satellites with different sensors. The evolution of the spectral response of an object is often more characteristic than a so-called 'spectral signature'.

From the point of view of data processing, this means that images covering the same region but which have been acquired by platforms with different orbit inclinations, different scan angles (small for high resolution land-use satellites and large for meteorological satellites), different pixel forms and sizes (from 10 by 10 metres for SPOT to 60 by 80 metres for LANDSAT MSS) or different viewing angles (for SPOT scenes), are used together. Even when only human visual interpretation is foreseen, working on images with different geometrical properties, different scales and which are rotated one to the other, involves horrible mental gymnastics and is, anyway, a source of error in the localization of observed phenomena.

When computer-aided interpretation is to be used, all images have to be registered with precision. The errors of registration should possibly be one order of magnitude smaller than the average size of the smallest objects of interest in the study. These remarks and what has been said about exogenous data integration explain why geometrical processing of digital images is unavoidable in most projects which use remote sensing techniques.

The problem of geometrical processing of images can be summarized in the two following statements:

- Where does this pixel of the image lie on the ground?
  and the inverse question:
- To which pixel of the image corresponds that point on the ground?

The processing techniques used are very different from those described in other lectures of this course. Such techniques as contrast enhancement, false colour coding, vegetation index, principal components, etc. treat the image pixel-by-pixel, working in channel spaces without explicit reference to the spatial coordinates (lines and elements) of the pixels. On the contrary, in geometrical processing, one works in geographical space and aims at moving the pixels (which means changing their coordinates) without modifying the grey level values.

2. SOME DEFINITIONS

2.1 Digital Image

This is the matrix of numbers, called pixels, grey levels, intensities, etc. The image is organized in lines and elements (also named columns) in order to allow to address single pixels; this is also called raster format. The value of the grey level of pixel x,y is noted Gr(x,y).

2.2 Master Image

This is the image on which the corrected image has to be superposed. It may exist as a digital image or not, it may be a map, an orthophoto, or any other document. The coordinates (lines, elements) will be noted x,y.

2.3 Input Image

This is the image which will be deformed in order to match the master image. It generally exists as a digital image. The coordinates (lines, elements) are noted u,v. It is often useful to split the coordinates into their integer and decimal parts writing

$$u = Iu+Fu, \quad v = Iv+Fv \quad \text{with } 0 \leq Fu < 1$$

2.4 Output Image

This is the digital image after transformation, the grey levels come from the input image, the geometry is that of the master image.

2.5 Rectification or Registration

When the master image is a geometrically well-defined document, i.e. a map, one speaks of image rectification or geometric correction. When the master image has no particular geometrical properties, i.e. when it is a satellite image of a particular date, one rather speaks of image registration. In such a case the choice of the master image is sometimes quite arbitrary.

3. GEOMETRIC DEFORMATION MODEL

The problem of geometric correction of an image may be split into two stages:
- the geometric deformation model definition and calculation;
- the actual manipulation of pixels for image transformation.

3.1 Direct or Inverse Functions

The geometric deformation model is defined by functions mapping one image on the other.

3.1.1 Direct functions are functions from

$$R*R \text{ to } R*R: (u,v), (x,y)$$

such as:

$$\begin{aligned} x &= f(u,v) \\ y &= g(u,v) \end{aligned} \qquad (1)$$

Given a pixel in the input image (u,v), these functions give its position in the master image after transformation.

3.1.2 Inverse functions are functions from

$$R*R \text{ to } R*R: (x,y), (u,v)$$

such as:

$$u = F(x,y) \qquad v = G(x,y) \tag{2}$$

Given a pixel (x,y) in the output image to be created, these functions indicate where to look for the corresponding grey level in the input image.

3.1.3 These functions, defined here in a quite general way, cannot always be expressed in a simple analytical form. It is sufficient that an algorithm exists to calculate them.

Direct and inverse functions are, of course, not independent; one can, theoretically, calculate functions (2) by solving the system of simultaneous equations (1) and vice-versa. Practically, this is possible only for very simple forms of the functions.

3.1.4 The geometric transformation of an image may use either direct or inverse functions. Each method has advantages and drawbacks.

Direct functions are often easier to be calculated but the image created will seldom be immediately usable; after round-off to integers, it may have holes when a pixel (x,y) has no antecedent (u,v), or several possible grey levels when several pixels (u,v) have the same image (x,y). Another step of processing, i.e. interpolation of missing pixels, will be necessary.

Inverse functions do not have this defect but may be more difficult to be determined.

3.2 Modellistic Approach

When the flight parameters (trajectory, speed, attitude of the platform) are known with good accuracy, an a priori model of geometric correction of the image may be built which will give either direct or inverse functions. This approach is, of course, specific for each platform.

3.2.1 Example: the panoramic correction of plane flight. This correction is applied to images from an airborne scanner with large scan angle (90° or more). As the altitude h of the plane is small as compared to the earth radius, the earth may be considered as being flat. The mirror of the scanner rotates with constant angular velocity and the sampling is done at constant time intervals. The result is that the angular element w is constant for each pixel but that the element of surface intercepted on the ground changes due to obliquity, increasing with the distance from nadir (point O in Figure 2). On the visualization screen, where the pixels are equidistant square points, the external

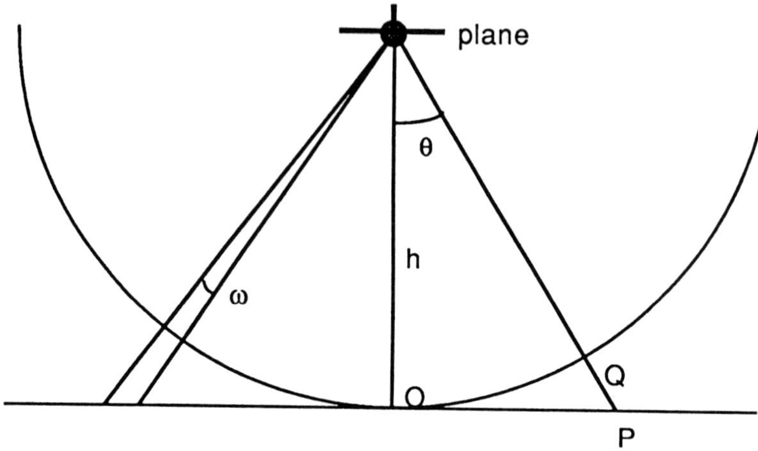

Figure 2. Panoramic correction scheme.

parts of the image appear squashed.

More precisely, the raw image is the development of a cylinder which has the flight line as the axis and on which the ground has been centrally projected (see Figure 2).

Inverse transformation functions will now be calculated. For simplicity, elements along the line are numbered starting from nadir.

Correction is made for each line independently so

$$u = x$$

s is the distance between two successive lines of the image; as one wants square pixels, s must also be the interval between elements along the line (in other words the pixel size). The pixel numbered y on the output image lies on the ground at a distance from nadir:

$$OP = y*s = h*tg\theta$$

The corresponding pixel on the cylinder (input image) is in Q at an angular distance q from nadir and is numbered:

$$v = q/\omega = artg(ys/h)/\omega$$

It is only necessary to know two variables, ω and s/h, which depend upon the speed of the plane, the rotation speed of the scanner mirror and the sampling time interval. It is not necessary to know accurately the flight altitude of the plane. As v will generally not be an integer, resampling of the image will be necessary; cubic convolution in one dimension will be used (see 4.2.3).

The panoramic correction as is described here supposes that the plane followed perfectly its ideal flight conditions (constant altitude, constant speed, no roll, no pitch, no yaw), so it is only a first approximation.

3.2.2 Correction of a satellite image. In the case of a satellite, the situation is more complicated. The earth curvature has to be taken into account and the earth rotates under the satellite between the registration of two successive lines.

3.2.2.1 Correction with nominal orbit parameters. The parameters involved in the model are:
- the geometric data of the orbit: altitude, excentricity, inclination on the equator;
- the cinematic parameters: speed of the satellite, angular velocity of scanning mirror, sampling time interval;
- earth parameters: curvature (sphere or geoid according to the type of satellite and the precision required), earth rotation under the satellite, type of map projection on which the corrected image must be superposed.

3.2.2.2 Correction with actual orbit parameters. There are always small deviations from the nominal values of orbit parameters, in particular: change of altitude and deviation of attitude of the satellite on the orbit. The axis of the scanner is not perfectly parallel to the velocity vector of the satellite (yaw and pitch angles) and the reference point of the scanner may not be directed to nadir (roll angle). These corrections of attitude cannot always be measured with sufficient precision so as to make a complete a priori model, but as the rate of variation of these parameters is slow they can be determined from a few control points of known position on the earth.

Models of this type are becoming available; this type of correction needs parameters which are known only by the receiving stations and image distribution centers; it uses huge computer software and the correction is made before image distribution; the resulting images are said to be "system corrected".

3.2.2.3 Example: SPOT pre-processing levels. SPOT images will exist at different pre-processing levels according to the degree of correction used:
- level 1: the images are corrected radiometrically and geometrically line by line but not from one line to another, taking into account earth' curvature and rotation, altitude and pointing angle of the satellite (in case of oblique vision);
- level 2: the images are completely corrected in both directions (lines and columns) and are rectified, using ground control points, to match a geographic graticule (UTM, Lambert or other). Relief is not taken into account;
- level 3: relief data are introduced in the correction model (they must be known from another source). A precision of 10 metres is claimed in flat or mountainous regions;
- level 4: a stereoscopic pair of oblique images is used to measure the relief and build a digital terrain model which may be used for correction of level 3 images.

3.3 Empirical Approach

In this approach, no a priori knowledge of orbit or attitude parameters is needed. The transformation functions contain undefined constants which are, in each case, calculated by fitting ground control points (GCP)

$$\begin{aligned} u &= F(x,y,a_1,a_2,...,a_n) = F(x,y,\mathbf{a}) \\ v &= G(x,y,b_1,b_2,...,b_n) = G(x,y,\mathbf{b}) \end{aligned} \qquad (3)$$

3.3.1 <u>Ground control points.</u>

3.3.1.1 Definition. A control point is a "point" which may be recognized and located with precision on both images (the master and the input image). Depending on resolution and scale, it may be a lake or an island (for small resolution images), a cross-roads, a building, a big isolated tree (for high resolution satellite or airborne scanner imagery). Selected GCPs should be stable in time; details on beaches and confluents of natural rivers which move with the season and with erosion, should be avoided despite their good contrast on image.

3.3.1.2 Search of GCPs on a digital image. The coordinates of ground control points are often searched visually on a visualization console. The point is located with the track-ball pointer; a program records the coordinates (lines and columns), and some comment is given, creating a file which may be edited later on.

Automatic search of GCPs is possible in particular cases (see paragraph 5.3).

3.3.1.3 Search of GCPs on paper documents. The coordinates are read using a digitizing table, already described in another course. In order to avoid further manipulations of coordinates, the pixel size of the output image should be defined at this stage. The factors to be taken into account are the pixel size(s) of the input image(s), the scale of the final printed document, together with the dimensions of the spot of the image restitution device.

3.3.2 <u>Fitting an empirical model.</u>

3.3.2.1 Least squares method. Each pair of GCPs (x_i,y_i), (u_i,v_i), introduced in functions (3) furnishes two equations in the unknown parameters **a**,**b**. As one needs as many equations as unknowns, the number of GCPs must be at least equal to half the number of parameters to be determined. Actually, in general, one prefers to have an over-determination of the problem, which means to have more equations than unknowns. The problem is then solved by the least squares method minimizing the sum:

$$S = \sum_{i=1}^{K} (u_i - F(x_i,y_i,\mathbf{a}))^2 + (v_i - G(x_i,y_i,\mathbf{b}))^2$$

extended to the K available GCPs.

From a geometrical point of view, in the input image the distance between a GCP and the image (calculated with the inverse functions) of the same point of the master image is minimized in the least squares sense. The values of these distances at the end of the minimization process are called residues of the least squares process.

When the F and G functions are linear functions of the parameters **a** and **b**, which is the case for polynomials of whatever degree in x,y, it is easy to show that the values of **a** and **b** which minimize S are solutions of a symmetrical system of linear equations often called "normal equations". The problem can then be solved in one step in a straightforward way (some care should be taken to control possible ill-conditioning of the linear system).

When the functions are not linear in the parameters, the problem is much more complicated and has to be solved by successive approximations. Efficient routines for non-linear fit are available in all mathematical packages of computer programs. An approximate value of the parameters, the "first guess", is then often requested.

3.3.2.2 Remarks. This least squares method avoids the amplification of random experimental errors on the coordinates of GCPs, errors of the order of 1 or 2 pixels, due to the subjective choice of the observer.

A good degree of overdetermination should be in the range of 2 to 3 equations per unknown.

The results are better if the GCPs are distributed on the whole image to be transformed (even outside the frame, if possible. In this case the coordinates are everywhere interpolated among the GCPs and not extrapolated).

3.3.3 Some empirical models.

3.3.3.1 Affine transformation. The transformation functions are bivariate polynomials of first degree

$$u = a_o + a_1x + a_2y$$
$$v = b_o + b_1x + b_2y$$

Six coefficients have to be determined, at least three control points are needed. Generally, about ten points are used.

Figure 3 shows how a square lattice is transformed. This transformation allows the translation of origin, change of scale variable for different directions, and/or different rotation angles for different directions. Straight lines are transformed into straight lines, a set of equidistant parallels is transformed into a set of equidistant parallels (generally with different direction). A circle is transformed into an ellipse. The transformation is homogeneous: a circle of constant radius is transformed into the same ellipse everywhere in the plane.

The affine transformation is practically the only case in which direct and inverse functions may be deduced easily one from the other.

This transformation is widely used and is generally sufficient to fit satellite images on UTM maps when the scan angle is small (case of LANDSAT and SPOT).

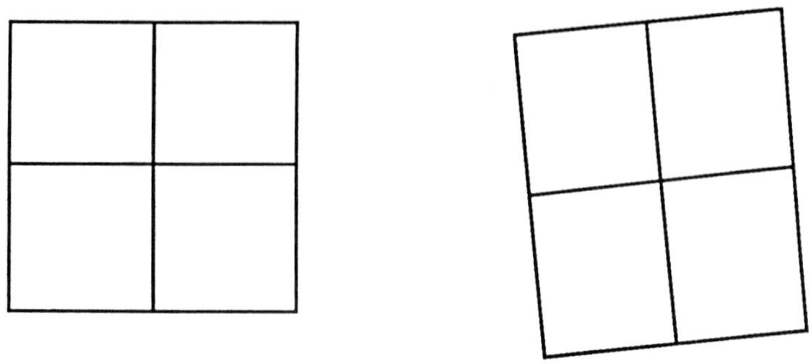

Figure 3. Affine transformation of a square lattice. $u = 1.1x-0.1y$; $v = 0.1x+0.9y$.

3.3.3.2 Higher degree polynomials. Bivariate polynomials of second or third degree offer more possibilities; straight lines are transformed into curves, the transformation is not homogeneous, change of scale is no longer constant at every point of the image. Figure 4 shows how a square lattice is transformed.

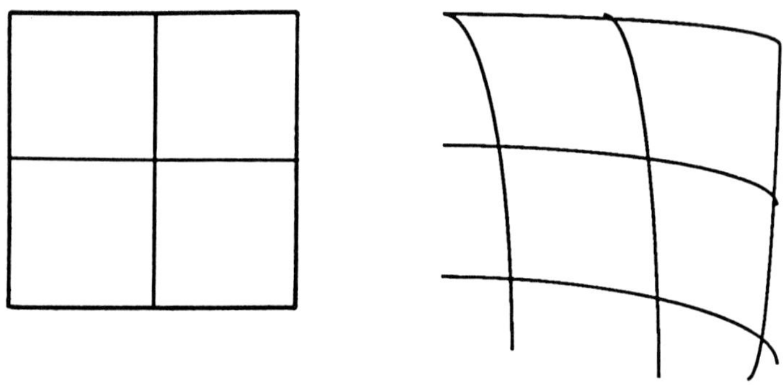

Figure 4. Transformation of a square lattice by high order polynomials.

When using more parameters, which means a greater number of degrees of freedom, these functions will fit the control points better than affine transformation and will give smaller residues. The drawback is that they may be worse when interpolating between the GCPs and they are unsuitable to transform an image by extrapolation outside the perimeter of control points.

3.3.3.3 Projective transformation. This is the well-known transformation used in photogrammetry. The transformation functions are:

$$u = (a_o+a_1x+a_2y)/(1+c_1x+c_2y)$$
$$v = (b_o+b_1x+b_2y)/(1+c_1x+c_2y)$$

They depend on 8 coefficients and are determined by at least 4 control points. Geometrically, it corresponds to the central projection of a plane on another plane from a point outside both. Straight lines are transformed into straight lines, a pencil of parallels is transformed into a convergent pencil. The transformation is not homogeneous. Affine transformation is a particular case of projective transformation. Figure 5 shows the effect of this transformation on a square lattice.

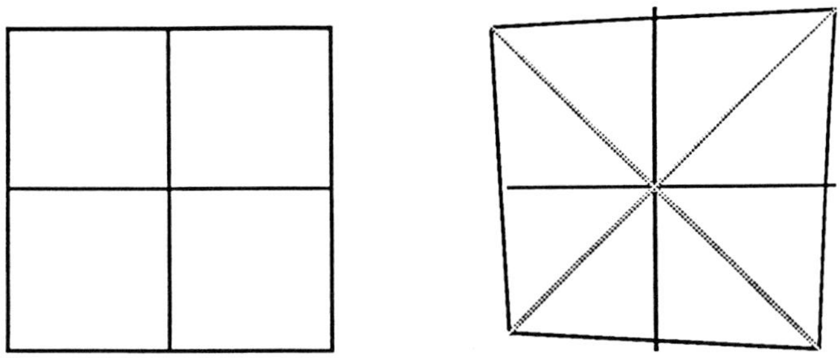

Figure 5. Transformation of a square lattice by projective transformation: $u = (1.1x-0.1y)/(1+0.1x-0.1y)$; $v = (0.1y+0.9y)/(1+0.1x-0.1y)$.

3.3.3.4 Rubber stretched triangles. Instead of using all available control points to define a unique transformation of the whole image, this approach defines several affine transformations, each defined by three control points and valid inside the triangle which has these points as vertices. Continuity is maintained along the sides of the triangles. This transformation is used for registration of images with unknown geometric characteristics but well-localized control points. It has been used for registration of night and day HCMM images years ago. Figure 6 shows the effect on the usual square lattice, straight lines are transformed into continuous broken lines.

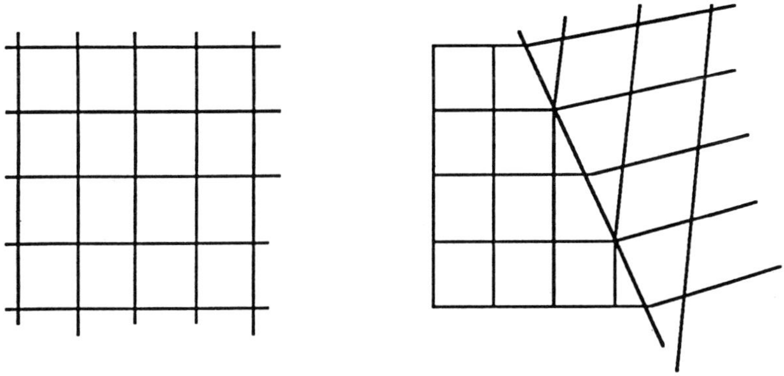

Figure 6. Effect of contiguous affine transformations on a square lattice.

3.4 Problems Due to High Resolution Satellites

With the new generation high spatial resolution satellites TM and SPOT, some new problems appeared in the geometric processing of images.

3.4.1 Local scale of maps. A topographic map is not a true scale document, roads, railways and buildings are represented bigger than the general scale of the map and this can introduce errors of several pixels in the coordinates of the control points. Geodesic points are perfectly located on a map but seldom visible on a satellite image. When orthophotographs, rectified to the same projection as the map, are available, they may bring a good help in choosing "good" control points. As these errors are random the least square model can partially correct them. A possible solution is to rectify a first image on the map and to use it later on as a master image.

3.4.2 Effect of relief. For LANDSAT MSS images, the effect of relief can generally be neglected; a difference in altitude of 1000 metres on the extreme side of an image brings a horizontal displacement of one pixel. With high resolution satellites, the effect is much bigger (see Figure 7) for nadir vision, and specially for oblique vision in the case of SPOT. This means that precision rectification of images is no longer possible using a "flat" map as a reference but needs a digital terrain model as a master image.

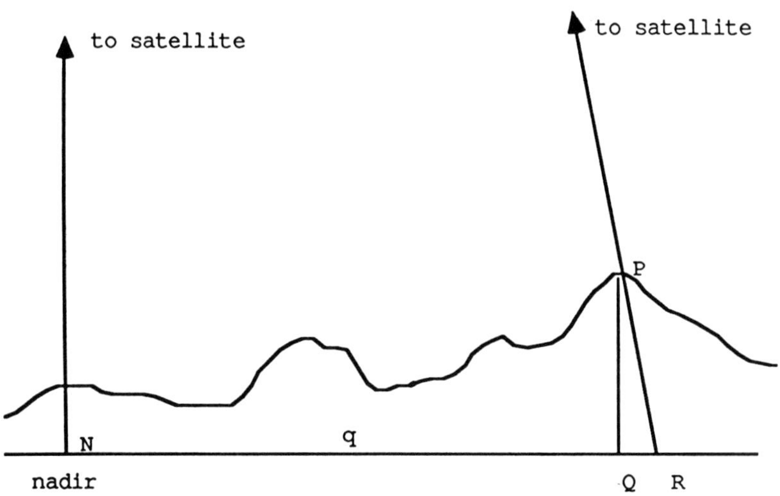

Figure 7. Point P on the ground is projected in Q on the map but seen in R by the satellite. If the height PQ is known from the DTM and the angle QPR calculated from the distance NR, the correction QR can be calculated.

3.5 Two-Step Geometric Correction

It is sometimes useful to carry out the correction process of an image in two steps, i.e. a first step with panoramic correction followed by a final fit with an affine transformation. Even if intermediate images have to be produced for the search of control points or for other reasons, it is important to organize the program in such a way that the final output image is produced directly from the input image avoiding several resamplings in chain.

4. IMAGE TRANSFORMATION

Once the geometric deformation model has been defined, the image must be actually transformed. Two cases apply:

4.1 Transfer of Lines

If lines, stored in vector mode (i.e. administrative limits) are to be transferred on a digital image, the direct functions are used, the coordinates of the line pixels are calculated, they are rounded off to integers, and a fix grey level is given to them; this process is mainly used for the presentation of final documents and it implies, of course, some loss of information from the original image.

4.2 Transfer of Digital Image, Resampling Problems

When a whole image has to be transferred to a master image, the easiest way is to scan the master image line by line and, using the inverse functions, to calculate where the corresponding grey level in the input image must be fetched. The coordinates (u,v) transforms of the pixel (x,y) will not be integers. As the input image is addressable in integer lines and elements, a rule must be defined for choosing the grey level which will be given to the pixel (x,y); the most often used rules are: nearest neighbour, bilinear interpolation, and cubic convolution.

4.2.1 <u>Nearest neighbour.</u> The grey level is simply that of the pixel in input image with coordinates (u,v) rounded off to the nearest integer.

4.2.2 <u>Bilinear interpolation.</u> The resulting grey level is a weighted average of the values of the four neighbours of point (u,v). Using the notation defined in paragraph 2.3:

$$Gr(x,y) = Fu*Fv*Gr(Iu+1,Iv+1) + (1-Fu)*(1-Fv)*Gr(Iu,Iv) + \\ + (1-Fu)*Fv*Gr(Iu+1,Iv) + Fu*(1-Fv)*Gr(Iu,Iv+1)$$

this formula is the result of two successive linear interpolations, in the line and in the column directions.

4.2.3 Cubic convolution. It is probably the most popular scheme of resampling used in image processing. It is justified by theoretical considerations of the signal theory.

The grey level Gr(x,y) is a linear combination (not a weighted average since some of the coefficients are negative) of the grey levels of the 16 pixels surrounding the point (u,v) in the input image. The coefficients are calculated using an approximation by polynomials of third degree of the function $\sin\pi d/\pi d$ where d is a local variable with origin at the point to resample.

It is generally applied in two successive one-dimensional steps to diminish the amount of numerical calculations. The following basic formula is used:

if u = U+t with U integer and $0 \leqq t < 1$ (instead of u = Iu+Fu), then:

$$Gr(u) = Gr(U-1)\sin\pi(1+t)/\pi(1+t) + Gr(U)\sin\pi t/\pi t +$$
$$+ Gr(U+1)\sin\pi(1-t)/\pi(1-t) + Gr(U+1)\sin\pi(2-t)/\pi(2-t)$$

where $\sin\pi d/\pi d$ is approximated by:

$$f(d) = 1-2d^2+d^3 \qquad \text{for } 0 \leqq |t| < 1$$
$$f'(d) = 4-8d+5d^2-d^3 \qquad \text{for } 1 \leqq |t| < 2$$

giving:

$$Gr(u) = Gr(U-1)f'(1+t) + Gr(U)f(t) +$$
$$+ Gr(U+1)f(1-t) + Gr(U+2)f'(2-t)$$

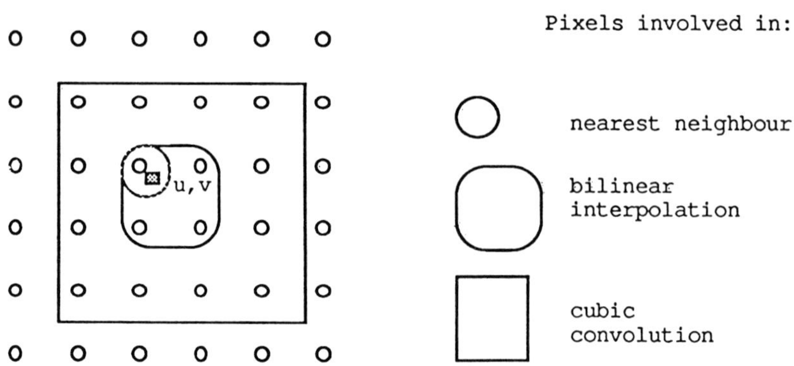

Figure 8. Surrounding pixels of point u,v involved in the different schemes of resampling.

4.2.3 Remarks on resampling. Other methods of resampling exist such as interpolation by polynomilas of degree 3, interpolation by spline functions, etc. Each of the three methods described here had advantages and drawbacks:

- the cost in computer time increases with the complexity of the scheme

used and the number of pixels involved in it. Cubic convolution is particularly heavy from that point of view;
- the nearest neighbour method keeps the original radiometry of the image, but may, due to round-off, introduce short-range deformation in the image;
- both other methods modify radiometry by a filtering effect (see paragraph 3 of Filtering), but give more regular images.

A complete discussion of effects of resampling on radiometry is given by Traizet (1981).

5. QUALITY CONTROL AFTER GEOMETRIC PROCESSING

5.1 Residues of the Least Squares Model

The values of the residues at the end of the least squares process (see 3.3.2.1) give two types of indication:
- in the first stages of the correction process, a large residue generally indicates a systematic error in the coordinates of the corresponding GCPs. This may be due to wrong identification of the point or to transcription errors of the coordinates;
- at the end of the process, the residues show the adequacy limit of the empirical model to take into account certain types of deformation present in the input image. In this case, there is a systematic drift in sign and direction of residues according to their position in the image. It is then possible to change the model, to segment the image or to take any ad hoc decision.

When calculating the deformation model, these types of control must be done before creating any output image.

Although large residues nearly always are an indication of bad results on the image, the reverse statement is not always true (see paragraph 3.3.3.2).

5.2 Visual Inspection of the Image

The aspect of the output image is an important first control; are straight roads still straight on the image?

Precise comparison of the two images (output and master) by flickering on the screen of the visualization console or on a light box if transparencies are available, is also important especially when high-degree polynomilas are used; it can give good match at the GCPs and bad interpolation between them.

5.3 Image Correlation

When two digital images are available, image correlation may be used as a control of superposition. The correlation of grey levels is calculated for different relative positions of the two images and the offset, which gives the maximum of correlation, is thus found. In fact, one must work on relatively small windows to obtain measures of local misfit in different regions of the image and to stay within reasonable computer time. A good order of magnitude of the window is 10*10 pixels.

This method has advantages:

- it is objective (independent of the observer);
- it appreciates the coincidence of images for surfaces and no longer for point details;

and disadvantages:

- a sharp maximum of correlation exists only if $\partial Gr/\partial x$, $\partial Gr/\partial y$, $\partial^2 Gr/\partial x \partial y$ are all large in the window; this means high and changing contrast in both directions, i.e. a cape, a highway cross-roads, etc.;
- there are spurious correlation maxima (sometimes even higher than the "true" maximum).

This correlation method, applied to the master and the input image, may be used before correction to retrieve the coordinates of the ground control points in one of the two images.

The last two remarks show the difficulties of implementing such a method in a completely automatic way either for control or for automatic GCP search.

6. IMPLEMENTATION OF GEOMETRIC PROCESSING PROGRAMS

6.1 Intermediate Numerical Files

Once the deformation model is completely defined, two approaches remain possible:

- the output image may be created immediately after the coordinates (u,v) are calculated for each pixel of the master image;
- alternatively, these coordinates (u,v) can be stored in an intermediate file on disc for the whole image and the output image can be created in a second step reading the intermediate file.

The advantages of the second method are evident when multichannel images have to be transformed, while it avoids the repetition of the calculation of the deformation model for each channel. It may also be advantageous when a two or several step transformation is used specially when only direct interpolation is implied. Inverse interpolation in such a bidimensional file might be expensive in computing time.

6.2 Lattice Transformation and Interpolation

When the calculation of (u,v) is complicated, i.e. for high-order polynomials, or when the direct functions have to be inverted by numerical methods, it is wise to exactly calculate (u,v) through the model, only for the nodes of a lattice with cells of a few tens of pixels and then calculate the coordinates of each pixel inside the cell by linear interpolation.

6.3 Memory Versus Input-Output

Very often disc access to digital images is possible only line-by-line. The output image will thus be created line after line. The transformation of a line of the master image generally cuts several (n_1 = some hundred) lines of the input image, or, from the other point of view, each line of input image contributes to several lines (n_2) of the output image. Two solutions are then possible:

- read each line of input image n_2 times and keep the minimum number of lines (4 in case of cubic convolution resampling) in the central memory;
- read each line only once and keep it in memory as long as it is useful, keeping then a block of images of max (n_1) lines in memory.

This choice depends on the capabilities of the host computer, memory size, possibility of asynchronous reading, etc.

7. CONCLUSIONS

Although the necessity of geometrically corrected images may now be obvious, the problem of when, where, and by whom (the satellite manager, the NPOC, the final user?) this pre-processing should be made remains open.

The quality of the system-corrected distributed images is now good and will still improve in the next years; images of the same frame for two dates will probably by superposable by simple translation and, as long as paper prints are used, match a map by simple rotation of the sheet.

Anyway, these images remain rectified to an "orbit" projection; the vertical is not oriented N-S but makes a different angle for each satellite (this is the price of sun-synchronous satellites).

So, if the user needs images from different platforms and does not like to tilt his head in front of the console screen to bring the North direction parallel to his cerebral vertical, he will have to process the data himself. When raw data are not available, the risk of "chain-resampling" exists, and after two cubic convolution resampling processes, each final pixel "feels" from about 144 original pixels in the neighbourhood for the best "health" of the radiometry.

So, the future of Commissions of Experts working on the standardization of distributed images for remote sensing is still very bright for some years to come.

REFERENCES

Anuta, P.E., 1977: 'Digital registration of topographic data and satellite MSS data for augmented spectral analysis', LARS Information Note 011277, The lab. for Appl. of Rem. Sensing, Purdue University, West Lafayette, Indiana.

Bernstein, R., 1976: 'Digital image processing of earth observation sensor data', IBM J. Res. Devel., 20, 40.

Traizet, M., 1981: 'Qualités radiométriques des images rectifiées, illustration des diverses méthodes de ré-échantillonnage', Note Technique No. 102, CNES, Toulouse 31055.

IMAGING RADAR FUNDAMENTALS

Daniel R. Nüesch
Remote Sensing Laboratories, Univ. of Zürich, Switzerland

Eric S. Kasischke
Environmental Research Institute of Michigan, ERIM,
Ann Arbor, USA

ABSTRACT. The purpose of this chapter is to give background material for a better understanding of remote sensing in the microwave region which differs quite substantially from techniques used in the optical region. Whereas those systems operating in the visible spectrum detect the reflected solar radiation, the radar systems provide their own sources of illumination and detect the reflected energy as it is scattered back from the earth surface. The fundamental concept on how an imaging radar works is explained. Special emphasis is given to the SAR image formation and parameters influencing the intensity of radar return from a given surface.

1. INTRODUCTION

Photographic cameras, TV cameras, scanning radiometers, multispectral scanners and imaging spectrometers cover the usual range of passive imaging devices which operate by detecting the solar radiation reflected by the earth or the emitted radiation of the earth. The reflected solar radiation is prominent in the 0.4 to 3 μm range and the emitted thermal radiation of the earth in the 4-30 μm range. In the microwave wavelength region (0.3 mm - 1 m), diffraction laws which relate resolution to the ratio of the antenna size (aperture size) to the wavelength, severely limit the practically achievable resolution. Thus, at microwave wavelengths, synthetic aperture radar (SAR) plays an increasingly significant role in spite of severe weight, power and data processing requirements. Radar's performance of being equally effective in day or night operation and providing consistent image quality in all weather conditions, i.e. the ability to observe the earth surface through most clouds, will possibly make SAR the dominant sensor for earth observation from satellites in the more distant future.

Although the transmission and reflection of electromagnetic waves were already demonstrated in 1887, radar (an acronym for RAdio Detection And Ranging) was not used for remote sensing purposes until World War II, for the detection of ships and aircraft. Imaging radars were developed as a consequence in the 1950s. In contrast to most other

F. Toselli (ed.), Applications of Remote Sensing to Agrometeorology, 157–180.

types of remote sensing systems, imaging radars provide their own source of illumination in the microwave portion of the electromagnetic spectrum. Because radars operate at wavelengths on the order of 10,000 times longer than those in the visible spectrum, they are different from the conventional visible and infrared spectrum imaging systems in the following ways:

a) radar does not detect the visible colour of the surface, but the moisture and the electrical properties of the surface;
b) radar systems record the phase and polarization (orientation of the electromagnetic wave) characteristics of the reflected microwave pulse;
c) radar is extremely sensitive to the surface roughness of the area being imaged;
d) radar produces images with speckle (image noise) due to the coherent nature of the system;
e) radar produces images with geometric distortions, due to relief displacements such as foreshortening, layover and shadowing.

The early imaging radars were flown in aircraft and are referred to as real aperture radar, or Side Looking Airborne Radars (SLAR). As shown in Figure 1, a SLAR sweeps a series of pulsed beams down and sideways from the sensor platform and records the reflected signal. Real aperture radar images are acquired by successively recording narrow side-by-side strips of the terrain.

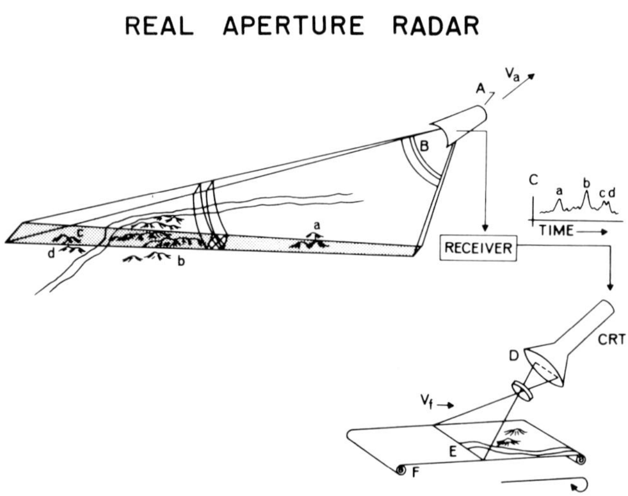

Figure 1. Geometry of a SLAR system (adopted from H. MacDonald).

The antenna first transmits a series of radar pulses and then receives them as they are reflected back from the earth's surface. When these pulses are detected by the receiver, it sends a signal to a cathode ray tube (CRT) which records the radar image on a photographic film. The film advances between each set of transmitted and received pulses, thus forming an image of the earth surface.

Since SLAR systems measure the distance to objects on the terrain by means of microwave pulses, the resolution of the image in the across track (range) direction is proportional to the length of the microwave pulse. Two objects can be resolved if their respective ranges are separated by at least half the pulse length. For example, a pulse length of 10^{-7} seconds is 30 m long, providing a resolution of 15 metres. By reducing the pulse length or using pulse compression techniques, finer resolution can be achieved.

The resolution in the along-track (azimuth) direction is proportional to the width of the beam sent out by the antenna. Since the angle of the beam remains essentially constant across the range, the width of the beam increases with range and the resolution in the along-track dimension deteriorates with range. The angular width of the antenna's beam is inversely proportional to the length of the antenna, so that the resolution could be improved by employing a larger antenna with a narrower beam width. Unfortunately, the physical length of an antenna that can be mounted on an airplane is limited by the plane itself.

Excellent imagery has been obtained from real aperture SLAR but the problem of the azimuth resolution degradation with distance makes real aperture radar incapable of imaging from space. However, it is possible to make a relatively short antenna with a wide beam "synthesize", or behave like, a very long antenna with a narrow beam, by taking advantage of the motion of the platform of the sensor. Synthetic Aperture Radar (SAR) was developed to solve this azimuth resolution problem (Figure 2). This is accomplished in the following way:

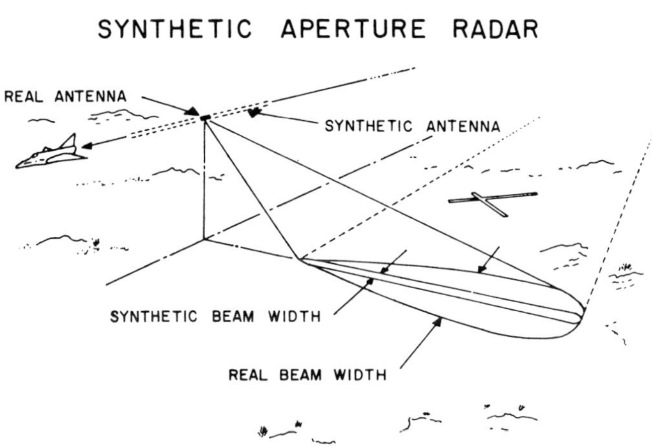

Figure 2. Geometry of a SAR system (adopted from H. MacDonald).

The short real antenna emits pulses at regular intervals along the flight path. As the platform flies past a feature on the earth surface below, the feature enters the antenna's beam, moves through the beam and finally leaves it. During the entire time the feature is in the beam, it reflects the series of transmitted microwave pulses.

Now, the greater the range of the feature to the sensor the longer it remains in the beam of the antenna. Seen from the features on the earth surface, the synthetic antenna is longer for more distant features and shorter for closer features. In fact, the effective length of the synthetic antenna is directly proportional to the length of the antenna but inversely proportional to the range of the feature. Since resolution is proportional to the length of the antenna but inversely proportional to the range, for synthetic aperture radar the two effects compensate for each other and the resolution of the image in the azimuth direction remains the same at all ranges.

Each feature (object) the radar illuminates is represented by a set of backscattered pulses, not just a single pulse. By using coherent energy, the phase histories of these multiple pulses can be recorded and processed into a SAR image. Representing each feature by a great number of backscattered pulses results in a tremendous amount of data for a given SAR scene.

At the beginning when SAR were first developed, there was a problem not only in storing the SAR phase histories, but also in processing these histories into an image. These problems were first solved by recording them on a signal film and processing this film, into an image film using an optical processor. Optical processing uses coherent laser beam illumination and the Fourier transform properties of lenses to compress the SAR signal histories simultaneously in two dimentions. Today digital techniques are being introduced more and more in SAR processing due to rapid advances in integrated circuit technologies. Hard-wired computers as well as Fast Fourier Transform (FFT) algorithms are being employed to speed up the computation time in order to achieve a throughput of several 100 million operations per second, the rate needed to guarantee a near real-time processing of today's SAR systems.

2. GENERAL RADAR CONCEPT

As mentioned in section 1, a radar system sends out energy in the form of a radio wave which strikes an object on the earth surface, causing the energy to be reflected back. The radar system, therefore, senses two things at the same time, namely:

- the intensity of the reflected wave; and
- the time taken for the round trip of the wave.

Sensing reflected intensity is the detection and measuring the time of reflection is the ranging. Radio detection and ranging (RADAR) is precisely what a radar system does.

Since the speed of radio waves is known (3×10^8 m/second), measurements of travel time of the reflected wave are easily converted to distance, which in radar usage is always called range. The range (R) is

$$R = ct/2 \qquad (1)$$

where c is the speed of light and t is the roundtrip time from the radar antenna to the reflecting object and back. Range is, therefore, used to indicate the location of an object in the cross-track dimension.

While discussing imaging theory, the resolution of a radar system is defined as the radar's ability to discriminate between two point (bright) targets in a particular scene. Two fundamentally different principles govern resolution in a radar system. Azimuth or along-track resolution is determined by the beamwidth of the radar. The beamwidth (β) is the angle by which a radar beam expands and is a function of the antenna size and the range as illustrated in Figure 3. The location of a reflecting object is somewhere within the cross-matched area, whose along-track dimension, or azimuth resolution (ρ_a), is

$$\rho_a = R\beta \tag{2}$$

The beamwidth of a radar is primarily a function of the antenna length, with a larger antenna producing a narrower beam. Therefore, the only way for a real-aperture radar to have fine azimuth resolution is to have a very large antenna and a very short range. Synthetic aperture radar overcomes this problem in a very sophisticated way, as will be discussed in section 6.

The range resolution (ρ_r) of a radar system is

$$\rho_r = c\tau_p/2 \tag{3}$$

where τ_p is the duration of a simple transmitted radar pulse. The range resolution is the distance by which two objects must be separated to be detected at two specific ranges by a radar (Figure 4).

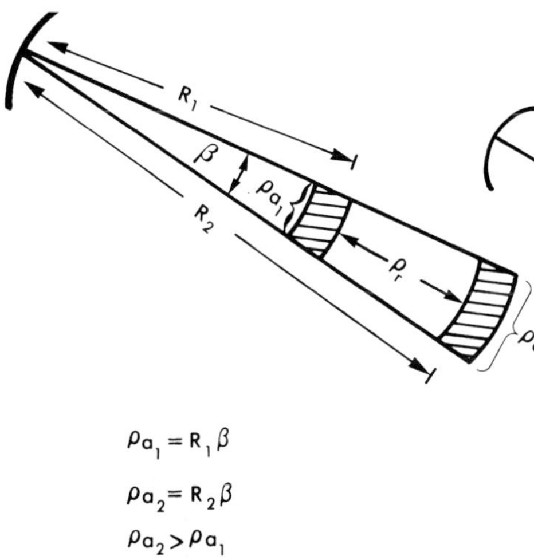

Figure 3. Dependence of azimuth resolution on range and antenna beamwidth (adopted from Kasischke et al., ERIM)

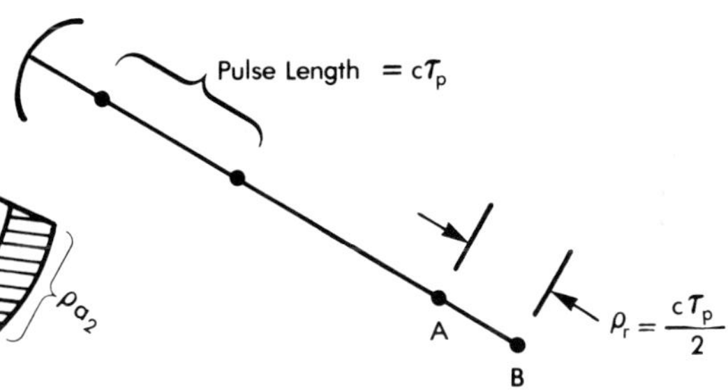

Figure 4. Dependence of range resolution on pulse duration (adopted from Kasischke et al., ERIM)

An important characteristic of imaging radars is that their transmitted pulses are polarized. Polarization is defined as the direction of the electric vector of an electromagnetic wave. The electric vector is the plane in which the electromagnetic energy is transmitted. While visible light is unpolarized because the direction of the electric vector is randomly distributed, imaging radars emit pulses either horizontally polarized, i.e. the electrical vector is parallel to the surface being illuminated, or vertically polarized, i.e. the electrical vector is perpendicular to the surface. The radar antenna then receives either the horizontally or vertically polarized return. In the near future, radars may have a circular polarization, where the plane of the electromagnetic waves actually rotates as the waves propagate. The effect of terrain on polarity of the radar return signal is still a complex research topic.

3. WAVELENGTH EFFECTS

Radar systems operate in microwave frequencies between 1 and 100 Gigahertz with wavelengths of 30 cm to 0.3 mm. Common imaging radar frequencies are in L-band at 1.25 GHz with 24 cm wavelength, in C-band at 5.3 GHz with 5.6 cm wavelength, in X-band at 9.4 GHz with 3.2 cm wavelength or in K-band at 30 GHz with 1 cm wavelength.

Since radar wavelengths are thousands of times longer than visible light, there are two basic differences between radar and visible, near infrared and infrared imaging systems. The first difference is in atmospheric penetration and the second is in interaction with the earth's surface. Radar waves are not appreciably attenuated by clouds, precipitation or the earth's atmosphere. Thus, radar can image at any time, including rainy nights. Because of this unique characteristic, perpetually cloud-covered regions such as Brazil were imaged for the first time by radar in 1970. This famous project; known as Project RADAM (RADar for the AMazon) was designed to produce imagery that could be made into semi-controlled 1:250,000 scale mosaic maps covering 1 degree in latitude and 1.5 degree in longitude (de Azevedo, 1971).

The second difference, the interaction of microwave energy with the earth's surface, enables a different remote sensing approach than those used in the visible and infrared wavelength regions. For smooth surfaces there is specular reflection of the incident radar waves and, therefore, no energy is reflected back towards the receiving antenna. As a surface becomes rougher, the longer radar wavelengths interact with the roughness of the surface in a way which causes diffuse reflection. The amount of target roughness compared to the wavelength controls the distribution of the reflected energy, and thus the intensity in the radar image. Whether or not a surface is rough when imaged by radar is a function of the radar wavelength (λ), the incidence angle of the radar beam (Θ_I) and the average vertical displacement of the surface Δh). These relations are shown in Figure 5.

The complex geometry of most natural surfaces makes it difficult to characterize them mathematically. One qualitative measure of roughness is the Rayleigh criterion, which relates surface roughness,

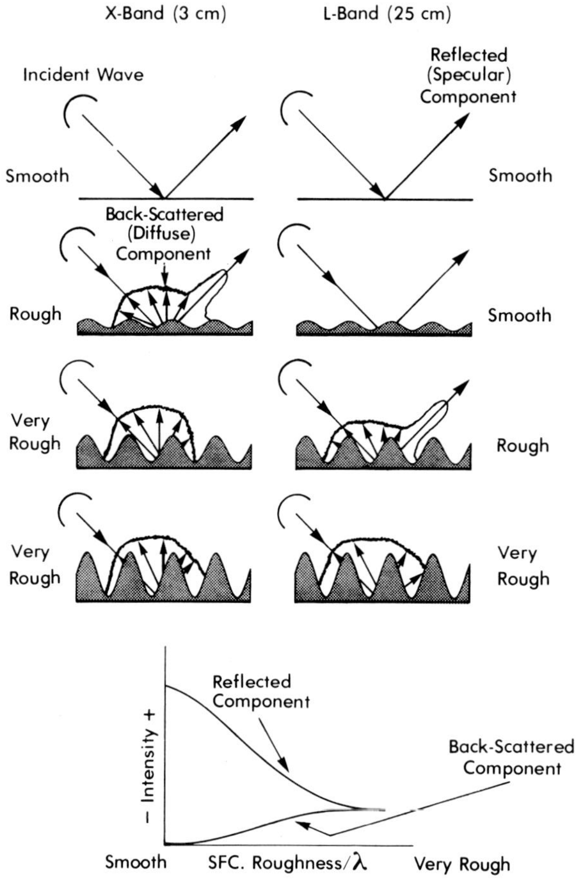

Figure 5. Theoretical scattering model of X- and L-band electromagnetic radiation from smooth to very rough surfaces (adopted from Kasischke et al., ERIM).

wavelength and incident angle, in order to define a theoretical boundary between smooth and rough surfaces. A surface is said to be rough if it meets the following condition:

$$\Delta h \cos \Theta_I > \lambda/8 \tag{4}$$

This equation is valid only for terrestrial targets but does not hold for water surfaces where interactions are even more complex. Because the Rayleigh criterion does not consider the intermediate range between smooth and rough, Peake and Oliver (1971) modified Rayleigh's approach to define the upper and lower values of the height of surface irregularities for surfaces of intermediate roughness.

4. RADAR EQUATION

All radar measurements are primarily a function of three parameters: a) transmitter power; b) antenna gain; and c) receiver sensitivity. Increasing the transmitter power will increase the radiated energy which, in turn, will result in a stronger target return. The power of the reflected signal at the radar is given by

$$P_R = P_T G \sigma A/(4\pi)^2 R^4 \tag{5}$$

where P_R is power received, P_T is power transmitted, G is the antenna gain, σ is the radar cross-section, and A is the antenna area (see Figure 6).

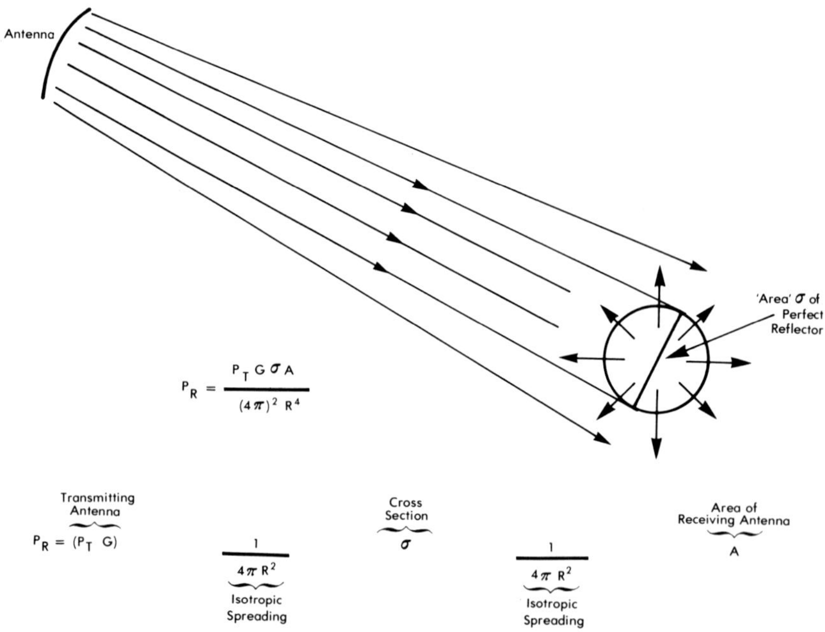

Figure 6. Schematic diagram of the components of the radar return equation (adopted from Kasischke et al., ERIM).

This equation is clarified by viewing the four parts of the radar equation separately:

$$P_R = \underset{(1)}{P_T/4\pi R^2} \quad \underset{(2)}{G} \quad \underset{(3)}{\sigma/4\pi R^2} \quad \underset{(4)}{A} \tag{6}$$

The energy on the outward propagating wave is given by term (1), where $4\pi R^2$ represents the area of a sphere with radius R.

This spherically expanding wave is focused down to an angular beamwidth by the antenna so that the flux becomes higher by a factor of G, term (2), over than of a spherically expanding wave.

This focused energy impinges on an object which has a cross-section σ. The cross-section is defined as the equivalent of a perfectly reflecting object of a given area which reflects isotropically (spherically), as shown by term (3).

Finally, the antenna area A, term (4), intercepts a portion of the reflected wave so that this portion of the flux produces the power received by the antenna.

If we consider the antenna gain at a particular angle Θ (typically $\Theta = 0$), the maximum gain of an antenna is related to its physical area A by the equation

$$G = 4\pi A/\lambda^2 \tag{7}$$

Solving Eq.(7) for A and substituting in Eq.(5) we obtain the power received at the radar as

$$P_R = P_T G^2 \lambda^2 \sigma/(3\pi)^3 R^4 \tag{8}$$

This basic radar return equation (8) is general and is known as "the radar equation".

Obviously the equation must be modified to include propagation losses caused by the atmosphere and the earth, e.g. effects of the earth's curvature and refraction as well as effects of interference on targets on the earth's surface. However, it can be applied to any object of any shape or composition and to any radar configuration.

Echo from land and sea is caused by the various scatterers within the resolution cell of a radar. In order to provide a normalized parameter that can be used to describe radar cross-section, a reflection coefficient (σ_o) has been introduced which is the radar cross-section (σ) per unit area and can be employed in the following equation (see also Figure 7):

$$\sigma_o = \sigma/a \tag{9}$$

where:

σ_o = backscatter coefficient;
σ = radar cross-section;
a = illuminated area on the earth surface.

5. RADAR RETURN PARAMETERS

Imaging sensors in the visible or infrared wavelength regions respond primarily to surface composition and with thermal IR to temperature. Radar, however, responds primarily to surface roughness and geometry and only secondarily to composition. This is not to say that radar is insensitive to composition changes, but rather that the radar wavelengths are comparable to the scale of naturally occurring variations

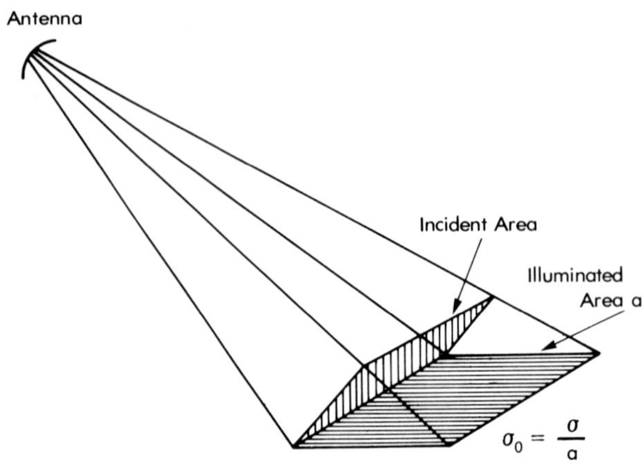

Figure 7. Illustration of the radar backscatter coefficient (σ_o) (adopted from Kasischke et al., ERIM).

in surface geometry. In addition the sidelooking configuration of SAR accentuates shape and geometry properties.

The intensity of the microwave return signal is dependent upon the interactive effect of major parameters of the radar pulse and the terrain, and also upon the mechanism of diffraction as well as reflection. These parameters can be separated into two groups:

a) radar system properties: frequency (wavelength), polarization and depression angle (incidence and aspect angle on surface);
b) terrain properties: surface roughness (surface scattering), volume scattering and dielectric constant.

The effect of these parameters is such that the radar returns cannot be attributed to a single parameter. The extreme complexity and interrelation of the first five parameters is the reason why the composition of the material, represented by the dielectric constant, is so difficult to determine with radar, particularly with the imaging radar such as SAR.

5.1 Frequency

The frequency determines the radar wavelength which, in turn, interacts with the roughness of the surface. A given surface can be rough with respect to a shorter radar wavelength, producing large radar backscatter, or it can be smooth with respect to a larger wavelength, producing little or no backscatter. The radar frequency thus directly influences the intensity of radar return and can give information on the surface roughness which, in turn, can be used for classification purposes.

5.2 Polarization

The polarization of the transmitted radar waves is important when considering the orientation of surface irregularities. Vertically polarized waves will interact with a surface geometry in different ways than horizontally polarized waves, resulting in a different radar backscatter between the two. The effect of terrain on the polarization of the reflected radar waves is probably the least well-understood and hence the least exploited radar parameter. Most of the energy returned to the antenna has the same polarization as the transmitted pulse. Depending on the surface roughness and the dielectric constant of the scattering material, some of the returned energy will be depolarized. Generally the more complex the surface the more depolarized will be the radar return; this effect is indicative of the amount of multiscatter and volume scatter. However, like and cross-polarized returns in SAR imagery have been found to differ for reasons which are not yet fully understood.

5.3 Depression Angle

The effect of slope which controls the incident angle of the transmitted wave to the illuminated surface becomes very important because of side-looking geometry, given by the depression angle of the radar system (Figure 8). A normal incidence angle gives a large reflection back to the antenna, an oblique angle a small reflection. The intensity of radar return is extremely sensitive to this parameter which makes imaging radar systems very effective in measuring changes in topographic relief.

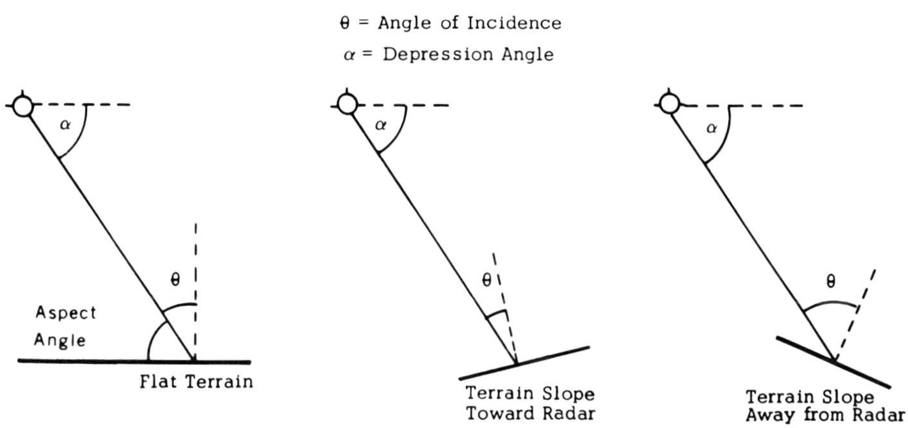

Figure 8. Effect of terrain slope (with constant depression angle) on incidence and aspect angle (adopted from MacDonald).

5.4 Surface Roughness and Volume Scattering

Surface roughness controls the intensity of radar return by controlling the amount of diffuse backscatter. As mentioned above, this backscatter is also dependent on radar wavelength. After Ulaby (1982), when an electromagnetic wave impinges upon the boundary surface between two media (i.e. the atmosphere and the earth surface), some of the incident energy is scattered backward and the rest is transmitted forward into the lower medium. If the lower medium is homogeneous the problem is a surface scattering problem. If the lower medium is inhomogeneous or is a mixture of materials of different dielectric properties, then a portion of the transmitted wave scattered backward may cross the boundary surface into the upper medium. In this case, scattering takes place within the volume of the lower medium and is referred to as volume scattering. An illustration of surface scattering and volume scattering is shown in Figures 9 and 10, respectively.

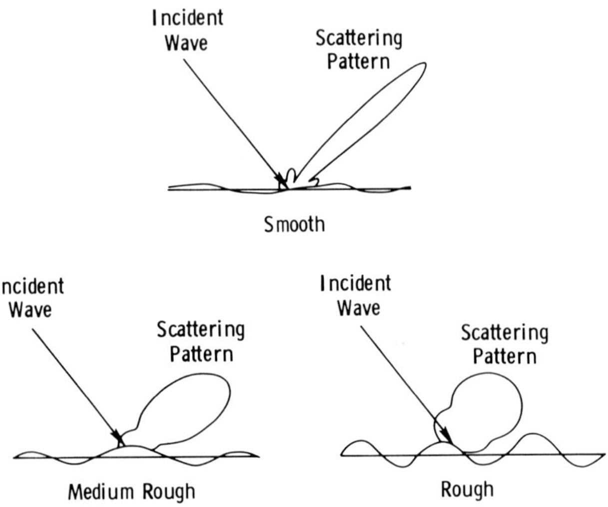

Figure 9. Example of surface-scattering (adopted from Ulaby et al., Vol. II, 1982).

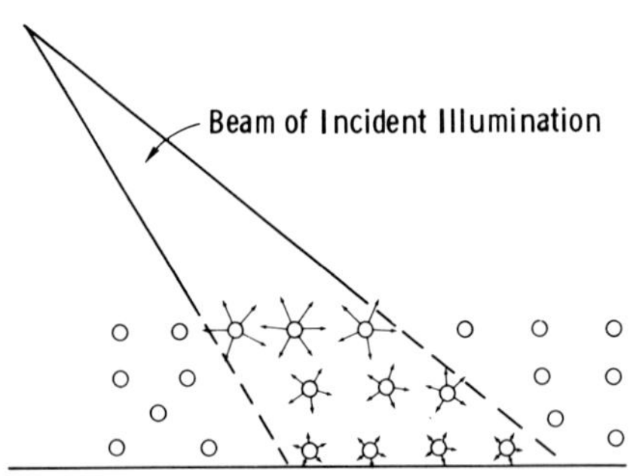

Figure 10. Illustration of volume scattering (adopted from Ulaby et al., Vol. II, 1982).

The mechanism of volume scattering causes a redistribution of the energy in the transmitted wave into other directions and results in a loss from the transmitted wave. The total loss is usually referred to as extinction and the extinction per unit length is called the extinction coefficient. The region in the inhomogeneous medium that is important for volume scattering can be ascertained by the depth of penetration, which is defined as the inverse of the extinction coefficient. A given surface that may appear very rough to short wavelengths, may appear very smooth to longer wavelengths. This is because the degree of roughness of a random surface is characterized in terms of statistical parameters that are measured in units of wavelength. An example has been given in section 3 with the Rayleigh criterion.

5.5 Dielectric Constant

The electrical properties of the surface material affect the radar return through permittivity and conductivity, i.e. the material's complex dielectric constant. The magnitude of this parameter directly affects the amount of an incident radar wave that is reflected, absorbed and dissipated. Without becoming too involved in the underlying physics, let us mention that it is difficult to measure the dielectric constant of natural materials and hence values for only a few materials have been investigated and published. Typical values for the dielectric constant at microwave frequencies range from 2.5 to 8 for dry sand, soil or rock, moist soils can have a value up to 15, while water can have a dielectric constant as high as 80. The dielectric constant increases almost linearly with increasing moisture content, corresponding to an increase in reflectivity. The reflectivity of a material varies as the square root of its dielectric constant. Because of the large variations in surface roughness and incidence angle of the radar waves on natural terrain, the dielectric constant (type and quality of material) is difficult to measure except under laboratory conditions, where surface roughness and incidence angle can be held constant. In nature the effects of these two parameters usually override effects due to changes in dielectric constant. Therefore, dielectric properties themselves are difficult to determine directly from microwave returns of imaging radar systems.

Since all these parameters affect the radar return simultaneously it is, in fact, difficult to assess the effect of each upon a given radar image. It will remain one of the major tasks of the radar image interpreter to determine which parameters cause brightness variations in the imaged scene.

6. SAR IMAGE FORMATION

The SAR image formation process consists of the following principal steps (Bayma et al., 1977):
1) signal recording; 2) range and azimuth correlation; 3) noncoherent integration (multiple look); 4) display.

The principal technologies applied to SAR processing have been (a) electronic, (b) optical, (c) digital, and (d) hybrid techniques. Analog electronic techniques were initially used to demonstrate that resolution finer than the real aperture beamwidth could be obtained by Doppler signal filtering. However, these processors did not have the capability of achieving the aperture limiting resolution of a fully focused SAR.

Early in the development of SAR techniques it was recognized that a coherently illuminated optical channel was well matched to the two-dimensional processing of suitably recorded SAR signals. In fact, most of the SAR images available for earth resources applications have been optically processed. However, digital techniques are now becoming operational although several thousand mathematical operations are required for each image pixel generated. The use of efficient signal processing algorithms such as Fast Fourier Transform (FFT) can reduce this computational load by orders of magnitude even if the computer memory requirements are enormous. But it is only due to rapid advances in integrated circuit technology that real time or near-real time digital processing of large-scene fine-resolution radar images has become technically and economically feasible.

6.1 Resolution

Unlike optical systems and real aperture radars where resolution degrades with distance, in SAR systems, resolution is independent of distance. Because a radar measures differences in range which can be equally sensed for objects which are very near as well as very far from the sensor, this is readily seen for the range dimension. This is not quite as obvious for resolution in the azimuth direction where a synthetic aperture is required to achieve this goal.

As long as an object is within the beam, a signal is returned and if the shape of this signal is retained and subsequently processed, the object can be imaged. Since the radar beam diverges and is, therefore, wider at longer ranges, an object is within the beam a longer time and thus can form a longer "synthetic" antenna. The increase in synthetic antenna length just offsets the decrease in resolution in real aperture imaging, with the result that resolution is also independent of distance in the azimuth direction.

Fine resolution in the azimuth direction is achieved by using the Doppler principle, where frequency is altered by motion toward or away from an illuminated object. Similar to acoustics, a Doppler shift is heard from approaching and receding vehicles, a higher pitch when the vehicle is approaching a lower when receding. Figure 11 illustrates the apparent motion of an object through the repeated radar beam from points A to C as a consequence of the forward motion of the sensor platform.

This same Doppler principle is used to record SAR signal histories. These signal histories can be recorded optically or digitally. Since the phase history generated by a point target is similar to a Fresnel-Zone Plate, it makes SAR data a natural candidate for optical processing. In addition, to simplify matters, we will use the optical case

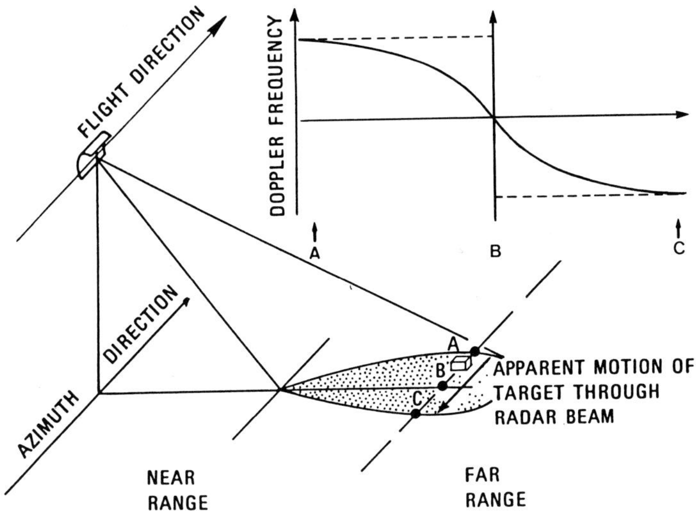

Figure 11. Doppler frequency shift due to relative motion of the target through the radar beam.

which is more readily visualized, to illustrate how this is accomplished.

A SAR transmitted pulse is linearly swept in frequency, i.e. the pulse wavelength decreases as the pulse proceeds. If a target on the ground reflects the transmitted pulse back towards the SAR, it is detected and mixed with a reference signal so that it is symmetric around zero frequency.

In a SAR system, an imaged target will be within the beamwidth of the radar for a specific length of time, called the integration time T_i ($T_i = T_N - T_S$). The integration time is dependent on the beamwidth of the system and the speed of the sensor platform (see Figure 12).

During the time the target is within the SAR beamwidth, it is continuously reflecting the linearly swept pulses being transmitted by the SAR antenna. As the platform advances (see Figure 13), the distance from the radar to the target decreases as the platform moves towards the target, i.e. $R_1 > R_2 > R_3$ and increases as the platform moves away from the target, i.e. $R_4 < R_5 < R_6$.

Figure 14 illustrates how a reflected pulse is detected and recorded by a SAR system. The SAR detects both the time between when a pulse is transmitted, and the intensity of the pulse. Therefore, as the SAR platform moves towards a target, consecutive pulses are detected earlier and earlier, i.e. $T_1 > T_2 > T_3$ and as the SAR moves away from the target the pulses are detected later and later ($T_4 < T_5 < T_6$).

The signal received is then used to modulate a cathode ray tube (CRT), which in turn exposes the SAR signal film (see Figure 15). The higher intensity portions of the linearly swept pulse result in an opaque film negative, where the lower intensity portions result in a clear film negative. Since the pulse is linearly swept, the distance

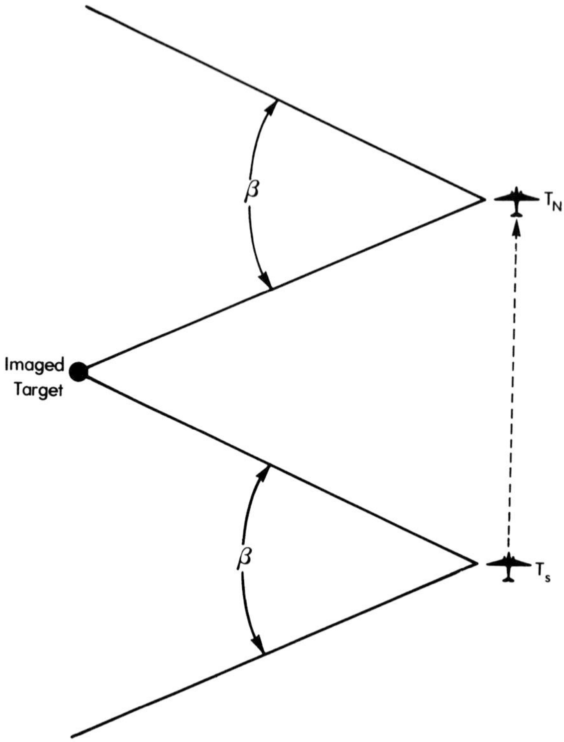

Figure 12. Illustration of integration time of target illumination (adopted from Kasischke et al., ERIM).

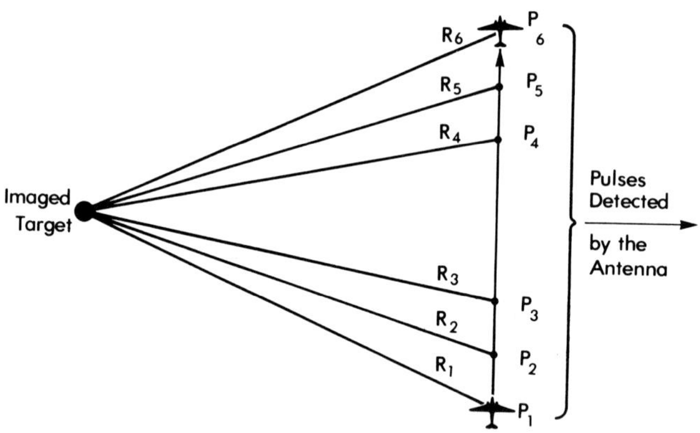

Figure 13. Illustration of variable target distance during integration time (adopted from Kasichke et al., ERIM).

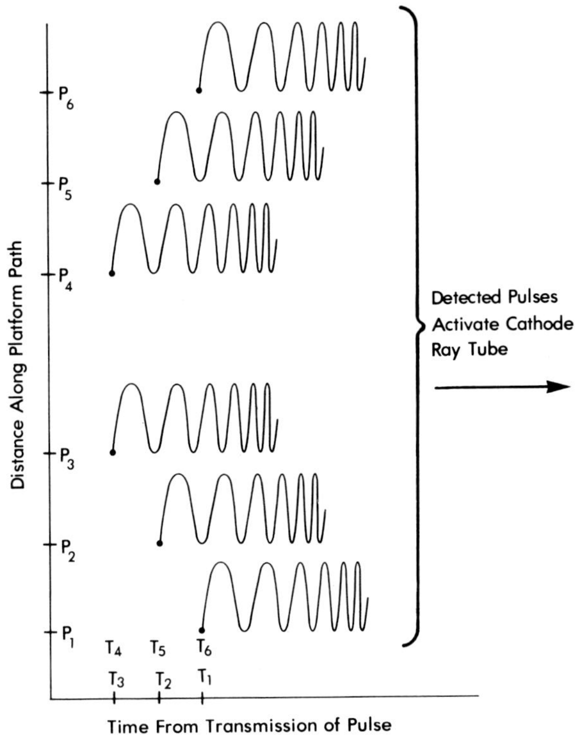

Figure 14. Schematic illustration of received signals from illuminated target (adopted from Kasischke et al., ERIM).

between opaque and clear areas on the signal film also varies.

The signal film is advanced proportionally to the SAR platform velocity. Therefore, a series of pulses are recorded. Because the SAR platform moves either towards or away from a target between pulses, the distance to the target either increases or decreases. Therefore, the modulation on the CRT will begin either earlier or later (see Figure 16).

All the CRT sweeps resulting from reflected pulses from a given target form a single Fresnel Zone Plate, which when placed in an optical processor, focusses to the original point target responsible for the formation of the zone plate.

Figure 17 is a computer plot of a schematic diagram of a Fresnel Zone Plate. Each line represents a recorded pulse backscattered from a single target. Each vertical line graphs the degree of opacity of the signal film; peaks of the lines are transparent and troughs are opaque. Each of the 30 range lines represents a reflected linearly swept pulse recorded side-by-side on film. The pulses are identical in terms of their frequency, frequency distribution and length. Only the phases differ. This can be seen by examining the ends of each vertical line. Near line 30 and line 1, the phase changes significantly from line to line. This change gradually diminishes to the centre of the

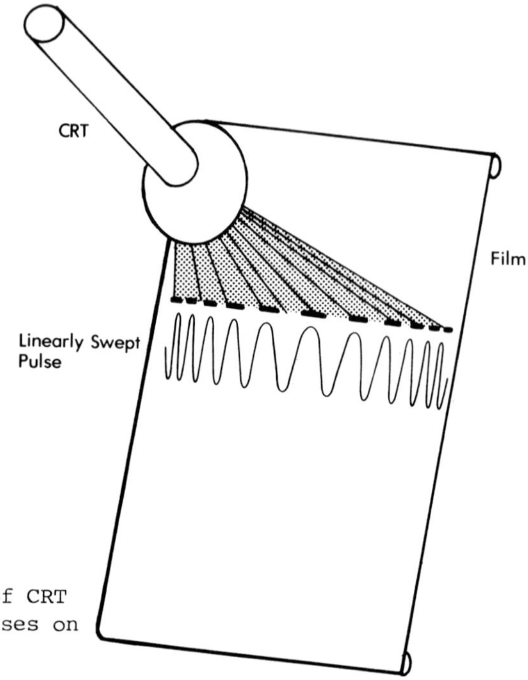

Figure 15. Illustration of CRT modulation by received pulses on signal film (adopted from Kasischke et al., ERIM).

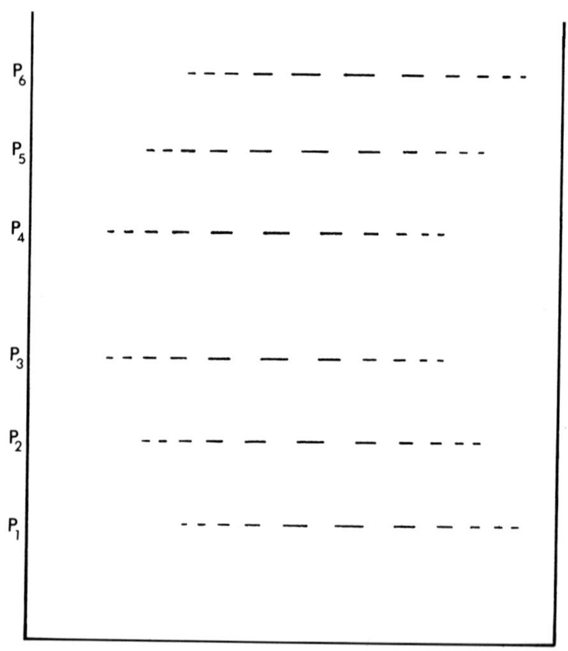

Figure 16. Schematic illustration of CRT modulated pulses from imaged target (adopted from Kasischke et al., ERIM).

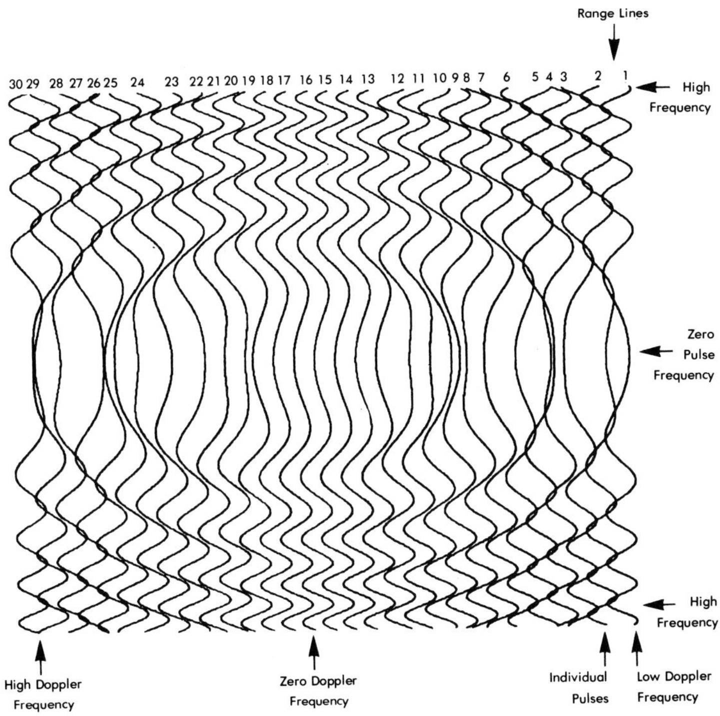

Figure 17. Schematic diagram of Fresnel Zone Plate (adopted from Kasischke et al., ERIM).

figure near lines 14 to 16. Lines 14 to 16 are near zero Doppler frequency.

The form of the zone plate is apparent in the elliptical alignments of the peaks and troughs of the lines. Thus, an actual Fresnel Zone Plate, shown in Figure 18, is formed by recording successively reflected pulses which change in phase from pulse to pulse.

The radar beamwidth limits the extent of the Doppler signature. The length of the antenna controls the angular width of the beam, which in turn controls the bandwidth of the Doppler signature. Because of this relationship, the resolution of SAR in the along-track (azimuth) direction is equal to one-half the length of the antenna. The antenna can, therefore, be small enough to be carried on a spacecraft and still be capable of attaining fine resolution at great distances.

6.2 Processing of SAR Signal Histories

As was mentioned at the beginning of this section, there are optical and digital techniques for recording and processing SAR signal histories into images. The areal representation of the signal histories must be compressed into the image scene.

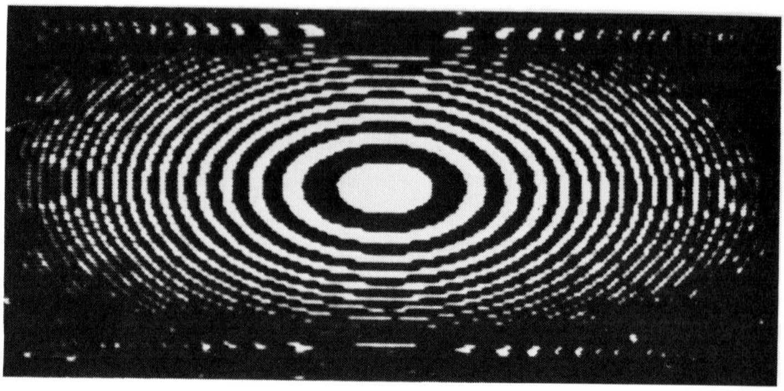

Figure 18. Image of an actual zone plate formed from a point (bright) target (adopted from Kasischke et al., ERIM).

6.2.1 Optical processing. A point target illuminated by a SAR sensor is represented over a large area on the SAR signal history in the form of a Fresnel Zone Plate. The SAR signal film (holographic record of the amplitude and phase history) contains literally thousands of these Fresnel zone plates, each one representing a point on the imaged area.

During optical processing the SAR signal film is placed in an optical processor, which consists of a laser beam and a series of lenses. Each zone plate is illuminated by a monochromatic beam of light and then focuses this light into a single point by diffraction. This focussed point represents the original illuminated point in the area illuminated by the SAR sensor. By passing the monochromatic light through the millions of zone plates which make up the SAR signal history recorded on signal film, the SAR image is reconstructed. A specialized processing system called the tilted plane optical processor (Figure 19) was developed to overcome the unique problems (Kozma et al., 1972)inherent in the SAR signal histories, producing an undistorted image at location F from the signal film at location E. The image formation is essentially instantaneous. The time required to process large amounts of data is equal to the time it takes to advance the signal film and to record the output image intensity, either on film or on magnetic tape.

One of the problems inherent in using optical processing is that of recovering the wide range of intensity information available from the SAR signal histories. Somewhere on the order of 10^4 to 10^5 grey levels of intensity information are available from a typical SAR scene which includes bright point targets (compared to 256 levels of intensity information recorded on Landsat data). SAR image intensities are recorded on film transparencies which usually only record between 50 to 100 levels of intensity. Obviously, intensity information is lost.

A method of making up for this lost information is called the hybrid concept, which combines the power and speed of coherent optical data processing with the flexibility and precision of digital methods. The key to the optical-digital interface is an image dissector camera which is positioned such that a small portion of the output image (F)

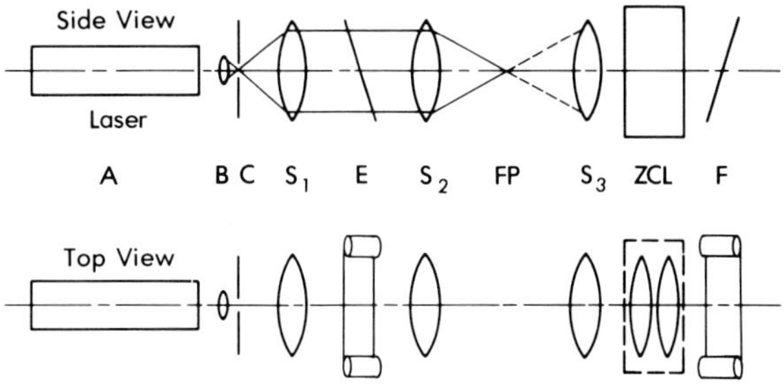

A — Laser
B — Lense to Focus Laser
C — Pinhole to Filter Light
S_1 — Spherical Lense
E — Location of Signal Film
S_2 — Spherical Lense
FP — Focal Plane, Position of Stop Filter
S_3 — Spherical Lense
ZCL — Magnification and Focusing Lenses
F — Location of Imaging Film

Figure 19. Schematic diagram of a tilted plane precision optical processor (adopted from Kasischke et al., ERIM).

of the optical processor is focused upon a photocathode (photomultiplier). As the input image is viewed by the photomultiplier, electrons are emitted with the current density being proportional to the intensity of the input image. The major advantage of the hybrid approach is that the medium used to record the output image can handle the large dynamic range present in a SAR image.

6.2.2 Digital processing. Digital processing of SAR signal histories represents a numerical evaluation of the synthetic-aperture beamformation process and is analogous to optical processing in the sense that the recorded signal histories must also be compressed in both the range and the azimuth directions. A large number of arithmetic computations are involved. Unlike optical processing, where the image is essentially instantaneously processed in two dimensions, digital processing is performed one-dimensionally on a line-by-line basis, first for all the range lines and then for all the azimuth lines. We all know by now that received signals from a point target have an azimuth encoding (Doppler frequency shift). In addition, the signals received have a similar range encoding as the transmitted pulses usually have a linear frequency modulation imposed on the carrier wave. This encoding allows the length of the pulse to be large in comparison with the desired range resolution so that the received signal to noise ratio can be made sufficiently large without a large peak power requirement. For every pulse that is transmitted, the signal received is first demodulated and stored as a horizontal line in a signal memory. Each successive pulse results in another line stored so that after a time signal memory corresponds to the synthetic aperture duration.

The principle of digital SAR processing is to form an image by compressing the point target information which is spread out in two dimensions in signal memory into a single, well-focussed point in the final image. This compression is achieved by cross-correlating the received signal with a reference function of the ideal transmitted waveform. The cross-correlation operation is equivalent to filtering the signal received with a matched filter whose transfer function is defined to be the complex conjugate of the spectrum of the ideal received signal. Whenever the reference function overlays a backscattered pulse in the signal history, a small pulse is formed which corresponds to an optically-focussed point. The intensity of the range compressed peak is controlled by the intensity of the backscatter from the target. Each range line throughout the scene is successively compressed in the above manner.

After range compression, the data in the signal memory are compressed line by line in the azimuth direction. The same cross-correlation technique is used, with the exception that the reference function must be changed because of range migration with increasing range. The image intensities resulting from the range and azimuth compressions are then recorded on digital tapes in a pixel format.

The arithmetic computation requirement for digital SAR image formation corresponds to the product of the number of computations involved to produce an output pixel and the pixel-throughput rate required for the system. For SAR image formation in real time, both the pixel rate and the per-pixel computation increase with the resolution capability. For a typical spaceborne SAR such as SEASAT, the computation rate required for real-time processing exceeds 10^9 operations/second (Elachi et al., 1982). Conventional digital computers work at only a rate of approximately 10^6 operations/second, therefore representing a major bottleneck in realizing an economical real-time digital SAR processor.

6.2.3 Optical versus digital processing. The advantages and disadvantages inherent in both optical and digital methods must be weighed carefully when choosing the one to use. There is no doubt that the trend today is towards full digital recording and processing of SAR data. The major advantage of optical processing is still efficiency in terms of both time and cost. It takes several hours of effort to process a small scene of SAR imagery digitally, while hundreds of scenes can be processed optically. The output format (film transparencies) allows for easy review of large quantities of data. The disadvantages of optical processing include the fact that unless the image intensities are recorded digitally, the full dynamic range of the received signals are not recovered. Also optically processed imagery is usually noisier because of the film grain, atmospheric refraction of the laser beam and dust. In addition, geometric distortions are introduced because of lens aberrations and film speed variations. Inaccuracies of the signal intensities may be introduced during development of the signal and the image films. Finally, optically-recorded imagery cannot be manipulated using digital processing techniques which limits the types of image analysis performed after the correlation process.

Digital processing has several advantages. Fewer processing artifacts (dust, lens, film) are present to degrade the geometric and radiometric accuracy of the data. Digital processing recovers the full dynamic range of the data. The output from digital processing allows a manipulation and analysis of the data using digital processing techniques similar to those used on Landsat data. The major disadvantage of digital processing is that today it is still slower than optical processing to achieve the same resolution and image product.

7. DYNAMIC RANGE

The most common ways of recording remotely-sensed data are photographic film and magnetic computer tapes. Photographic film can record up to around 100 grey levels or intensity levels of information even though the human eye can only discern up to 40 grey levels. Landsat data are typically recorded on magnetic tapes, with 256 grey levels. Compared to this more conventional sensor, SAR data are recorded at a greater dynamic range. Dynamic range refers to the ratio between the maximum measurable signal (I_1) and the minimum detectable signal (I_2) for the scene resolution cells represented by the image pixels. SAR systems can detect between 10^6 and 10^7 grey levels of intensity from a SAR scene, with approximately 10^4 to 10^5 grey levels being recorded in the SAR signal history. Because of these large intensity levels, radar scene intensity is usually discussed using decibels (dB) as the unit. A decibel is defined in terms or ratio

$$dB = 10 \log I_1/I_2 \tag{10}$$

where I_1 and I_2 are the pixel intensities or grey levels. A SAR system, therefore, senses between 60 and 70 dB and records approximately 40 to 50 dB.

Conversion table of dB values and power ratio

Power ratio	dB
1.0233	0.1
1.2589	1.0
1.5849	2.0
1.9953	3.0
3.9811	6.0
10.0000	10.0
100.0000	20.0
1000.0000	30.0
⋮	⋮
1000000.0000	60.0

8. SPECKLE EFFECTS

The speckle, a grainy pattern observed in all radar images, is a result of the interaction of the coherent electromagnetic wave with a rough surface. It is basically the interference pattern resulting from the combination of the returns of a scattering point. Speckle arises from the coherent nature of the radar wave, causing constructive and destructive interference and, therefore, random bright and dark areas in the image. The speckle effect can be reduced by processing separate portions of an aperture and recombining these portions so that interference does not occur. This process, called "multiple-look", or non-coherent integration, produces a more pleasing appearence to the eye and also in some cases may help in the interpretation of the image but, unfortunately, at a cost of degraded resolution.

RADAR POST-PROCESSING

Daniel R. Nüesch and Erich H. Meier
Remote Sensing Laboratories
University of Zürich
Switzerland

ABSTRACT. The major goal of this chapter is to discuss various problems involved in radar image analysis. Although the information contained in analogue radar imagery will satisfy the needs of a majority of users, more sophisticated analysis of the data may be required for specific tasks. Without going deeply into the mathematics, this chapter covers two areas of radar image processing: geometric rectification for co-registering the image to a map in order to locate the position of a feature in an absolute sense, and digital image processing techniques to support the information extraction procedure.

After the SAR image formation process, also called "image correlation", the radar data are in a two-dimensional image format, such as a regular photograph or an image stored on magnetic tape. However, a number of processing steps are still required to make the images more easily interpretable for specific applications. For example, the use of high or low contrast paper and the over and underexposing of film are two techniques which may be used to highlight certain features. Similarly, numerous techniques can be used to enhance digital remotely sensed data on high speed computers. In this chapter we will discuss procedures which are used to achieve radiometric and geometric calibration. Radiometric calibration involves corrections for antenna pattern shading and other nonlinear processes in the sensor or the correlator, in order to relate the intensity of each pixel to the surface backscatter cross section directly. The geometric correction also serves the same purpose in the sense that it involves the adjustment of the precise spatial location of each pixel in the image such that it could be superimposed on a well-defined cartographic reference.

1. RADIOMETRIC CORRECTION

Radiometric correction of SAR imagery removes effects which are the result of the SAR sensor system and the SAR viewing geometry.

We know from the radar equation (Eq. (8) of Imaging Radar Fundamentals) that the power returned is inversely proportional to the fourth power of the range (R). This results in less radar return signal

F. Toselli (ed.), Applications of Remote Sensing to Agrometeorology, 181–197.

at increasing range, hence less intense imagery. This phenomenon is referred to as the R^4 power loss. The SEASAT satellite, for instance, contained an electronic device called a sensitivity time control which adjusted the recorded radar signal to produce a uniform SAR image. Unfortunately this system failed upon the launch of the satellite so that the image intensities are brighter at the near edge than at the far edge.

The range power loss today is usually corrected automatically during the digital correlation process. If not, when older data are used, a correction can be made by identifying areas having the same surface state at various range locations in the image. The average signal is then computed for each area and plotted versus range and a smooth curve is fitted through these points. A multiplicative correction factor for each range sample is then obtained by dividing the maximum value on this curve by the value at a specific range.

Each SAR system has a specific antenna pattern. This antenna pattern occurs because more power is transmitted from the centre of the antenna than from its edges. The correction of the antenna pattern is usually performed after the digital correlation process in a way similar to that mentioned before by computing intensity profiles along columns and rows over the whole image. These profiles are then used to describe a function which, in turn, can be used in the correction algorithm.

2. CONTRAST STRETCHING

As described in chapter 7 of Imaging Radar Fundamentals, SAR has a very wide dynamic range with grey levels from 0 to 10^5. Depending upon the specific application, the intensity levels of the processed SAR image recorded on digital tapes are stored between 0 (dark) and 255 (bright). Usually the square root of intensity is recorded, after the very bright targets are truncated and all put into the top signal level (i.e. 255). Therefore, a very few pixels occur at the 255 level (bright targets such as houses, bridges, ships, etc.) with the remaining pixels grouped at digital levels between 0 and 100. The result is that 100 to 150 grey levels are not used and greatly reducing the contrast of the image generated from the digital data.

In order to utilize the full brightness range of the digital tape, contrast stretching techniques are applied. This is done usually through mathematical manipulations of the data which evaluate the distortion of the pixels and then redistributes the pixels based upon pre-determined criteria. One of the most common procedures performed is a linear contrast stretch, where a specific percentage of the pixels is assigned to the highest (255) and lowest (0) pixel level, while distributing the remaining pixels between the other values.

Numerous interactive processing techniques exist which allow the operator to designate the mathematical or statistical distribution he wishes the image to have. An example of a nonlinear contrast stretch algorithm is the Gaussian distribution, but there are quite a few more listed in the literature (Pratt, 1978; Castleman, 1979).

3. SPECKLE REDUCTION (FILTERING)

A fact which plays an important role in the SAR calibration is the presence of speckle. As discussed earlier, the coherent noise adds uncertainty to the measured intensity of a specific image pixel. By using a large number of looks (noncoherent integration), the ratio of the standard deviation to the mean intensity can be minimized. This, in turn, will degrade the spatial resolution, so that a trade-off must be made between spatial and radiometric resolution.

If one must work with one-look images, different data processing algorithms have been developed which reduce the speckle within a SAR image. The simplest way is to apply an average filter to the data. In this operation, a window of n rows and m columns is selected. The intensity of the centre pixel of this window is then replaced by the average intensity of all the pixels in the window. Thus, the whole image is smoothed by sliding the window along the rows and down the columns, reducing the speckle in an image, because the intensity of each single pixel is now replaced by an average of the surrounding pixels.

One of the best known smoothing algorithms is the median filter, which works in the same way as the average filter, except that the intensity of the centre pixel is replaced by the median intensity of all pixels within a window (where median is defined as the middle value in a set of observations ordered by size). The most important disadvantage of the average and median smoothing algorithms is that they are more or less insensitive to edges and sometimes smooth out the patterns of interest within the image if the window size is too large.

A variance filter smooths digital data by replacing the intensity of the centre pixel by the statistical standard deviation of all pixel intensities within the window. Thus, it becomes more sensitive to sharp changes in image intensity which are associated with edges of patterns.

Another smoothing algorithm is the convex hull filter, which is a sophisticated geometric filter operating in three-dimensional space (Crimmins and Brown, 1985). It is an iterative filter which evaluates the intensities of neighbouring pixels in its smoothing operation. Essentially it smoothes out small-scale intensity variations such as speckle while retaining the larger-scale features in the imagery.

4. GEOMETRIC ASPECTS

Once the SAR phase histories have been processed either optically or digitally, the resulting image will be recorded in analogue format on film or digitally on computer compatible tapes (CCT). Regardless of how it is stored, the SAR image will display unique distortions due to sensor platform motion, earth rotation and earth curvature (spaceborne SAR) as well as geometric displacement errors which are inherent to any SAR system due to the side-looking geometry.

The radar image format basically represents the slant range or distance from the radar to each of the respective surface elements in a scene. The effect is illustrated in Figure 1.

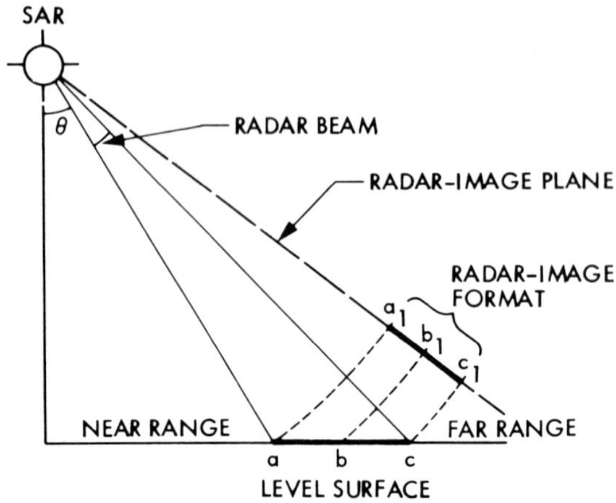

Figure 1. Radar image format showing slant range representation a,b,c, relative to ground range abc, assuming a level surface (adopted from Ford et al., 1980).

The radar range-dependence results in a relief displacement in the image. Relief displacement is the change in position of a terrain feature due to its elevation. Relief displacement in a radar image is always toward the sensor platform compared to optical images where relief displacement is away from the position of the imaging system. This relief displacement can sometimes be so severe that it results in a reversal of positions of two terrain features. But let us now go systematically through the geometric relief displacement in the SAR images which is the apparent variation in length of equal terrain slopes when the respective slope lengths are imaged at different incidence angles. The displacements result in: foreshortening, layover, and shadowing.

a) Foreshortening

Slopes inclined toward the radar appear compressed relative to slopes inclined away from the radar. This situation is illustrated in Figure 2. The slope ab is projected as a_1b_1.

b) Layover

Layover is an extreme case of foreshortening that occurs when the look angle Θ is smaller than the foreslope α^+ ($\Theta < \alpha^+$). This effect is illustrated in Figure 3. In this case a mountain appears laid over on its side. Whereas foreshortening effects can be removed from radar imagery with digital processing algorithms, this layover distortion cannot be corrected, even when the surface topography is known, because return signals from spatially separated locations in the range direction arriving at the radar receiver simultaneously, appear superimposed on the image and cannot longer be separated.

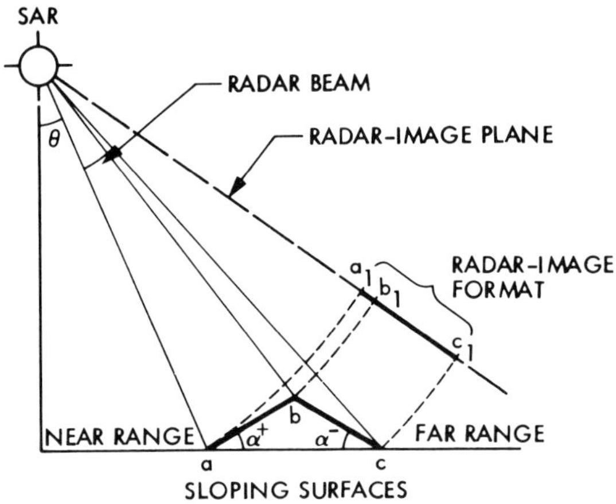

Figure 2. Radar foreshortening of slope ab (adopted from Ford et al., 1980).

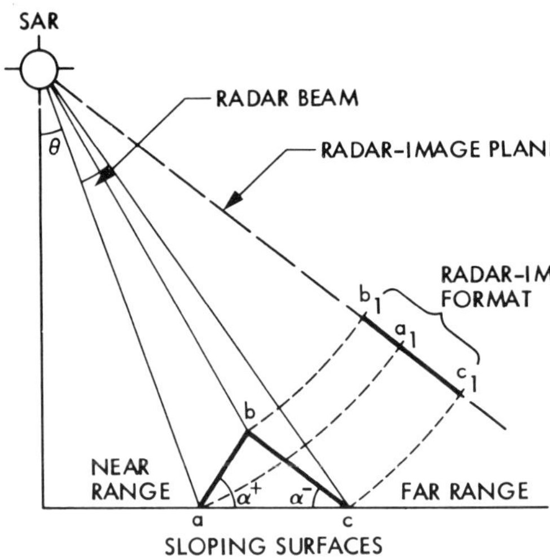

Figure 3. Radar layover of slope ab (adopted from Ford et al., 1980).

c) Shadowing

Slopes inclined away from the radar are in shadow when the look angle θ plus the backslope angle α^- are greater than 90°. Therefore, shadowing represents essentially the opposite of foreshortening. Shadows are caused by ground features that obstruct the radar beam and prevent illumination of the area behind them. This effect is shown in Figure 4.

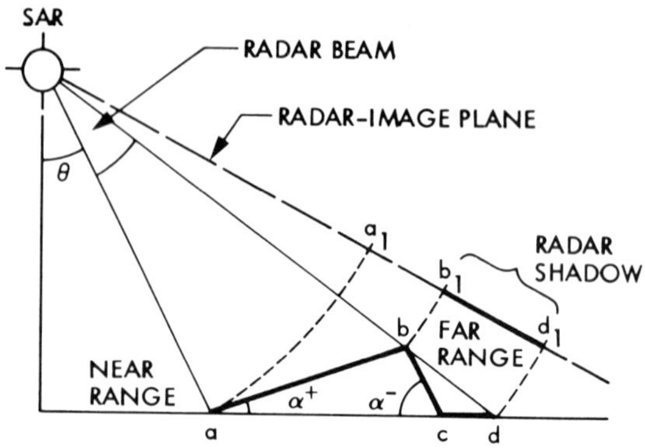

Figure 4. Radar shadow of surface bcd (adopted from Ford et al.,1980).

4.1 Geometric Rectification

It has been recognized for a long time that the inherent geometric distortions in SAR imagery represent a limiting factor of the data product because the side-looking view is only capable of measuring the absolute distance between the sensor and the objects on the surface and sampling the echo data at uniform delay intervals in the slant range plane. In order to obtain a higher accuracy in the position of surface pattern in a SAR image, geometric rectification algorithms must be applied to the data. We can distinguish on the one hand between internal distortions due to the side-looking view of the earth surface or variations in the SAR system operation, and on the other hand external distortions due to altitude changes, velocity variations and attitude changes of the sensor platform as well as earth rotation in the case of spaceborne SAR systems. Distortions due to internal causes can be removed through theoretical calculations or through monitoring of the SAR system operational characteristics. External distortions are much more complex.

Two fundamentally different aspects must be considered for an understanding of SAR image distortions. These are differently defined geometric conditions in the range and azimuth direction, i.e. perpendicular and parallel to the direction of flight.

The largest perceptible distortions are due to the definition of the geometry in the range direction which is principally caused by earth curvature in flat terrain or the topography in rugged terrain and, therefore, relatively easy to correct for. In contrast, the more complex but quantitatively much smaller deviations in the azimuth direction are much harder to deal with.

At this point we must distinguish in addition the airborne from the spaceborne SAR system data collection configuration, which must be treated in a different way when geometric rectification is concerned. In actuality the earth's surface is curved, but from aircraft altitudes, the earth curvature is usually small enough to be ignored.

However, from spacecraft altitudes, the earth curvature and the earth's rotation during the imaging period must be accounted for.

While a high precision rectification (similar to orthophotos in the visible wavelength region) of SAR data collected from airborne platforms has not been achieved up to this date, techniques have been established to register spaceborne SAR data to large topographic maps (Meier and Nüesch, 1985). The reason why this has not yet been done for airborne SAR data lies in the enormous difficulty in accounting for an accurate registration of the platform's flight path where the slightest changes due to atmospheric turbulence will affect the oppler frequency shift of the illuminated target. Some techniques are still at an early stage of development but potentially will produce positive results.

4.2 Rectification in the Range Direction

A large number of SAR image producers offer their products in so-called ground range presentation. Figure 5 shows the relationship between a slant range and a ground range image. Wu (1981) provides a derivation for the calculation of the horizontal range x as a function of the slant range r.

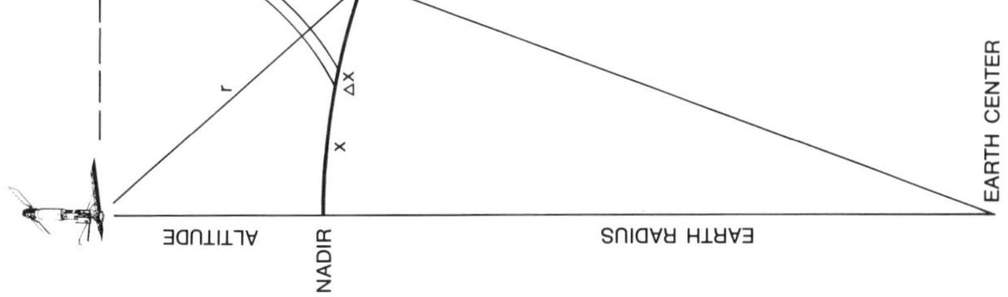

Figure 5. Relationship between slant range and ground range presentation.

Transformation of slant range to ground range presentation produces no basic improvement in the image geometry. This is because the topographic effects are not taken into account. In particular the shortened or extended representation of mountain slopes which are inclined towards or away from the sensor remains (see section 4). Where the precise process of the slant to ground range conversion or the associated variables such as sensor altitude and curvature radii are not known, an accurate rectification of a SAR image data set becomes impossible, because the absolutely imperative information on the initial situation during data collection is then not available for the geometric correction. The use of slant range images as a working basis for processing geometric problems is, therefore, recommended.

4.3 Rectification in the Azimuth Direction

A highly computer-bound frequency analysis of the reflected signals is required for attaining satisfactory ground resolution parallel to the sensor flight direction. The frequency of these signals recorded by the receiver is dependent on the relative velocity between the sensor and the backscatter points on the ground. Radar pulses which are reflected from areas in front of the sensor reach it with a higher frequency than the one transmitted, because the backscatter points are moving towards the instrument. Similarly, the frequency of pulses reflected from areas behind the sensor is lower than the transmitted frequency because of the so-called Doppler shift. Figure 6 illustrates the simplified situation with an airborne SAR over flat terrain. The lines connect all points from which the backscatter signals are received at the same frequency. These are, therefore, termed iso-Doppler lines.

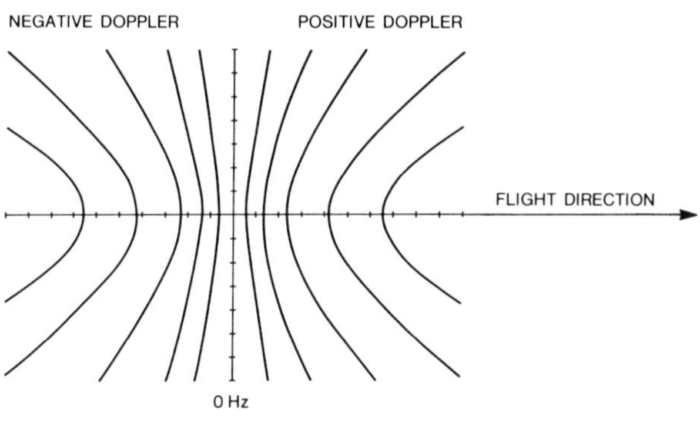

Figure 6. Iso-Doppler lines generated from an airborne SAR (horizontal flight, flat terrain).

If, in correlating the received phase histories, we now use a reference function corresponding to 0 Hz Doppler shift, range lines are produced which follow the 0 Hz iso-Doppler line. In the example of

Figure 6 this runs perpendicular to the flight axis and exhibits no distortions in the azimuth direction.

The situation is much more complex in the case of the spaceborne SAR system. The earth's rotation, together with the earth's curvature, creates Doppler shifts which are additionally dependent on the location of the existing backscatter centres within the swath (see Figure 7). On ascending orbits, targets in the far range exhibit considerably higher relative velocities away from the sensor than do points at near range, therefore causing large negative Doppler shifts. At a latitude of 46 North, a Doppler bandwidth in excess of 200 Hz must be taken into account within the swath width of 100 km.

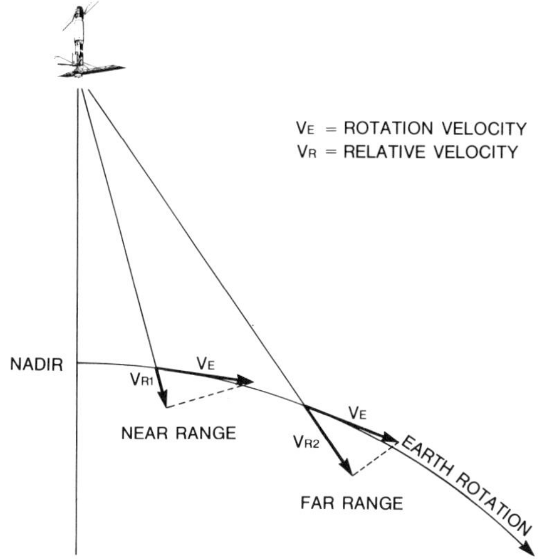

Figure 7. Representation of the different relative velocities in the near and far range of the swath width.

Figure 8 provides a very good impression of the geometric situation of a spaceborne SAR. It shows the isodoppler and isorange lines calculated for a particular satellite position, as well as the theoretical antenna axis (cross track line) associated with this position. The numerous layover regions in the Savoy Alps area, which permit no clear geometric correction, are particularly striking (see also Figure 10).

Because it is very difficult to develop a correlator which will take these different frequencies into account, various processors operate with only one mean Doppler shift as a reference. The consequence is the so-called azimuth skew, an increasing tilting of the range lines in the azimuth direction with increasing slant range. In the case of ascending orbits these shifts run with the satellite motion, on descending orbits they run towards it. In both cases, however, they are approximately linear over the whole swath width.

The SAR processor installed with DFVLR in Oberpfaffenhofen (FRG) allows the subdivision of the image-acquired slant range into several range segments. Processing with separate reference functions is

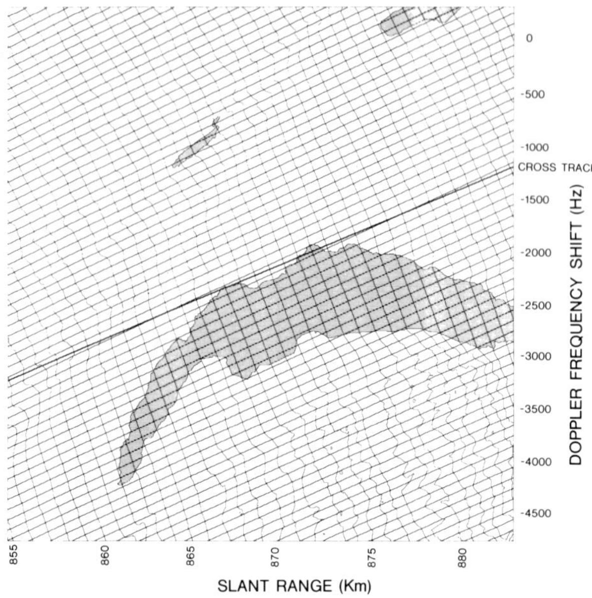

Figure 8. Calculated isodoppler and isorange lines with reference to a satellite position.

possible within these segments, enabling the correlation of a range line to take place in step fashion, across the swath. With an optimal calculation of the different reference functions, it is therefore possible for range lines to be processed which vary only slightly from the theoretical antenna axis which is vertical to the flight direction. The consideration of roll, pitch and yaw angles is not longer necessary. In any case these values are very inaccurate and have tolerances of ± 0.3.

4.4 Rectification Procedure

Considering the different geometric conditions in the range and azimuth directions, the conclusion was reached that the methods previously used for rectifying radar images like rubber-sheeting techniques are not suitable. Satellite radar images in particular, which cannot be correlated with reference functions corresponding to 0 Hz Doppler shift, require a more specific procedure which gives special consideration to the azimuth deviations. Traditional concepts from photogrammetry, such as interior and exterior orientation, are to be applied with caution. Although the attitude of the sensor in space, and thus the look direction of the antenna, is an important variable for image correlation, in order to obtain optimal signal to noise ratio and optimal azimuth ambiguity, it has no influence on the geometry of the processed image. It is likewise unsatisfactory to replace these attitude parameters by the satellite motion components, since the actual look direction of the sensor is only determined in the correlation procedure. It is found from the direction of reception of those signals whose frequency is within the Doppler bandwidth used in processing and which is situated within the pulse transit times allowed for.

It is far more practical to speak of the orientation of the complete system, defined by the position and velocity vector of the sensor, and by the pulse transit times and reference functions used in image correlation. In this way it is possible to reconstruct the image geometry as defined by the correlator.

The position of a particular backscatter element within the swath width is obtained from the slant range between the sensor and this element, together with the Doppler frequency used in processing. Curlander (1982) demonstrates how, with these two specifications, together with the satellite position data and a geoid model, it is possible to localize individual pixels. The accuracy of the procedure described is of the order of mean deviations of approximately 60 m in range and azimuth direction on a flat earth surface. This value was achieved with the aid of high precision ephemeris data from the Goddard Space Flight Center (GSFC) and applies only to individual, accurately measured points which can be easily localized in the image (corner reflectors). If this degree of accuracy is to be extended over an entire image, it is necessary to calculate slant range and relative velocity to the sensor for each backscatter element on the earth's surface. Therefore, the geoid model values are to be replaced by those from a digital terrain model which is exactly defined in spatial terms.

Figure 9 shows a SAR image recorded over the French/Swiss border on 19 August 1978 by Seasat. Lake Geneva can be seen in the centre with the city at its lower end. The image topography includes altitude differences up to 2000 m in the Savoy Alps. The image is a slant range representation processed by DFVLR Oberpfaffenhofen with a nominal pixel spacing of 12.5 m. It comprises 3000 azimuth lines each of 6000 pixels and covers a ground surface of 75 km in azimuth direction and 100 km in range direction.

Figure 10 shows the rectified image comprising 1560x1520 pixels which are oriented according to the map projection system used by the Swiss Ordnance Survey (oblique Mercator). The coarse resolution of 50 m is due to the grid size of the digital elevation model used.

To achieve a radar ortho-image from a SAR precision product corresponding to a well defined map projection system used by the national topographic map agencies represents an ambitious goal. It enables the user of remote sensing data to superimpose multispectral and multitemporal images taken of the same geographical area. This, therefore, implies the possibility of combining the information content of data from various sensors and land information systems for a complete, continuous monitoring of the complex processes on the earth's surface. In order to be able to succeed in this matter we must at least consider briefly the problem of the extraction of information from radar images in the next section.

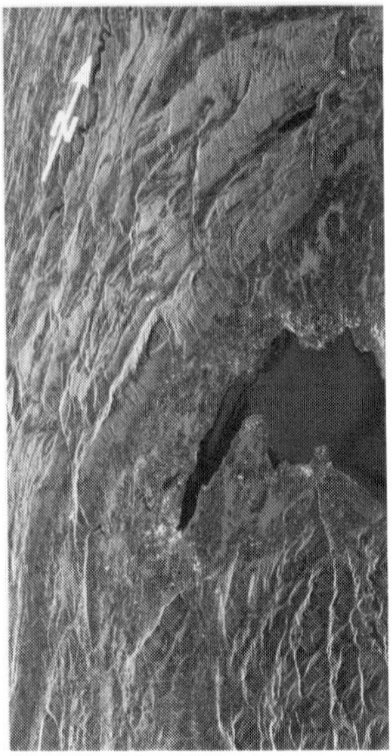

Figure 9. SEASAT SAR slant range image (pixel spacing 12.5 m).

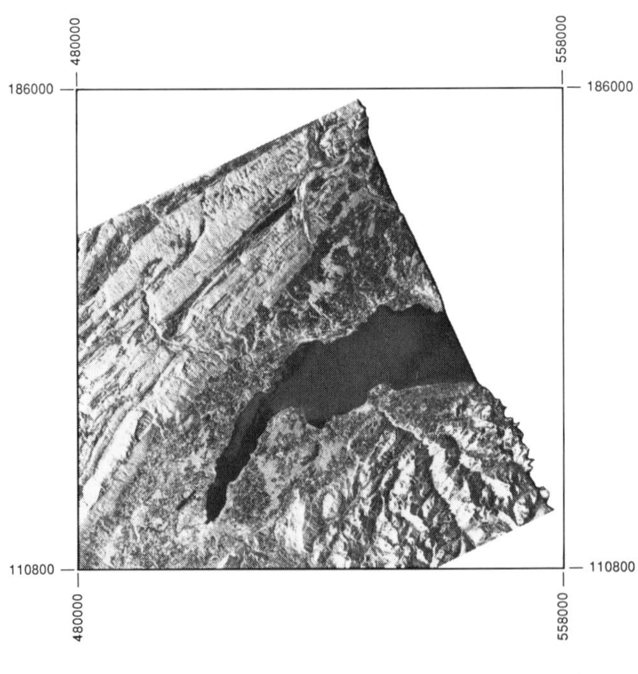

Figure 10. Geocoded radar image (pixel spacing 50 m) satellite position.

5. RADAR IMAGE ANALYSIS

Because SAR imaging is done by a coherent sensor, images of relatively homogeneous areas, such as single crop fields, appear "speckled" although black and white photographs of the same area would appear uniform in intensity. Many objects which scatter diffusely in the visible region become specular at radar wavelengths which are on the order of $5x10^4$ longer than visible wavelengths. The radar cross sections of these specular scatterers may change by several orders of magnitude as the viewing angle changes by only a few milliradians. This coherent speckle gives rise to the very large contrasts observed in radar imagery.

For an interpreter of radar imagery the tone, texture and context represent the major recognition elements in an image which contributes to the information extraction. In addition, shape, pattern, size and shadow are important features used in image classification. While image tone refers to the local average intensity for a given image area for which the radar backscattering properties may be considered spatially uniform, image texture is defined as the spatial distribution of image intensity for a given scene. Image context then refers to the relationships between a given image area and the area surrounding it. Going into detail of all radar imaging features and analysis techniques (analog and digital) would go far beyond the scope of this review. Nevertheless I would like to draw attention to one specific problem which will play an important role in the future, namely automatic analysis of SAR data.

Because SAR imagery is based on an enormous amount of data, digital analysis of imagery from synthetic aperture radar has been a long sought after goal. The coherent nature of the data makes it difficult to adapt successfully statistical pattern recognition methods like those developed for use on Landsat data, for example. The presence of coherent speckle precludes a pixel-by-pixel classification unless the data have been highly smoothed. In this situation textural analysis opens a new dimension for a potential solution.

5.1 Textural Analysis

It is widely accepted that image texture and tone are used in a complementary fashion by human interpreters to recognise objects of interest. In particular, when the range of tones in an area of interest is comparable with the range of tones in the entire image, human interpretation draws heavily on textural appearance. SAR imagery seems to be a good example of this fact. Although texture, which we will define as the frequency of tonal change within a given limited area, is a very important characteristic used by humans to identify objects of interest in an image, machine classification of digital images is frequently based solely upon spectral features which describe the overall tonal variations within an image. Textural features, however, describe the local spatial distribution of tonal values within an image (Haralick, 1973). In spite of tremendous advances in the field of computer technology during the last decade, the cost of characterizing digital image texture on a general-purpose computer is still very high. The application of classification procedures involving image texture to remote sensing surveys has remained primarily a research topic because of the magnitude of the computational resources required to achieve a reasonable throughput. With the recent advances in digital computation devices, there is a hope that texture information can be used in support of SAR image information extraction.

Reviewing the literature on textural models, Haralick (1979) listed eight statistical approaches to the measurement and characterization of image texture:

- autocorrelation functions;
- optical transforms (Fourier transform);
- digital transforms (Hadamard transform);
- textural edgeness (edgeness per unit area);
- structural elements (mathematical morphology);
- spatial grey tone dependence (co-occurrence);
- grey tone run lengths;
- autoregression models.

In a comparative study of texture measures for terrain classification by Weszka et al. (1976) it was concluded that the spatial frequency features (Fourier features) are less effective measures of texture than second-order grey level statistics (co-occurrence matrices) or first-order statistics of grey level differences (means, variances). Conners and Harlow (1980) made a theoretical comparison of texture algorithms using a set of textures generated by Markov chains. Their results showed that the spatial grey level dependence method (SGLDM) is more powerful than the power spectral method, the grey level run length method or the grey level difference method. While the SGLDM has not been conclusively proven to be the most powerful statistical texture analysis algorithm, there is strong supporting evidence that it is representative of the best texture algorithms currently available.

Under the SGLDM, as described in Haralick (1973) it is assumed that the texture information in an image is contained in the overall or "average" spatial relationships which the grey tones in the image have to one another.

This relationship can be characterized by a set of co-occurrence matrices $P(i,j)d,\Theta$, whose i,j^{th} element is the relative frequency with which two picture elements separated by distance d in a direction Θ occur in the image, one with grey tone i and the other with grey tone j, respectively. Such co-occurrence matrices are usually computed over an image subregion which is to be discriminated from other subregions. The shape of this region (square, circular, rectangular, etc.) and its size must be chosen carefully with regard to the application.

Haralick (1973) proposed a variety of 14 textural features (scalar functions) which can be employed to extract useful textural information from these co-occurrence matrices. Some of these scalar functions relate to specific textural characteristics of the image such as homogeneity, contrast and the presence of organized structure within the image. Two scalar functions which can easily be implemented in a computational process are given below:

$$\text{ENTROPY}(d,\theta) = \sum_{i=1}^{N_G} \sum_{j=1}^{N_G} P(i,j)\log P(i,j) \qquad (1)$$

where N_G is the number of grey levels. This function is largest for equal P(i,j) and small when they are very unequal. This feature is a measure of the disorder in the subregion over which the co-occurrence matrix was computed. The analogous definition for inertia is:

$$\text{INERTIA}(d,\theta) = \sum_{i=1}^{N_G} \sum_{j=1}^{N_G} (i-j)^2 P(i,j) \qquad (2)$$

where again N_G is the number of grey levels. This function is a difference moment of the co-occurrence matrix and represents a measure of the contrast or the amount of local variation present in the subregion over which the co-occurrence matrix was computed.

What makes the co-occurrence approach valid is its characterization of the spatial interrelationship of the grey tones in a textural pattern in a manner which is invariant under monotonic grey tone transformations. This is a very strong point for its use in remote sensing applications where data from different dates or flight parameter conditions often require special preprocessing to make them commensurate. Moreover, no a priori knowledge of the location of boundaries of areas of interest is required. Unfortunately the scalar functions do not capture the shape aspects of tonal information as well.

In spite of the tremendous computational burden of computing texture measures from local co-occurrence matrices, experience gained during investigations (Holmes et al., 1984) and (Nüesch, 1985) confirmed that considerable information is contained in the local texture of SAR images. One approach to reduce the computational burden associated with texture measures based on co-occurrence is to work with co-occurrence data rather than with co-occurrence matrices or only with difference images. Co-occurrence data are data where each pixel contains information from another pixel located at a fixed displacement from the first pixel. A difference image is closely related to a co-occurrence data image. These data images (co-occurrence or difference) might provide additional texture information especially when processed by nearest neighbour type processors (Young and Fu, 1986).

REFERENCES

Bayma, R.W., R.L. Jordan and B.N. Manning, 1977: 'A survey of SAR image-formation processing for earth resources applications', Proc. of the 11th Int. Symp. on Remote Sensing of the Environment, Ann Arbor, Michigan, USA, 137-159.

Bernstein, R., 1983: 'Imaging geometry and rectification', in: Manual of Remote Sensing, Second edition, edited by R.N. Colwell, Sheridan

Press, 873-922.

Castleman, K.R., 1979: Digital Image Processing, Prentice-Hall Inc., Englewood, N.J., USA.

Connors, R.W. and C.A. Harlow, 1980, 'A theoretical comparison of texture algorithms', IEEE Transactions on Pattern Analysis and Machine Intelligence, Vol. PAMI-2, No.3, 204-222.

Crimmins, T. and W. Brown, 1985: 'Image algebra and automatic shape recognition', IEEE Trans. on Aerospace and Electronics Systems, January, 60-69.

Curlander, J.C., 1982, 'Location of spaceborne SAR imagery', IEEE Transactions on Geoscience and Remote Sensing, Vol.GE-20, No.3

De Azevedo, L.H.A., 1971: 'Radar in the Amazon, Projecto RADAM', Proc. of the 7th Int. Symp. on Remote Sensing of the Environment, University of Michigan, Ann Arbor, 2303-2306.

Elachi, C., T. Bickness, R.L. Jordan and C. Wu, 1982: 'Spaceborne synthetic-aperture imaging radars; applications, techniques and technology', Proc. of the IEEE, October 1982, 1174-1209.

Ford, J.P. et al., 1980: Seasat Views North America, the Carribean and Western Europe with Imaging Radar, JPL Publication, 80-67.

Haralick, R.M., K.S. Shanmugam and I. Dinstein, 1973: 'Textural features for image classification', IEEE Transactions on Systems, Man and Cybernetics, Vol. SMC-3, 610-621.

Haralick, R.M., 1979: 'Statistical and structural approach to texture', Proc. of the IEEE, Vol.67, 786-804.

Holmes, Q.A., D.R. Nüesch and R.A. Shuchman, 1984: 'Textural analysis and real time classification of sea ice types using digital SAR data', IEEE Transactions on Geoscience and Remote Sensing, GE-22, 113-120.

Hovanessian, S.A., 1984: Radar System Design and Analysis, Artech House, Inc., Dedham, MA, USA.

Hovanessian, S.A., 1980: Introduction to Synthetic Array and Imaging Radar, Artech House, Inc., Dedham, MA, USA.

Kasischke, E.S., G.A. Meadows and P.L. Jackson, 1985: 'The use of synthetic aperture radar imagery to detect hazards to navigation', ERIM-Report No.169200-2F, Ann Arbor, Michigan, USA.

Kennie, T.J.M. and M.C. Matthews, 1985: Remote Sensing in Civil Engineering, Survey University Press, Halsted Press, London.

Kozma, A., E.N. Leith and N.G. Massey, 1972: 'Tilted plane optical processor', Applied Optics, 11, 1766-1777.

Long, M.W., 1983: Radar Reflectivity of Land and Sea, Artech House, Inc., Dedham, MA, USA.

MacDonard, H., 1980: 'Techniques and applications of imaging radars', in Remote Sensing in Geology, edited by Siegal B. and Gillespie A., John Wiley and Sons, New York.

Meier, E.H. and D.R. Nüesch, 1985: 'Registration of spaceborne SAR data to large scale topographic maps', Proc. of the 19th Int. Symp. on Remote Sensing of the Environment, Ann Arbor, Michigan, USA, 581-593.

Nüesch, D.R., 1985: 'Classification of SAR imagery from an agricultural region using digital textural analysis', Investigators Final Report of the Eur. SAR-580 experiment, SA/1.04.E2.85.12/2.

Nüesch, D.R., 1982: 'Augmentation of Landsat MSS data by Seasat SAR imagery for agricultural inventories', Reprint of NASA Report IT-E2-04233, in Remote Sensing Series, 7, Dept. of Geography, University of Zürich.

Peake, W.H. and T.L. Oliver, 1971: 'The response of terrestrial surfaces at microwave frequencies', Ohio State University Electronics Laboratory, 2440-7, Technical Report AFAL-TR-70-301, Columbus, Ohio.

Pratt, W., 1978: Digital Image Processing, John Wiley and Sons, New York.

Siegal, B.S. and A.R. Gillespie, 1980: Remote Sensing in Geology, John Wiley and Sons, New York.

Ulaby, F.T., R.K. Moore and A.K. Fung, 1981, 1982, 1986: Microwave Remote Sensing, I,II,III; Addison-Wesley Publishing Co., Reading, MA, USA.

Weszka, J.S., C.R. Dyer and A. Rosenfeld, 1976: 'A comparative study of texture measures for terrain classification', IEEE Transactions of Systems, Man and Cybernetics, SMC-6, 269-285.

Wu, C., B. Barkan, B. Huneycutt, C. Leang and S. Pang, 1981: 'An introduction to the interim digital SAR processor and the characteristic of the associated Seasat SAR imagery', JPL-Publication 81-26.

Young, T.Y. and K.S. Fu, 1986: Handbook of Pattern Recognition and Image Processing, Academic Press, Inc., Orlando.

EXTRACTION OF SURFACE TEMPERATURE FROM SATELLITE DATA

Hartmut Grassl
GKSS-Forschungszentrum Geesthacht GmbH
Max-Planck-Strasse
D-2054 Geesthacht - Fed. Rep. of Germany

ABSTRACT. The surface temperature figures in nearly all equations for energy fluxes through a surface element. It is routinely derived from satellite radiances over the oceans. However, over land the changing surface emissivity, strong daytime heating as well as nighttime cooling and the difficulty of defining surface temperature for a canopy, have prevented a routine application. This paper shows the possible influences of water vapour content, near surface thermal structure, aerosol particles, thin clouds, slant paths and surface emissivity on the correction of atmospheric masking. A procedure for land surface temperature determination for the NOAA 7, 9 satellite data comprising four steps is put forward:

- water surface temperature derivation;
- interpolation of the water surface correction;
- additional correction due to heated or cooled surfaces;
- choice of the highest temperature from the three thermal IR channels.

The procedure is applied to a daytime and a nighttime scene covering the Alps.

1. INTRODUCTION

All fluxes determining the energy budget of a surface element, except solar radiation flux, involve - if parameterized - the surface temperature, either directly like longwave net flux or indirectly via a temperature gradient like sensible heat flux and latent heat flux into the atmosphere as well as the sensible heat flux into the soil. Therefore, the two-dimensional surface temperature distribution is needed, whenever the total energy budget or parts of it are sought for any area. Since the direct measurement of surface skin temperature is impossible (due to the disturbance by any thermometer) an ideal means is the radiometric determination of temperature via surface emission within the thermal infrared spectrum from an airplane or a satellite.

Immediately two error sources come into play:

- the non-blackness of all natural surfaces;
- the emission of the atmosphere between the measuring platform and the

F. Toselli (ed.), Applications of Remote Sensing to Agrometeorology, 199–220.

surface at temperatures differing from the surface temperature.

Both errors vary with time for a distinct surface element. In both cases this is mainly caused by changes in water content either by a change from a dry to a wet surface or by changes in atmospheric water vapour.

Since the atmosphere has no perfect spectral window in the thermal infrared, a one-channel radiometer for the remote measurement of surface temperature may at best apply a climatological mean correction of the atmosphere's masking, if no calibration points are available.

This lecture will give - after a short section on theory - an account of surface parameters, gases and aerosol particle characteristics influencing the remote measurement of surface temperature, will then show algorithms to establish a stand-alone remote surface temperature measurement, and will finally discuss a variety of applications of different algorithms also including the daily temperature amplitude determination.

2. ESSENTIAL PHYSICAL BACKGROUND

2.1 Governing Equations

Remote sensing by electromagnetic waves in the visible and infrared part of the spectrum normally applies time averages over many waves. The appropriate equation governing the time averaged squares of amplitudes of electromagnetic waves is the radiative transfer equation.

Written for spectral radiance I_λ, the energy flowing through a unit surface from a unit space angle per unit time and unit spectral interval (here wavelength λ; thus the dimension is $Wm^{-2}sr^{-1}\mu m^{-1}$)

$$\frac{dI_\lambda}{d\tau_\lambda} = - I_\lambda + J_\lambda$$

or (1)

$$\frac{dI_\lambda}{ds} = (-I_\lambda + J_\lambda) \cdot e_\lambda$$

It simply states that firstly the incoming spectral radiance I_λ reduces more for a greater spectral optical depth increment $d\tau_\lambda$ of a medium, and that secondly I_λ may be increased by a source J_λ. The spectral optical depth τ_λ is defined here as the integral over spectral volume extingtion coefficient e_λ along a distinct direction s

$$\tau_\lambda = \int_0^{s^*} e_\lambda \, ds \tag{2}$$

For thermal infrared remote sensing the integration of Eq.(1) (an integro-differential equation in scattering media) may be simplified considerably by neglecting scattering in the atmosphere, thus reducing the spectral source J_λ to Planck's function B_λ, which merely depends on temperature T. This simplification is a good approximation for all cloudless areas in the thermal infrared outside very strong atmospheric turbidity layers. As a rule of thumb the scattering optical depth caused by aerosols is by a factor of 5 smaller in the main thermal infrared window from 10-30 μm wavelength as compared to the visible wavelength, where $\tau_{aerosol} < 0.2$ is mostly satisfied.

After a formal integration along a distinct path in a non-scattering atmosphere, here from $\tau_\lambda = 0$ at the top of the atmosphere - where as a boundary condition no emission from space is assumed, i.e. $I_\lambda^- = 0$ - to the ground with total optical depth τ_λ^*, we obtain for the spectral radiance I_λ^+ emerging at the top of the atmosphere

$$I_\lambda^+ = \varepsilon_\lambda B_\lambda(T_s) e^{-\tau_\lambda^*} \quad \text{surface contribution}$$

$$+ \int_0^{\tau_\lambda^*} B_\lambda T(\tau_\lambda) e^{-\tau_\lambda} d\tau_\lambda \quad \text{atmospheric emission} \qquad (3)$$

$$+ (1-\varepsilon_\lambda) \int_{\tau_\lambda^*}^{0} B_\lambda(T(\tau_\lambda)) e^{-\tau_\lambda}\, d\tau_\lambda e^{-\tau_\lambda^*} \quad \text{surface reflection of downwelling emission of the atmosphere (here written for a smooth surface only)}$$

Only the first term on the right side contains information on surface temperature T_s via Planck's function $B_\lambda(T_s)$; ideally the spectral emissivity ε_λ reaches unity for a blackbody. Fortunately, most natural surfaces, especially water and vegetated surfaces, reach $\varepsilon_\lambda > 0.95$, even $\varepsilon_\lambda \cong 0.99$ in thermal infrared windows. In order to obtain the spectral surface emission $\varepsilon_\lambda B_\lambda(T_s)$ to the satellite spectral transmission $t_\lambda = e^{-\tau_\lambda}$ of the atmosphere should be as small as possible, i.e. only atmospheric windows are suitable wavelength positions.

The second term, atmospheric emission, masks the surface contribution. If near surface layers of the atmosphere with $B_\lambda T(\tau_\lambda)$ only slightly smaller than $B_\lambda(T_s)$ constitute the main part of the integral of the second term, the transmission loss of surface emission in term number one is to a large extent compensated. However, this is not at all the case for high aerosol particle layers or subvisible cirrus clouds both emitting at low tempeatures. In this case even very small spectral optical depth values with $\tau_\lambda \leq 0.01$ may cause up to one percent error in spectral upward radiance I_λ^+, which at ambient temperatures of $20^\circ C$ is approximately equivalent to a one degree temperature error if measuring in the 10-13 μm central infrared window. Since vertical temperature and humidity structure (water vapour absorption is the dominant absorption in the windows) vary considerably depending on weather conditions and geographical position the degree of compensation mentioned above also varies, leaving a variety of necessary corrections for the atmosphere's masking ranging from no correction to up to 10 K.

The third term accounts for the reflection of downwelling thermal infrared radiances by the surface. Its importance decreases with increasing ε_λ and extinction by the atmosphere; thus it is nearly negligible over water surfaces in the tropics and it becomes considerable in moderately humid atmospheres over rock surfaces. If handled accurately for a rough surface, either composed of surface elements with different slopes or vegetated, this term needs a proper description of spectral emissivity ε_λ as a function of both the angle of incidence of downwelling radiance and the surface element slope distribution. Its effect is a partial compensation of the blackbody emissivity deficit in term number one, with growing compensation for increasing surface roughness.

2.2 Relation to Temperature

The quantity desired is the surface temperature T_s and it is hidden in the spectral Planck's function $B_\lambda(T_s)$. Any upwelling spectral radiance I_λ^+, the satellite signal, may be expressed in a so-called brightness temperature T_B, which merely is a temperature a blackbody emitting the same spectral radiance would have. The simple inversion of Planck's law

$$B_\lambda(T)d_\lambda = \frac{2hc^2}{\lambda^5} \cdot \frac{1}{e^{hc/\lambda T} - 1} d\lambda \tag{4a}$$

$$B_\nu(T)d_\nu = \frac{2h\nu^3}{c^2} \cdot \frac{1}{e^{\nu h/kT} - 1} d\lambda \tag{4b}$$

(if written for wave-number $\nu = 1/\lambda$ in spectroscopy)
h = Planck's constant;
c = velocity of light;
k = Boltzmann's constant;

gives brightness temperature T_B

$$T_B = \frac{a}{\ln B_\nu(T) - b}; \quad a = c_2\nu, b = \ln c_1\nu^3, \quad c_1 = 2h/c^2, c_2 = h/k \tag{5}$$

The conversion of satellite signals I_λ^+ into brightness temperatures T_B is generally accepted. However, these values are often erroneously used directly as surface temperatures before the atmospheric masking has been corrected. Before these corrections we will introduce Eq. (3) and also Eq. (5) which, when written for the monochromatic idealization, must be adapted to the finite spectral width of a radiometer channel.

2.3 Account of the Finite Spectral Width

The width of satellite radiometer channels causes a modification of the idealized equations (3) and (5). If the normalized spectral transmission $\varphi^*(\nu)$ or $\varphi^*(\lambda)$, for which we have

$$\int_{\nu_1}^{\nu_2} \varphi^*(\nu)\,d\nu = \int_{\lambda_1}^{\lambda_2} \varphi^*(\lambda)\,d\lambda = 1 \tag{6}$$

is introduced into Eq.(3) the real satellite signal becomes

$$I^+_{\Delta\nu} = \int_{\nu_1}^{\nu_2} I^+_\nu \cdot \varphi^*(\nu)\,d\nu = \int_{\nu_1}^{\nu_2} B_\nu(T_B)\cdot\varphi^*(\nu)\,d\nu$$

$$I^+_{\Delta\lambda} = \int_{\lambda_1}^{\lambda_2} I^+_\lambda \cdot \varphi^*(\lambda)\,d\lambda = \int_{\lambda_1}^{\lambda_2} B_\lambda(T_B)\cdot\varphi^*(\lambda)\,d\lambda \tag{7}$$

with:
ν_1, ν_2 = wavenumber for which $\varphi^*(\nu) = 0$ if $\nu < \nu_1$ and $\nu > \nu_2$;
λ_1, λ_2 = wavelength for which $\varphi^*(\lambda) = 0$ if $\lambda < \lambda_1$ and $\lambda > \lambda_2$.

Since $\varphi^*(\nu)$ cannot be simply expressed as an analytical function, Eq.(7) cannot be exactly - as Eq.(5) - converted into a brightness temperature T_B. Following an empirical approach by Singh (1984) the brightness temperature T_B of existing satellite radiometer channels is sufficiently accurate if described by

$$T_B = a_1 + b_1 \cdot T'_B \tag{8}$$

where the so-called preliminary brightness temperature T'_B is related to $I^+_{\Delta\nu}$ as follows

$$T'_B = \frac{b_2}{\ln I^+_{\Delta\nu} - a_2} \quad \text{(equivalent to Eq.(5))} \tag{9}$$

This approach also accounts - in Eq.(8) - for any occurring non-linear calibration curves of a satellite radiometer.

As an example, Table I displays the coefficients for all three thermal infrared channels with central wavelengths $\bar{\lambda}$ at 3.7, 11 and 12 μm of the NOAA 7 polar orbiting satellite, the first to contain the split-window at 11 and 12 μm.

2.4 ΔT for Different Atmospheres

The dependence of the temperature correction ΔT on different situations is best explained by showing ΔT for variations of spectral emissivity ε_λ, water vapour content, temperature structure, viewing angle, aerosol particle optical depth and optically thin, undetected clouds.

The best way of obtaining an overview of possible corrections is

TABLE I
Coefficients in Eqs.(8) and (9) needed for the determination of T_B for the Advanced Very High Resolution Radiometer (AVHRR) infrared channels onboard the NOAA 7 satellite

Channel	$\bar{\lambda}$	a_1 K	b_1	a_2	b_2 K
3	3.7 μm	0	1	12.2554	-3821.046
4	11 μm	-12.920	1.045	9.2058	-1344.832
5	12 μm	- 7.717	1.027	8.9373	-1226.189

by numerical simulations with a radiative transfer model. The following results were all available before a procedure for land surface temperature determination had been devised.

2.4.1 Emissivity dependence. Within the $0.95 < \varepsilon_{\Delta\lambda} < 1.0$ range of mean surface emissivity $\varepsilon_{\Delta\lambda}$ for the wavelength band $\Delta\lambda$ of a radiometer channel, the temperature correction ΔT used to infer surface temperature T_S by

$$T_S = T_B + \Delta T \tag{10}$$

is nearly linearly dependent on the deviation from the blackbody emission $1-\varepsilon_{\Delta\lambda}$ with $\Delta T/\Delta\varepsilon_{\Delta\lambda} = -0.45$ K for one percent change in mean emissivity $\varepsilon_{\Delta\lambda}$. Moreover, $\Delta T/\Delta\varepsilon_{\Delta\lambda}$ is only slightly dependent on temperature structure close to the ground and on spectral position within the main thermal infrared window (Figure 1).

2.4.2 Water vapour influence. Water vapour is the dominant absorber within the thermal infrared window from 10 to 13 μm wavelength. It is not possible to reduce its influence strongly by choosing narrow spectral intervals, since the main absorption is not due to single water vapour lines but to the so-called water vapour continuum, often termed e-type absorption with reference to the vapour pressure e dependence. The origin of this absorption, either wings of distant lines or water vapour polymers, is still not known.

As both absorption within the atmosphere and the different emission temperatures determine ΔT we should expect a strong dependence of the water vapour contribution to ΔT on temperature structure. This is clearly shown in Figure 2 (modified after Stork, 1984), where ΔT for strong inversions nearly vanishes and also loses water vapour content dependence while water vapour dependence is very strong for heated surfaces (daytime dry land surface). Another fact from Figure 2 should be stressed: even in midlatitude atmospheres with water vapour content < 2.5 $g.m^{-2}$ ΔT may vary between daytime and nighttime conditions by up to 8 K. Figure 2 also indicates the advantage of a two channel or split-window approach. The different water vapour content dependence allows an approximate calculation of ΔT without ground truth or knowledge of climatology.

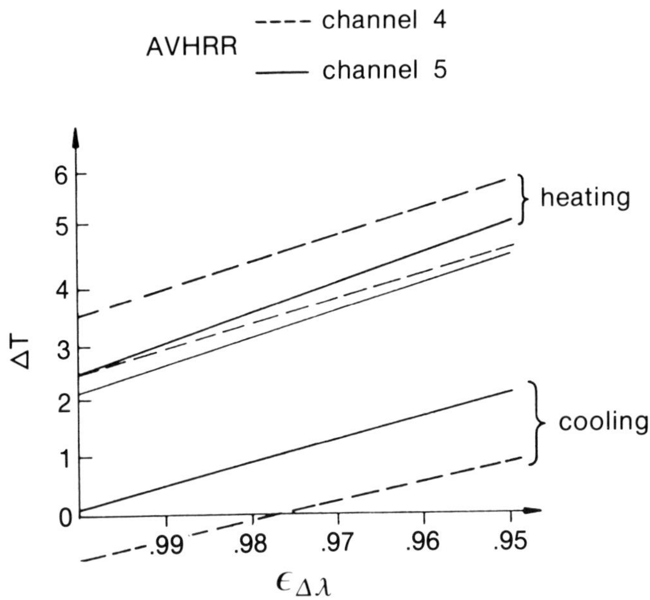

Figure 1. Temperature correction ΔT in K in a $45^{O}N$ standard atmosphere as a function of mean spectral surface emissivity $\varepsilon_{\Delta\lambda}$ for the wavelength band of a radiometer channel. Extreme temperature structure variations at the ground (10 K heating, 10 K cooling) do not change slopes appreciably (Stork, 1984).

2.4.3 Temperature structure influence. The lower the temperature at which an atmospheric absorber emits (if compared to T_s), the higher the ΔT. Therefore, whenever atmospheric absorbers mask the surface, we need information on the temperature structure of the atmosphere in order to estimate the absorber's influence on remote surface temperature determination. Since water vapour content is strongly concentrated near to the ground and since it is the most important absorber in this context, the near surface temperature structure should be investigated first.

The main results of systematic simulations of lower tropospheric temperature variations are:

- temperature inversions at the surface reduce ΔT, eventually giving $\Delta T \leq 0$ for strong inversions (see also Figure 2);
- heated surfaces lead to a strong enhancement of ΔT, if compared to corresponding atmospheric temperature structure at adiabatic lapse rates;
- both above effects increase with increasing absorption, i.e. in channel 5 of the AVHRR, $\Delta T < 0$ is reached earlier at increasing inversion strength;
- in many cases ΔT varies by more than 2 K between daytime and nighttime vertical temperature profiles, thus rendering the application of

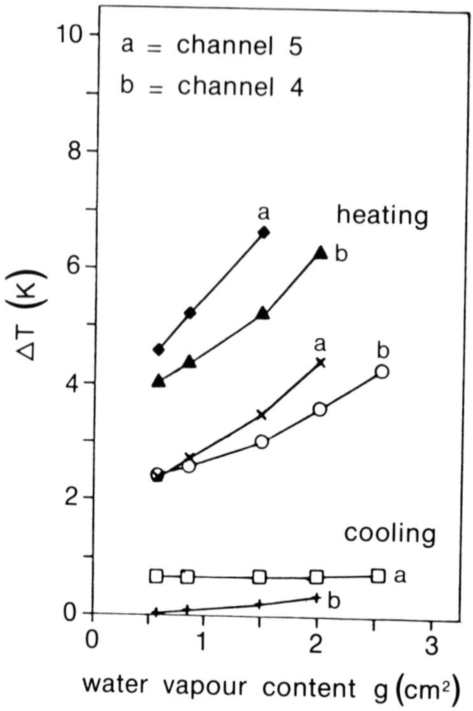

Figure 2. Temperature correction ΔT as a function of water vapour column content for a 45°N standard atmosphere, AVHRR channels 4 and 5, as well as surface heating and surface cooling by 10 K; nadir view; water surface emissivity ε_λ.

a mean ΔT to a one-channel brightness temperature T_B of little use;
- determination of remote surface temperature is most difficult for daytime situations over heated land surfaces.

The second effect of temperature structure, if ordered corresponding to importance for surface temperature determination, is the low temperature in the upper troposphere and lower stratosphere. Optically thin cirrus clouds, often not detected, and stratopheric aerosol layers, therefore, become most important. They reduce the first term in Eq.(1), but however do not add much to the second term ($B_\lambda(T) \sim T^4$ near 10 μm wavelength and thus very small at $-50^{o}C$), at the low temperatures mentioned.

2.4.4 Off-nadir view. Imaging infrared radiometers like those onboard METEOSAT or NOAA-N look onto the Earth's surface at different angles, changing from pixel to pixel. Therefore, ΔT must be different for all pixels even with a fixed vertical structure of the atmosphere and fixed surface properties. While all results up to now were for the nadir view, Figure 3 demonstrates the ΔT increase for the off-nadir

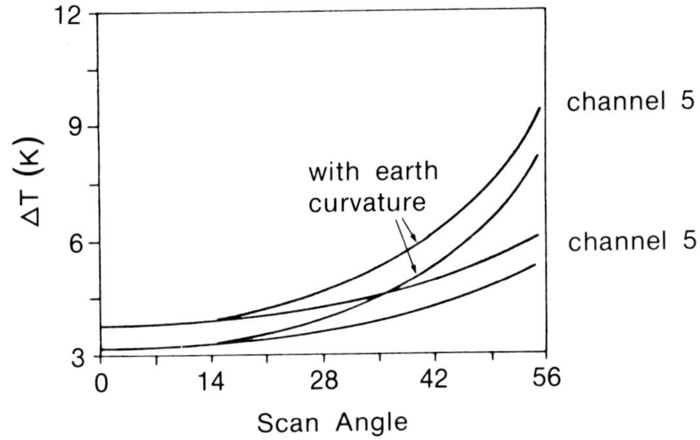

Figure 3. ΔT as a function of the scan angle (off-nadir) for the split-window channels of the AVHRR both for a flat surface and the Earth's surface.

view also.

The curves, shown for the split-window channels of NOAA-7, also account for the sphericity of the Earth. If the curves were plotted as a function of relative airmass m, they would not show a linear dependence. This is for the most part not due to the temperature decrease with height in the troposphere but to the compensational effect of the second term in Eq.(3). $\Delta T \sim m^n$ with $n < 1$ leads to a doubling of ΔT (nadir) at $m > 2$, which is however, because of the sphericity of the Earth, already reached at about 50° scan angle, an angle inside the $\pm 54^o$ cross-track scan of the AVHRR.

2.4.5 Aerosol particle effects. Aerosol particles scatter and absorb; thus it is more difficult to understand their impact on ΔT than that of gases. Even as pure scatterers they would increase ΔT, because the compensating second term in Eq.(3) then vanishes. The typical value of total aerosol optical depth τ_a (see Eq.(2)) in the visible (0.55 µm wavelength normally used as a reference wavelength) is 0.2. However, $\tau_{a,0.55}$ may reach 1.0 during Saharan dust storms or in metropolitan areas. At thermal infrared window wavelengths $\tau_{a,0.55}/\tau_a \approx 5$, the exact relation depending mainly on the size of the particles. Also the partitioning of extinction into scattering and absorption, and thus the chemical composition, play an important role since the wavelength dependence of $\Delta T/\Delta\tau_{a,0.55}$ is not well known in the thermal infrared window. We may expect the failure of the split-window technique for aerosol effect correction and thus need additional channels for a satellite-alone retrieval of T_s.

Since the vertical distribution of aerosol particle number density is even more variable than that of water vapour density, the aerosol particle contribution to ΔT may vanish as for clear polar air or may be as important as water vapour in cases with high lying aerosol layers. For cases where most of the aerosol load is confined to the planetary boundary layer - the standard case - $\Delta T/\Delta\tau_{a,0.55} \cong 5$ K per unit optical depth; i.e. typically .5 K must be added to satellite brightness temperatures to compensate for the aerosol effect. The exact value of $\Delta T/\Delta\tau_{a,0.55}$ additionally depends rather strongly on relative humidity and water vapour column content.

2.4.6 Optically thin clouds. Any surface temperature algorithm applied so far must rely on a cloud detection procedure applied before the temperature algorithm. Because of a frequent failure of cloud detection algorithms for optically thin or sub-pixel sized clouds, the possible influence of thin cirrus clouds on brightness temperatures is shown as an example. Figure 4 from Manschke (1985) also includes aerosol effects. It is clearly seen that

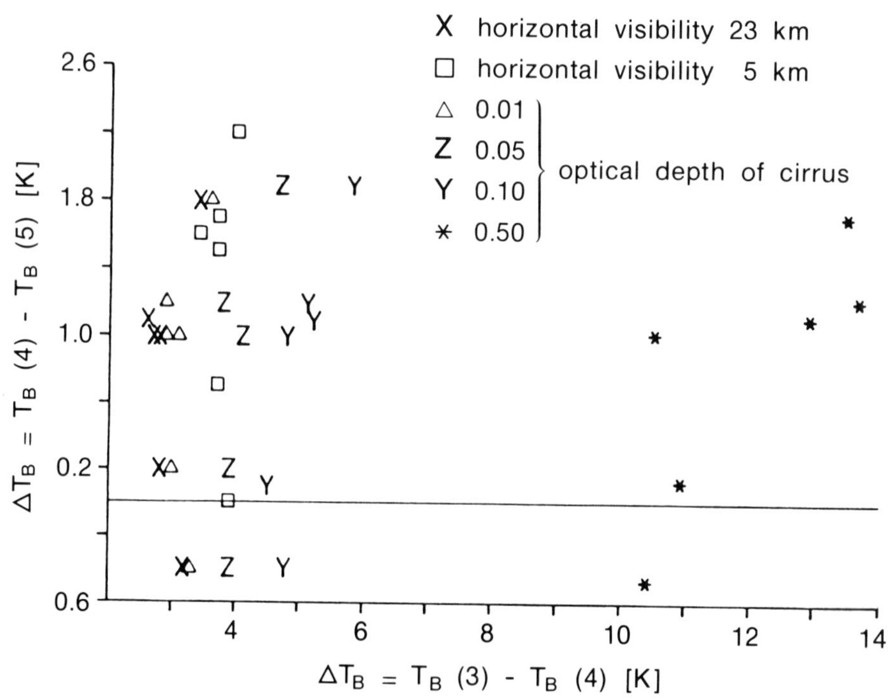

Figure 4. Brightness temperature differences $\Delta T_B = T_B(3) - T_B(4)$ versus $\Delta T_B = T_B(4) - T_B(5)$ for 6 atmospheres (45°N summer, autumn, winter; subarctic summer and winter; tropics 10°N) and at 6 cirrus or aerosol disturbances as indicated.

- the difference $\Delta T_B = T_B(3) - T_B(4)$ of channels 3 at 3.7 µm and 4 at 11 µm of the AVHRR is well suited for a cloud detection;
- aerosol effects at a horizontal visibility change from 23 to 5 km are smaller than those for 0.05 optical depth of a cirrus at 10 km height;
- aerosol effects also have more impact on $T_B(3) - T_B(4)$ than on $T_B(4) - T_B(5)$;
- differences among atmospheres dominate ΔT_B for $\tau_{cirrus} \leqq 0.1$, thus offering T_s determination through thin cirrus as shown by Manschke (1985) for water surfaces.

Different procedures for detecting thin and sub-pixel sized clouds as well will be presented in section 3.5.

2.4.7 Other influences. The thermal infrared window from 10-13 µm as well as other smaller thermal infrared windows are not free of absorption by other gases than water vapour. For well-mixed gases like carbon dioxide (CO_2), methane (CH_4) and nitrous oxide (N_2O), the influence has been taken into account for the simulations presented. They also pose no severe problems when applying T_s retrieval algorithms to satellite data. Depending on the satellite radiometer channel, ozone may also be important for the exact ΔT (Grassl and Koepke, 1980).

3. ALGORITHMS

It is now common practice to call an equation coupling satellite radiances and atmospheric parameters via an empirical or semi-empirical though simple relation an algorithm. Despite a possible success when "playing" with brightness temperatures or their differences during the search for a retrieval procedure or an algorithm, one should normally start with the basic equation, simplify it and then adjust some parameters by validating remote measurements with in-situ data whenever possible. This way will be chosen for the multi-channel approach, but one-channel surface temperature estimation will be considered first.

3.1 T_s from a Single Channel

From the variability of the temperature correction ΔT, shown in section 2.4, it follows that the surface temperature T_s from a single channel will be least accurate over heated land surfaces at high relative humidity and high level turbidity, if only climatological corrections ΔT are applied to the satellite data. Under these circumstances and for the outer parts of the swath $\Delta T \cong +10$ K is not rare.

On the other hand, for a nearby lake in spring, as shown by McClain (1986), $\Delta T \cong 0$ K may be correct. Thus the satellite brightness temperature map often smoothes horizontal surface temperature differences. How should one then proceed, if only one channel is available? The answer is: Use as much additional information as is available. This additional information may be
- radiosonde data;

- short-term temperature profile forecasts;
- routine air temperature (2 m height) observations;
- bulk temperature observations in lakes and rivers;
- physical models of relative near surface temperature profiles as a function of season and daytime;
- calibration by a multi-channel satellite radiometer (for example AVHRR for METEOSAT as shown by Singh et al., 1985).

In this context the impact of the emissivity difference $\Delta\varepsilon_\lambda = 1-\varepsilon_\lambda$ over land is often exaggerated, since over vegetated surfaces it is small; even if for a tiny surface element (dry leaf) $\Delta\varepsilon_\lambda$ may reach 0.1, the roughness of all vegetation causes the mean emission of a satellite pixel to approach the blackbody emissivity: a wheat field is composed of many "black" holes. However, this discussion leads us to the difficulty of defining the mean surface temperature of rough surfaces. While we have no difficulties for smooth water surfaces (it is the skin temperature with radiation originating from the uppermost micrometers), we may still call it the surface temperature of a forest, although all portions of trees, the grass and the ground below contribute in varying amounts.

ΔT is more strongly influenced by heating or cooling of surfaces relative to the atmospheric temperature above the surface layer than by the emissivity deficit $\Delta\varepsilon_\lambda$.

3.2 T_s from at Least Two Channels

Firstly, by simplifying Eq.(3) the basic two-channel or split-window algorithm will be derived, following McMillin (1975). Neglecting the third term in Eq.(3), the contribution by reflecting atmospheric emission, thus setting $\varepsilon_\lambda = 1$ and applying the mean value theorem, we obtain:

$$I_\lambda^+ = B_\lambda(T_s)e^{-\tau_\lambda^*} + \bar{B}_\lambda(\bar{T})(1-e^{-\tau_\lambda^*}) \tag{11}$$

with $\bar{B}_\lambda(\bar{T})$ = mean spectral radiance of the atmosphere, Planck's function at an appropriate temperature $\bar{T}$ for a given transmission $t = e^{-\tau_\lambda^*}$.

For two channels, i.e. two different wavelength intervals, the spectral dependence of upward radiance I_λ^+ and the source function, Planck's function $B_\lambda(T)$, must be introduced, here by a Taylor expansion truncated after the first expansion term

$$B(\lambda_r,T) \cong B_\lambda(T) + \frac{\partial B_\lambda(T)}{\partial\lambda}(\lambda_r-\lambda) \tag{12a}$$

and

$$I^+(\lambda_r) \cong I_\lambda^+ + \frac{\partial I_\lambda^+}{\partial\lambda}(\lambda_r-\lambda) \tag{12b}$$

where λ_r is a reference wavelength within the 10-13 μm window. Since T_s, $\bar{T}$ and T_I (reached by inversion of I_λ^+) are close together for the

10-13 μm range and thus will not change appreciably with λ, we put

$$\frac{\partial I_\lambda^+}{\partial \lambda} \cong \frac{\partial B_\lambda(\bar{T})}{\partial \lambda} \cong \frac{\partial B_\lambda(T_s)}{\partial \lambda} \cong \frac{\partial B_\lambda(T_I)}{\partial \lambda} \tag{13}$$

Eq.(11) for any radiometer channel wavelength interval $\Delta\lambda_i$ within the 10-13 μm range is then simplified into (Eq.(11) into (12b); Eq.(12a) into the result)

$$I_{\Delta\lambda_i}^+ = B_{\lambda_r}(T_s)e^{-\tau_{\lambda_i}^*} + \bar{B}_{\lambda_r}(\bar{T})(1-e^{-\tau_{\lambda_i}^*}) \tag{14}$$

Writing Eq.(14) for two channels using $B_{\lambda_{1r}}(\bar{T}) \cong B_{\lambda_{2r}}(\bar{T})$, which is proved by radiative transfer calculations, subsequent subtraction and resolving for $B_{\lambda_r}(T_s)$, gives

$$B_{\lambda_r}(T_s) = I_{\lambda_{1r}}^+ + \frac{1-e^{-\tau_{\lambda_1}^*}}{e^{-\tau_{\lambda_1}^*} - e^{-\tau_{\lambda_2}^*}}(I_{\lambda_{1r}}^+ - I_{\lambda_{2r}}^+) =$$

$$= I_1^+ + \gamma(I_1^+ - I_2^+) \tag{15}$$

If this equation is converted into an equation for brightness temperatures T_B (see Eq.(5)), the basic version of the split-window algorithm is found

$$T_s = T_B(1) + b\ T_B(1) - T_B(2) + a \tag{16}$$

where the constant a accounts for errors due to the simplifications (Eqs.(11), (12), (13) and $B_{\lambda_{1r}} = B_{\lambda_{2r}}$) used to derive Eq.(16) from Eq.(3).

3.3 Achievements of the Split-Window Algorithm over Water

The algorithm (16) has been used primarily for sea surface temperature determination and numerous coefficients have been derived by different groups (examples are Strong and McClain (1984; Llewelyn-Jones et al. (1984); Schlüssel et al. (1987)). The differences in coefficients a and b mainly stem from the set of atmospheres and in-situ data used, and the accuracy and the quantity with which a comparison has been made, for instance bulk or skin temperature measured from a buoy or from a ship. It has also been shown that a regional algorithm is superior to a global, if only two channels are available (Llewelyn-Jones et al., 1984). The accuracies achieved by the two NOAA satellites with a split-window (NOAA 7 and NOAA 9) are best documented by showing the increased accuracy during the buoy-satellite matchups in Figure 5 and the accuracies after classification into daytime, nighttime and humidity

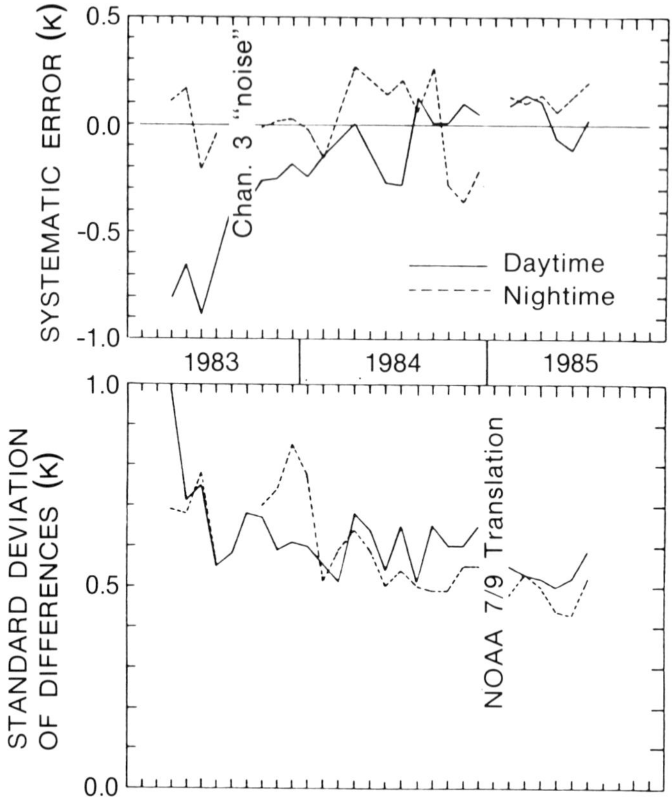

Figure 5. Systematic error and standard deviation monthly means between satellite derived and buoy bulk temperatures as a function of time for the split-window technique. Nighttime temperatures also use channel 3 and thus a modified algorithm.

intervals in Table II (taken from WCP-110, 1986).

3.4 Algorithms over Land?

Knowing the main result for water surfaces, namely a temperature nearly free of systematic errors and with only 0.5 K rms error, we ask whether the split-window algorithms may be applied to land surfaces. The answer is no, since the following facts must be taken into account:

- spectral emissivity changes from one surface type to the next, causing a ΔT_B of the same magnitude as humidity (see Table IIb) and thus leading to a false ΔT;
- cloud detection during daytime is exceptionally difficult over bright surfaces like snow or sand dunes;
- strong near surface heating or cooling is only partly accounted for by two spectral channels.

TABLE IIa
February - August 1985: matchup statistics for day and night and for different areas

REGION	DAY			NIGHT		
	No.	Bias	RMSD	No.	Bias	RMSD
World	1625	+0.01	0.55	1042	+0.12	0.48
W. Trop. Pac.	189	-0.02	0.47	102	+0.22	0.40
E. Trop. Pac.	433	-0.15	0.55	290	+0.08	0.41
Trop. Atl.	424	+0.08	0.51	277	+0.12	0.51
Trop. Ind.	277	-0.03	0.50	163	+0.11	0.46
N. Atl. & Pac.	208	+0.18	0.60	115	+0.18	0.54
N. Pac.	112	+0.25	0.65	74	+0.16	0.59
N. Atl.	96	+0.10	0.53	41	+0.23	0.46
Southern Ocean	139	+0.08	0.63	113	+0.11	0.44

TABLE IIb
February - August 1985: matchup statistics by moisture class, indicated by the brightness temperature difference classes

MOISTURE (Tropical Oceans)	DAY			NIGHT		
	No.	Bias	RMSD	No.	Bias	RMSD
$0^o < T_4 - T_5 < 1^o$	500	-0.08	0.47	218	+0.16	0.47
$1^o < T_4 - T_5 < 2^o$	694	+0.00	0.56	553	+0.09	0.46
$2^o < T_4 - T_5 < 3^o$	69	-0.01	0.61	38	-0.01	0.52

My proposal for a procedure for land surface temperature determination from two or more satellite brightness temperatures, therefore, was:

1) determine cloud-free parts;
2) use the split-window algorithm (16) over water surfaces within the area envisaged;
3) inter- and extrapolate the "water surface" atmospheric correction;
4) derive an additional correction from the brightness temperature difference between water surfaces and a land surface pixel for the channels available;
5) choose after adding the second correction the highest temperature to reach a good estimate of the surface temperature.

3.5 Application of the Proposed Procedure to ALPEX SOP Satellite Data

During one of the special observing periods (SOP) within the ALPEX (Alpex Experiment) rather cloudfree satellite data during day- and nighttime could be acquired on the 25th and 26th March 1982. For this time the procedure outlined above is applied to NOAA 7 data by Stork (1984). Since the images cover a major portion of the Alps, this was a hard test as far as cloud clearing and interpolation were concerned.

Step 1: cloud clearing. Four different procedures were applied simultaneously to daytime images. They are shown in Table III.

TABLE III
Cloud clearing thresholds (cloudy if above this threshold)[1]

Water surfaces	Snowfree land surfaces
Channel 1: $A_s \geqq 0.15$[2]	$A_s > 0.15$
Channel 2: $A_s \geqq 0.15$	$A_s > 0.24$
$T_B(3) - T_B(4) > 3$ K	$T_B(3) - T_B(4) > 5$ K
$T_B(4) - T_B(5) > 1.7$ K	$T_B(4) - T_B(5) > 1.9$ K

[1]all thresholds were chosen interactively at an image processing system.
[2]A_s = albedo.

Nighttime cloud clearing, rather easy over water, as shown by Olesen and Grassl (1985), is more difficult over land. Therefore, no table similar to Table III can be shown.

Step 2: surface temperature over water surfaces. For midlatitude atmospheres and the NOAA 7 satellite the split-window algorithm (16) for water surfaces is, following Jürgensen (1984)

$$T_s = 3.345\ T_B(4) - 2.363\ T_B(5) + 5.74\ \text{K} \tag{17}$$

This algorithm was applied to all big lakes within the scene and to the Adriatic Sea.

Step 3: inter- and extrapolation. Before an interpolation the atmospheric corrections over water ranging from 2 to 3 K for all big lakes and the Adriatic Sea were spatially averaged. Thus the noise in single ΔT_s due to the 0.12 K noise equivalent $\Delta T_s'$ (NEΔT) of the satellite radiometer was strongly suppressed. The rather small number of lakes does not allow a polynomial of higher order to be fitted to the water surface $\Delta T_s'$. Therefore, only a first and second order polynomial were chosen. Even the second order polynomial led to strong errors within the extrapolation area. Whenever extrapolation is necessary we have chosen as atmospheric correction ΔT for the image coordinates x,y and for channel i

$$\Delta T_i(x,y) = a_{1_i}x + a_{2_i}y + a_{3_i} \tag{18}$$

leading to a first approximation of surface temperature

$$T_{s_1}(x,y) = T_B(i) + \Delta T_i(x,y) \tag{19}$$

T_{s_1} still contains errors due to the near surface heating or cooling as well as to emissivity changes.

Step 4: additional correction derived from the water-land contrast. As shown in section 2.4 the heating or cooling of land surfaces increases or decreases the correction ΔT. For a midlatitude atmosphere (45°N) the additional correction may reach 3 K for a +20 K change in surface temperature. From model simulations we found for channels 4 and 5:

$$\Delta T_4' = 0.14\ \Delta T_{s_1}(4)$$
$$\Delta T_5' = 0.22\ \Delta T_{s_1}(5) \qquad (20)$$

These relations also hold for cooling, i.e. for negative temperature differences ΔT_{s_1}, if ΔT_{s_1} is the difference between T_{s_1} and the interpolated water surface temperature. After this correction the second approximation for T_s is reached

$$T_{s_2}(i) = T_B(i) + \Delta T_i(x,y) + \Delta T_i' \qquad (21)$$

$\Delta T_4'$ is shown in Figure 6.

Step 5: correction of emissivity changes. The effects of unknown surface emissivity and thus of unknown spectral dependence can only be removed partially. After applying corrections (19) and (21) to channel 4 and 5 one should choose the channel with higher temperature, since this channel should be the channel nearer to a blackbody emission. Choosing the higher temperature we arrive at T_s, the quantity desired. However, whenever the difference of T_s between both channels exceeds 1 K, pointing to $\Delta\varepsilon_\lambda > 2\%$, we discard this value and classify the pixel as cloud contaminated.

T_s from Figures 7 and 8a,b contains a wealth of information for climatologists.

3.6 Error Analysis

The following error sources should be discussed:
- conversion from counts to brightness temperatures;
- split-window algorithm for water surfaces (Eq.(17));
- interpolation errors (Eq.(18));
- water-land differences (Eq.(20));
- remaining emissivity difference.

The conversion from counts with a 10 bit resolution to brightness temperature is, according to Singh (1984), accurate to 0.01 K and thus errors are negligibly small. The radiometric noise expressed by the noise equivalent temperature difference (NEΔT = 0.12 K for channels 4 and 5 of AVHRR and for the 270 to 300 K range) is easily suppressed by spatial averaging. The split-window algorithm combining errors due to changing atmospheric structure not resolved and the NEΔT after spatial averaging is affected by a 0.5 K rms error (see Table II). If - as done here - T_s is determined on a pixel by pixel basis the combination leads to a slightly higher standard error for water surface temperature determination.

The interpolation error strongly depends on the number of water surfaces found in a scene. For 23 lakes in the chosen scene and the first order polynomial (18) the error of the atmospheric correction ΔT for water surfaces reaches 0.6 and 0.85 K for channel 4 and 5, respectively.

The adoption of this interpolated correction for synthetic water surfaces, $\Delta T_i(x,y)$, as a first approximation (Eq.(19)) for land surface temperature, enhances the rms error from 0.51 K for lakes to 0.8 K over land if the more transparent channel 4 is chosen. This error would increase to 1.0 K for channel 5. One should, however, be aware of the large systematic errors left because of heated or cooled surfaces and emissivity changes. These systematic errors are reduced by steps 4 and 5 (section 3.4). On the other hand, step 4 increases the rms error to 0.9 K for channel 4 and 1.1 K for channel 5.

The discussion of errors due to surface emissivity changes is more difficult. Since $\Delta\varepsilon_{\Delta\lambda} = 0.02$ is the upper limit for differences within the 10-13 µm window for smooth rock surfaces and since this would cause a systematic underestimate of T_s by about 1 K, we expect over vegetated surfaces with $\varepsilon_{\Delta\lambda} > 0.95$ and $\Delta\varepsilon_{\Delta\lambda} < 0.02$ an additional systematic error $\leqq$ 1 K which, however, is reduced by step 5 where the higher temperature of both channels is chosen.

An accuracy of 1-2 K for T_s of land surfaces is a step forward at least for daytime situations, where the atmospheric masking may reach 10 K.

A verification of the approach described is only possible by radiometric measurements from masts and airplanes. This has been shown to be successful for ocean surfaces (Schlüssel et al., 1987) by measurements from a ship, where by a combination of satellite radiometers a 0.3 K rms error was also reached.

3.7 The Daily Temperature Amplitude

The climatologist is sometimes even more interested in daily temperature amplitudes ΔT_D than in T_s, because they are an indication for evaporation and thus help in sensing latent heat fluxes remotely. Already before an application of the approach discussed above for remote land surface temperature determination we had derived ΔT_D for the ALPEX scenes used so far (Grassl et al., 1985).

In the inner Inn Valley, at Schwaz in the Tyrol, we could show by using ground observations and radio soundings that ΔT_D (Figure 9) is accurate to at least 2 K without any atmospheric correction in a valley wind system. The time of the NOAA-7 satellite overpasses at approximately 2.30 a.m. and p.m. guarantees observations at the daily temperature maximum and - best for valleys with a valley wind system -

very near to the temperature minimum. Under calm conditions, however, with strong heating during daytime, the procedure outlined in section 3.4 must be applied especially for the daytime overpass (see Figure 2).

3.8 Further Improvements

Recently, Schlüssel (1987) has demonstrated that a combination of two radiometers onboard the NOAA satellite series, namely the AVHRR and the HIRS (High Resolution Infrared Sounder) improves not only temperature and humidity profiles but also surface temperature over water surfaces. A 0.3 K rms error was found during a verification cruise in the North-East Atlantic for some case studies. The additional information comes from a 6.3 μm band water vapour channel whose weighting function peaks in the lowest troposphere and from three 15 μm-CO_2 channels peaking in the lower troposphere and at the ground. Especially the algorithm over big lakes and marginal seas would gain in accuracy by introducing the combined algorithm which also accounts for changing coefficients at different viewing angles. However, the latter improvement is also possible for the split-window algorithm (16). Despite the higher accuracy demonstrated for marginal seas there are still extreme conditions (McClain, 1986) in spring over the American big lakes where the improved algorithm failed. Therefore, it might be dangerous to take 0.5 K rms error for granted over lakes as done in section 3.6.

REFERENCES

Grassl, H. and Koepke, P., 1981: 'Corrections for atmospheric attenuation and surface reflectivity in satellite-borne SST measurements', in Oceanography from Space, Plenum Press, 97-107.

Grassle, H., Stork, T. and v. Rüsten, C., 1985: 'Daily temperature amplitude at the surface from NOAA-7', Zbornik Meteoroloskih i Hydroloskih Radova, **11**, 36-39.

Jürgensen, A., 1984: 'Bestimmung des Wasserdampfgehaltes der Atmosphäre über dem Ozean aus Satellitendaten', Diploma Thesis, University of Kiel, FRG.

Llewellyn-Jones, S.T., Minnett, P.J., Saunders, R.W. and Zavody, A.M., 1984: 'Satellite multichannel infrared measurements of sea surface temperature of the North-East Atlantic ocean using AVHRR/2', Quart. J. Roy. Met. Soc., **110**, 613-631.

Manschke, A., 1985: 'Einfluss dünner Cirren auf den Strahlungshaushalt', Diploma Thesis, University of Kiel, FRG.

McClain, P., 1986: private communitation.

McMillin, L.M., 1975: 'Estimation of sea surface temperatures from two infrared window measurements with different absorption', J. Geophys. Res., **80**, 5113-5117.

Olesen, F.-S. and Grassl, H., 1985: 'Cloud detection and classification over oceans at night with NOAA-7', Int. J. Rem. Sens., **6**, 1435-1444.

Schlüssel, P., 1987: 'Infrarotfernerkundung von Oberflächentemperaturen sowie atmosphärischen Temperatur- und Wasserdampfstrukturen',

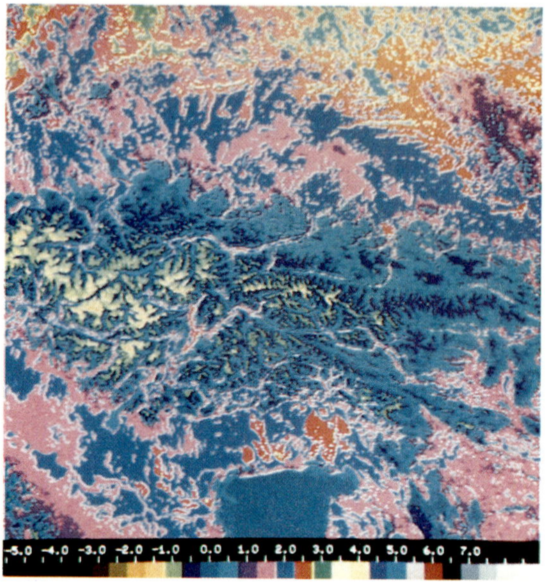

Figure 6. $\Delta T_4'$ for the daytime overpass in steps of 0.5 K indicated by different colours (see scale below image).

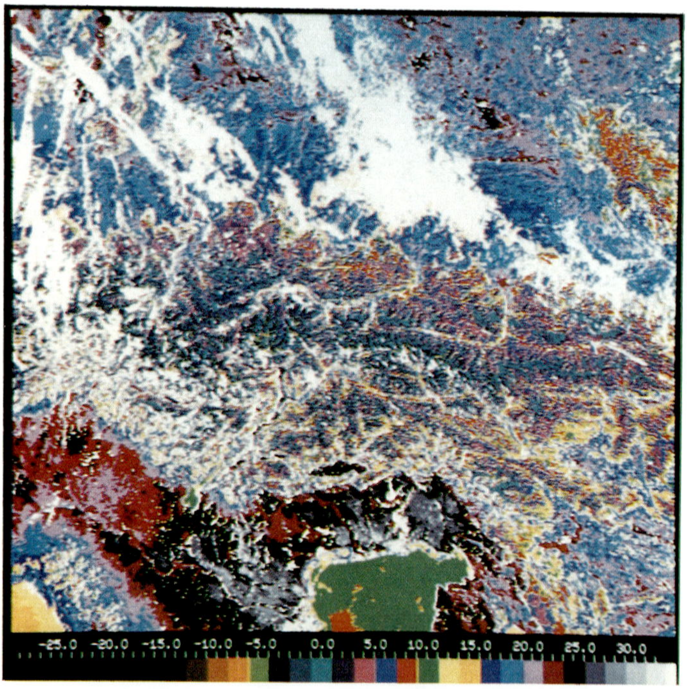

Figure 7. Surface temperature for the daytime overpass in steps of 2 K (25 March 1982, early afternoon); clouds are white.

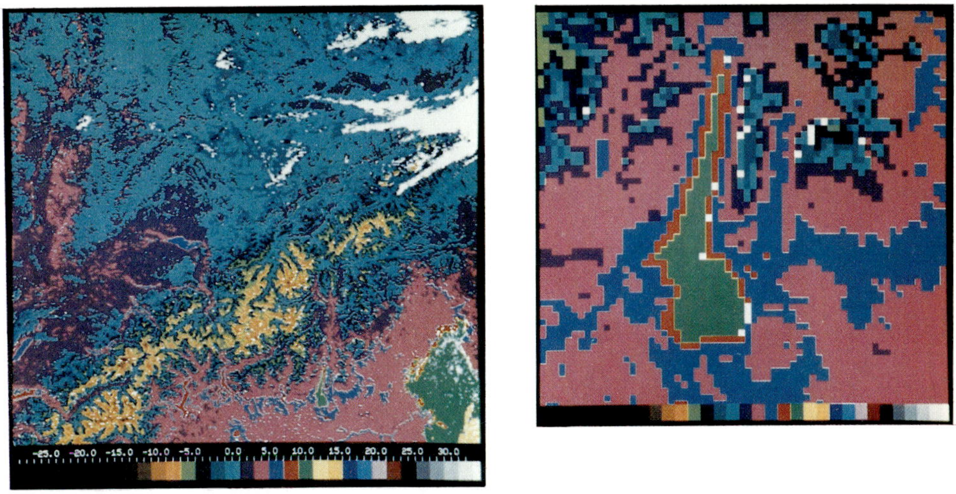

a) b)

Figure 8. Surface temperature for the nighttime overpass in steps of 2 K (26 March 1982, early morning).
a) for the full image;
b) zoom around Lago di Garda, Northern Italy.

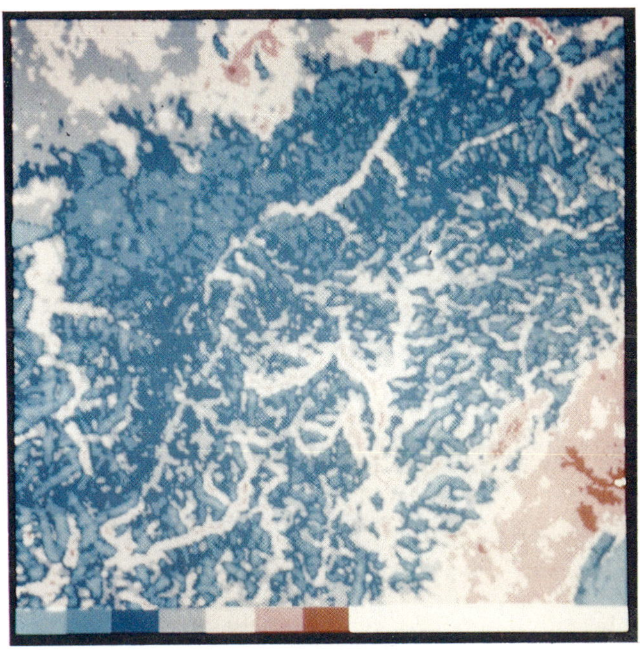

Figure 9. Daily temperature amplitude for parts of the Alps derived solely from brightness temperature differences between 25 March, ∿ 2.30 p.m. and 26 March, ∿ 2.30 a.m.

Ph.D. Thesis, University of Kiel. Berichte aus dem Inst. für Meereskunde, Nr.161, 116 pages.

Schlüssel, P., Shin, H.-Y., Emery, W.J. and Grassl, H., 1986: 'Comparison of satellite derived sea surface temperatures with in-situ skin measurements', J. Geophys. Res., March issue.

Singh, S.M., 1984: 'Removal of atmospheric effects on a pixel basis from the thermal infrared data from instruments on satellites. The advanced very high resolution radiometer (AVHRR)', Int. J. Rem. Sens., 5, 161-183.

Stork, T., 1985: 'Bestimmung der Oberflächentemperatur über Land aus Satellitendaten', Diploma Thesis, University of Kiel, FRG.

Strong, A.E. and McClain, E.P., 1984: 'Improved ocean surface temperatures from space: comparison with drifting buoys', Bull. Am. Met. Soc., 65, 138-141.

WCP-110, 1986: 'Satellite-derived sea surface temperatures for global climate applications', Report of the COSPAR int. workshop, Camp Springs, Maryland, USA, 28-31 May 1985; World Climate Programme Series, WMO, Geneva.

USE OF SURFACE TEMPERATURE IN AGROMETEOROLOGY

Bernard Seguin
INRA - Station de Bioclimatologie
BP 91
F 84140 Montfavet - France

ABSTRACT. After a short introduction indicating the interest of surface temperatures derived from satellite thermal IR data as compared to ground climatological measurements, a first chapter describes their physical bases resulting from the surface energy balance equation. Possible applications in agrometeorology are then listed, and the specific application of computing evapotranspiration is finally described and discussed.

1. INTRODUCTION

The recent development of thermal IR data has allowed us to obtain frequent and accurate measurements of the surface temperature of land surfaces. If ground experiments with portable radiothermometers may give a new tool for estimating crop water stress and then scheduling irrigation (see Jackson, 1984, for a comprehensive review), the most important applications come from the thermal IR data of satellites.

The satellite derived data of surface temperature T_S (see lectures by Raschke and Grassl) appear as new measurements which are useful for agrometeorology in two ways, compared to conventional measurements of screen air temperature in a climatological network:

1) The spatial scale, from 100 m to 5 km depending on the satellite resolution, is far more precise than ground climatic data (typically near 100 km for synoptic stations and 10-50 km for local less regular stations) (Figure 1).
   Although LANDSAT TM (120 m) and HCMM (500 m) give access to a high spatial resolution, they only allow us to perform detailed studies on specific days, because of their low temporal resolution (every 18 days for TM, archives for HCMM which operated for 3 years, 1978 to 1980). On the other hand, NOAA-AVHRR and geostationary satellites like METEOSAT (5 km) are fully available for agrometeorological monitoring (4 passes per day with two satellites for AVHRR, theoretical possibility of every half hour for METEOSAT).

F. Toselli (ed.), Applications of Remote Sensing to Agrometeorology, 221–240.

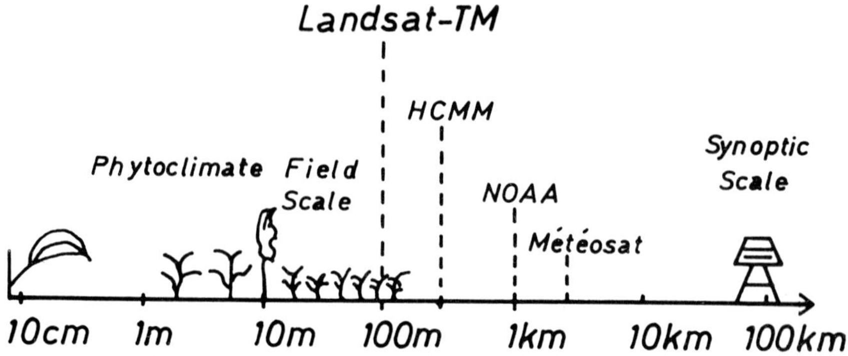

Figure 1. Typical scales for climatic action on crop production, with possible inputs from satellites.

2) The access to surface properties, as opposed to screen air temperatures which attempt to characterize the air mass properties as independently as possible of underlying surface properties (Figure 2).

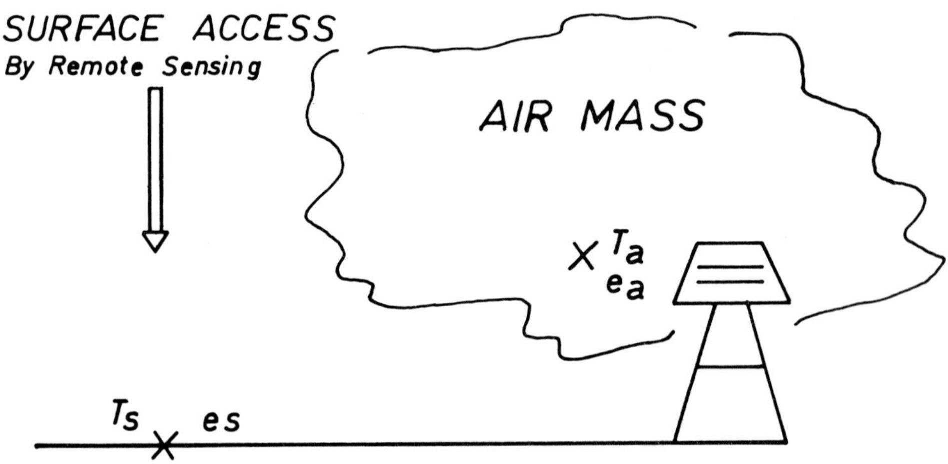

Figure 2. Schematic description of surface characteristics compared to screen measurements.

Access to surface properties (albedo, surface temperature, surface moisture) is a determinant input for the local functioning of the soil-vegetation interface, in relation to the thermal equilibrium governed by the combined processes of energy balance and soil water budget. A short analysis of this equilibrium, which governs the significance of the surface temperature, is necessary before developing its possible use in agrometeorology.

2. THE SURFACE ENERGY BALANCE. RESULTING RELATIONSHIP BETWEEN T_s AND EVAPOTRANSPIRATION

2.1 Energy Balance of a Thin Surface

If we consider the daytime case of a thin surface (bare soil or low crop less than 10-20 cm high), the classical equation of energy balance, at a given time,

$$R_n = G + H + LE$$

expresses the division of net available energy (net radiation R_n) between ground conductive heat flux G and convective fluxes, sensible heat H and latent heat LE (corresponding to evaporation for a bare soil and evapotranspiration for a vegetation canopy) (Figure 3).

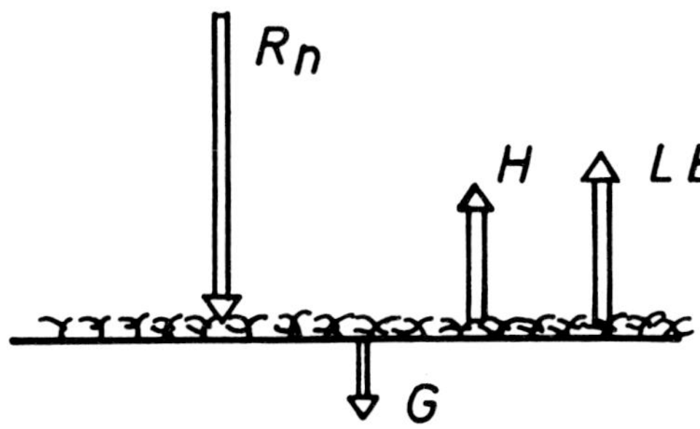

Figure 3. Energy balance components.

As the surface is thin, this equation may be applied directly at the surface level where radiative absorption and emission take place.

The net radiation R_n is found from the various components of radiation exchanges:

$$R_n = (1-a)R_g + R_a - \sigma T_s^4 \tag{1}$$

where the solar global radiation R_g is partly reflected depending on the surface albedo a, σT_s^4 corresponds to the long-wave radiation emitted by the surface (assuming the emissivity being equal to 1.0) and R_a by the atmosphere down to the surface (long-wave atmospheric radiation).

In order to obtain an independent expression for T_s, it is usual to define a "climatic net radiation" R_n^* where T_s is replaced by T_a

$$R_n^* = (1-a)R_g + R_a - \sigma T_a^4 \tag{2}$$

the difference $R_n^* - R_n$ being equivalent to $\sigma T_s^4 - \sigma T_a^4 \approx 4\sigma T^3 (T_s - T_a)$.

Using the expression for H

$$H = \rho C p h (T_s - T_a) \tag{3}$$

where the sensible heat flux is related to (T_s-T_a) by the turbulent exchange coefficient h (ρ and Cp being constants, air density and specific heat, respectively), one obtains

$$R_n^* = R_n + 4\sigma T^3 (T_s-T_a) = G + H + 4\sigma T^3 (T_s-T_a) + LE$$

and then

$$R_n^* - G = (\rho Cph + 4\sigma T^3)(T_s-T_a) \tag{4}$$

so that the deviation T_s-T_a may be expressed by

$$T_s-T_a = \frac{(R_n^* - G) - LE}{\rho Cph + 4\sigma T^3} \tag{5}$$

2.2 Relationship between T_s and Evaporation (or Evapotranspiration)

Taking into account the usual assumption that the ground heat flux G is, either negligible (which is the case in wet situations or with a dense canopy), or a fixed proportion of R_n^* (0.1 for example, whilst it may increase up to 0.2 - 0.3 in the specific case of dry bare soil), we can state from Eq. (5) that, for given atmospheric conditions (R_n^* and T_a), T_s is a linear decreasing function of LE. Knowing that $4\sigma T^3 \approx 6$ to 8 $W.m^{-2}.K^{-1}$ and that ρCph between 10 and 100 in the same units, it can be said that the slope of this relationship depends mainly upon the exchange coefficient h.

From the laws of turbulent transfer in the atmospheric surface layer, h can be written as (see Seguin, 1981)

$$h = h_N \cdot F(z/L) \tag{6}$$

where the exchange coefficient in neutral conditions h_N is

$$h_N = \frac{ku^*}{\mathrm{Log}(z/z_{o\ t})} \tag{7}$$

with the friction velocity u^* derived from the logarithmic wind profile

$$u^* = \frac{k\ u(z)}{\mathrm{Log}(z/z_o)} \tag{8}$$

(z_o and $z_{o\ t}$ being the aerodynamic roughness and the heat exchange roughness, respectively, which are equivalent for thin surfaces)

and F(z/L) the stability effect correction depending on the ratio z/L (L Monin Obukhov length).

So that, on the whole, h depends upon:

- the wind velocity u(z);
- the stability effects (L being related to u^* and T_s-T_a);
- the surface aerodynamic roughness z_o (which is related to the mean height of surface elements h, z_o being of the order of 0.10-0.15 h).

The relationship between T_s and LE is then globally dependent on:
- the atmospheric variables: R_n^*, T_a, u;
- the surface parameters: a and z_o.

2.3 The Extension to Tall Canopies

When applied to tall canopies, the concept of energy balance and the linked equations are still valid, but problems arise with the definition of the surface level and its properties.

A simplification may occur with the ground heat flux G, because the interception of radiation inhibits the input of energy directly to the soil surface, thus giving a ratio G/R_n less than 0.05 for dense canopies.

Two main differences arise compared to a thin surface:

- The roughness length for heat transfer $z_{o\ t}$ becomes significantly different from the aerodynamic roughness z_o, corresponding to an excess resistance for heat transfer in the case of rough surfaces (Figure 4) (see Thom, 1975; Garatt, 1978; Brutsaert, 1982). The ratio $z_{o\ t}/z_o$ may be estimated as from 0.2 to 0.1 for short to medium-size crops (grass, soybeans, wheat). There is some uncertainty for taller canopies (like corn, fruit trees, forest, etc.).

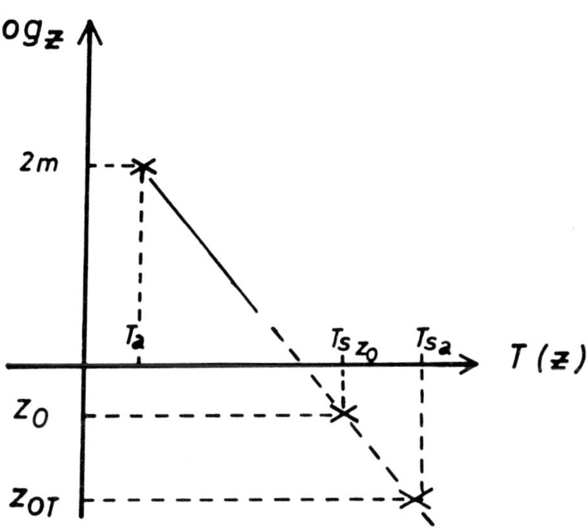

Figure 4. Semi-logarithmic temperature profiles, allowing the separation of $z_{o\ t}$ and z_o.

- The corresponding "aerodynamic surface temperature" $T_{s\ a}$ thus defined, by extrapolation of air temperature profiles down to the level $z_{o\ t}$, may differ from the "radiative surface temperature" $T_{s\ r}$ as measured from the thermal IR radiation. This last parameter results from the radiation emitted by various layers inside the canopy, and pay incorporate a significant contribution from the generally colder lower layers (Figure 5). $T_{s\ r}$ is, therefore, in general lower than $T_{s\ a}$. The deviation has been measured as 1° in the case of a wheat canopy (Huband and Monteith, 1986). It is not well known for taller canopies.

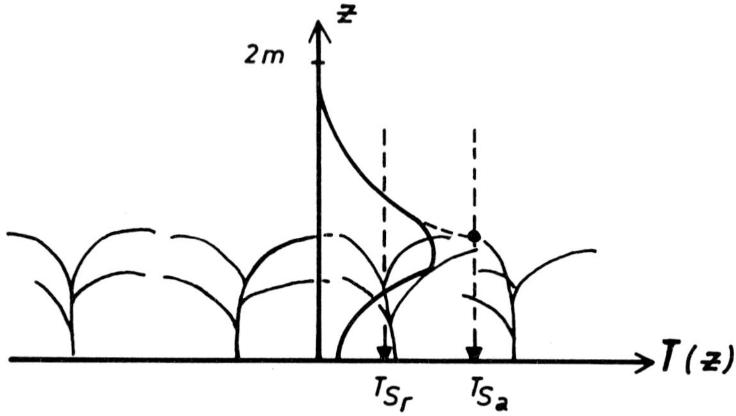

Figure 5. Scheme indicating the case of a developed canopy with $T_{s\ r}$ significantly different from $T_{s\ a}$.

These two points ($z_{o\ t} = z_o$ and $T_{s\ r} = T_{s\ a}$) which are closely linked (for instance if $z_{o\ t}/z_o$ is computed from experiments where $T_{s\ a}$ is measured by $T_{s\ r}$, as in Garratt (1978)), lead to large uncertainties on the precise form of the relations between T_s and LE for elevated canopies.

3. THE VARIOUS APPLICATIONS OF SURFACE TEMPERATURES FROM SATELLITE DATA

Starting from the knowledge of the determination of T_s-T_a expressed by Eq.(5), and considering the trivial expression

$$T_s = T_a + (T_s - T_a) \qquad (9)$$

we may consider that three types of application are derivable from T_s data from satellite thermal IR.

3.1 Agrometeorological Applications with T_s Itself

Up to now agrometeorological studies have mainly been based on air temperature data, generally because it is the only thermal parameter accessible on the spatial scale needed. Regular access to T_s by satellite data allows us to define some specific applications where it appears to be better adapted than air temperature. For instance:

- Nighttime maps of minimum temperatures for frost episodes: the air temperature is known to be an imperfect characterization of spring frost conditions, because surface temperatures may differ by some 3 to 5° for the same screen minimum temperature (depending upon cloud coverage or wind velocity). Maps of surface temperatures may ensure a better assessment of frost conditions, either for the monitoring of cold evolution during the night (Sutherland et al., 1979), or for the mapping of frost effects upon crops (Byrne et al., 1984; Caselles et al., 1983; Chew et al., 1982; Kalma et al., 1983). Experiments in the Rhone valley (Southeastern France) displayed a good overall agreement between NOAA nighttime data and ground measurements and the superiority of satellite mapping in terms of precision on a spatial scale (Kerr et al., 1983) as shown in Figure 6, which displays two contrasted nights.

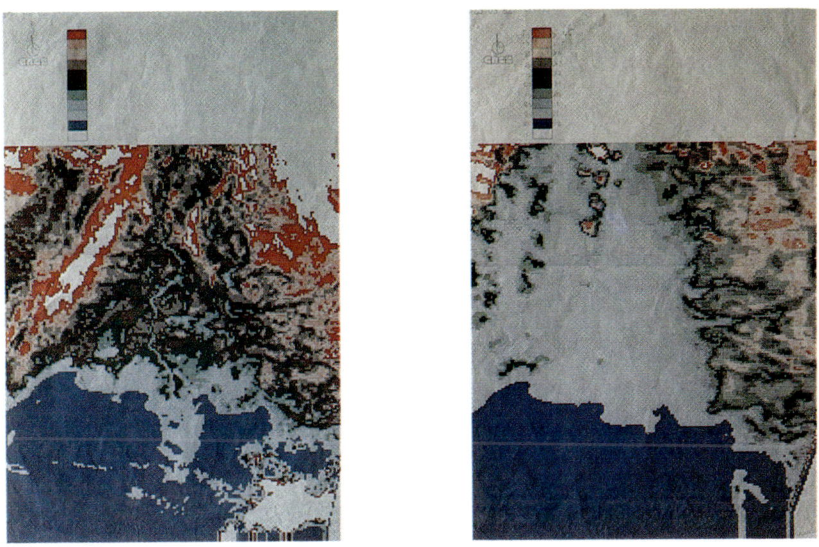

a) b)

Figure 6. Two night frost situations in the lower Rhone valley from NOAA. a) calm night with a high variability; b) windy night displaying a uniform situation.

- The use of surface temperature instead of air temperature to monitor the summation of temperatures in growth and development agrometeorological applications. For a large number of crops, the duration of phenological phases (for instance, between sowing and maturation) is calibrated against ΣT_a, which allows us to characterize a given year against normal climatological values or to delineate the potential of agricultural regions for cropping specific ranges of cultivars. The use of ΣT_s instead of ΣT_a may give a new tool for this range of agrometeorological applications.

3.2 Mapping of Air Temperature with Satellite Thermal IR Data

In other cases we may think of using T_s data to derive a better characterization of T_a spatial variations for zones where the climatic ground network is insufficient, for instance in mountainous regions.

This use of satellite data for topoclimatic studies supposes that the spatial variation of T_s is significantly related to that of T_a, which needs a fairly constant value of T_s-T_a. Knowing the main sources of variation of this parameter (water availability in the first place, then canopy structure), this condition may be fulfilled if vegetation is spatially homogeneous and well-watered.

Studies by Lagouarde (1983), based on HCMM data, established the potential use of thermal IR in the description of several mountainous regions of France, the altitudinal gradients of T_s obtained being quite similar to the well-known climatological values of T_a gradients in mountainous regions.

3.3 Estimation of Evapotranspiration

The third application consists in the analysis of T_s in relation to the surface water status and its rate of evaporation (or evapotranspiration) LE, thus using more or less directly the expression (5) for T_s-T_a. It will be developed in the following chapter.

4. USE OF SURFACE TEMPERATURE FOR THE ESTIMATION OF EVAPOTRANSPIRATION

This is the most studied and promising use of surface temperature from satellite data,

- Firstly because T_s-T_a is primarily governed by water availability. Its range of variation extends from near 0° when LE approaches the potential rate of PET (LE being of the order of $R_n^* - G$) to some 20 to 25° in fully dry conditions (Eq.(5) gives T_s-T_a = 20° for $R_n^* - G$ = 600 W/m^2, ρCph = 23 W.m^{-2}.K^{-1} and $4\sigma T_s^3$ = 7 in the same unit). The influence of the canopy structure is far lower, of the order of 2 to 3° for vegetation under the same water status.
- Secondly because it is difficult to assess spatial variations of evapotranspiration ET by conventional methods. Most ET measurement methods are usable on a local scale (some m^2) and a few possibilities exist on the larger regional scale (see Seguin et al., 1982, for a global review). Remote sensing then appears as the only available system for characterizing the latter.

This field of work has been extensively studied in the last ten years. The setting-up of a methodological basis had started, before that date, using ground radiothermometers, but the availability of real satellite thermal IR data with HCMM gave an effective impulse to this field of work.

4.1 Methodological Basis

The basic equation upon which the common methodology relies may be derived from Eq.(4) slightly modified

$$LE = R_n^* - G - (\rho Cph + 4\sigma T^3)(T_s - T_a) \tag{4'}$$

where LE is indirectly computed from $(R_n^* - G)$, using the measurement of T_s-T_a and the estimation of h (which justifies the appelation of combination equation, because it combines the energy balance concept and the aerodynamic expression for the sensible heat flux H).

We should mention here that a direct approach could also be proposed by the aerodynamic expression for LE

$$LE = \frac{\rho Cph}{\gamma}(e_s - e_a) \tag{9}$$

involving the vapour pressure (with the psychrometric constant γ = 0.66 $mb.deg^{-1}$) instead of the temperature gradient. It requires the measurement of surface moisture, for instance by using microwave wavelengths, as described by Schmugge in another lecture.

A first step in building methods to infer evapotranspiration from surface temperature is the validation of Eq.(4'), using small-scale experiments. Studies performed by Fuchs et al (1969), Brown and Rosenberg (1973), Brown (1974), Stone and Horton (1974), Heilman and Kanemasu (1976), and more recently by Huband and Monteith (1986) for a wheat canopy have established the generally good performance of this approach, leading to an overall precision of 10 to 15% compared to reference methods. It is important to mention, however, that these tests were applied to well-watered canopies. In this situation, the sensible heat flux H is low, and resulting values of LE are less sensitive to errors in the determination of H. When applied to dry conditions, LE values appear to be very sensitive to the adopted value for the heat exchange coefficient h, as displayed on Figure 7, derived from experiments in the lower Rhone valley (Seguin et al., 1982).

4.2 Simulation Models

As the combination equation supposes the quasi-permanent measurement of T_s (at least on a hourly basis), it is hardly applicable to the operational use of satellite data. Several authors have then developed simulation models, allowing a simultaneous computation of LE and T_s values, which are then compared to measured T_s data to infer resulting value for LE.

These models were developed for bare soils (Rosema, 1975; Vauclin,

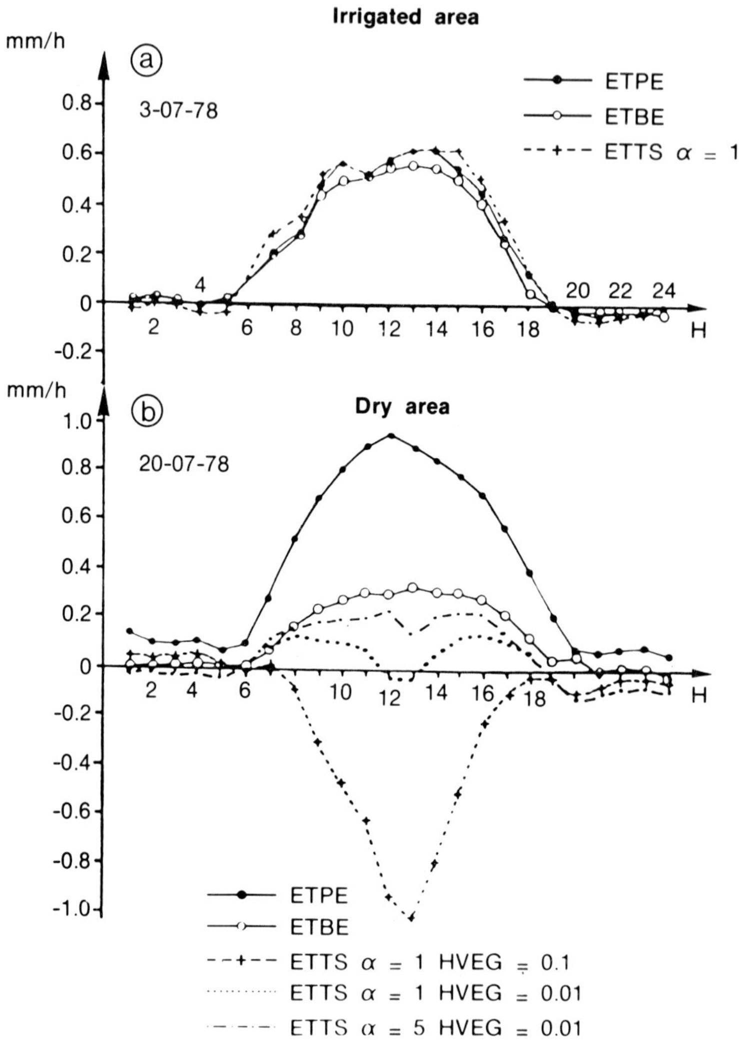

Figure 7. Hourly variation of evaporation fluxes for irrigated site (a) and dry zones (b) from the Crau experiment. In the last case, LE values using (4'), called ET_{TS}, are highly variable depending upon the roughness estimate and the ratio $z_o/Z_{o\ t} = \alpha$.

1978; Recan, 1982; Brunet, 1984) and extended to the case of vegetated surfaces (Deardorff, 1978; Soer, 1980; Carlson et al., 1981; Camillo et al., 1983). As they require the simultaneous estimation of numerous parameters (both for the lower atmosphere, the vegetation and the soil sublayer), however, they rarely are applied to satellite data (see one example in Figure 8, derived from Bernard et al. (1986)).

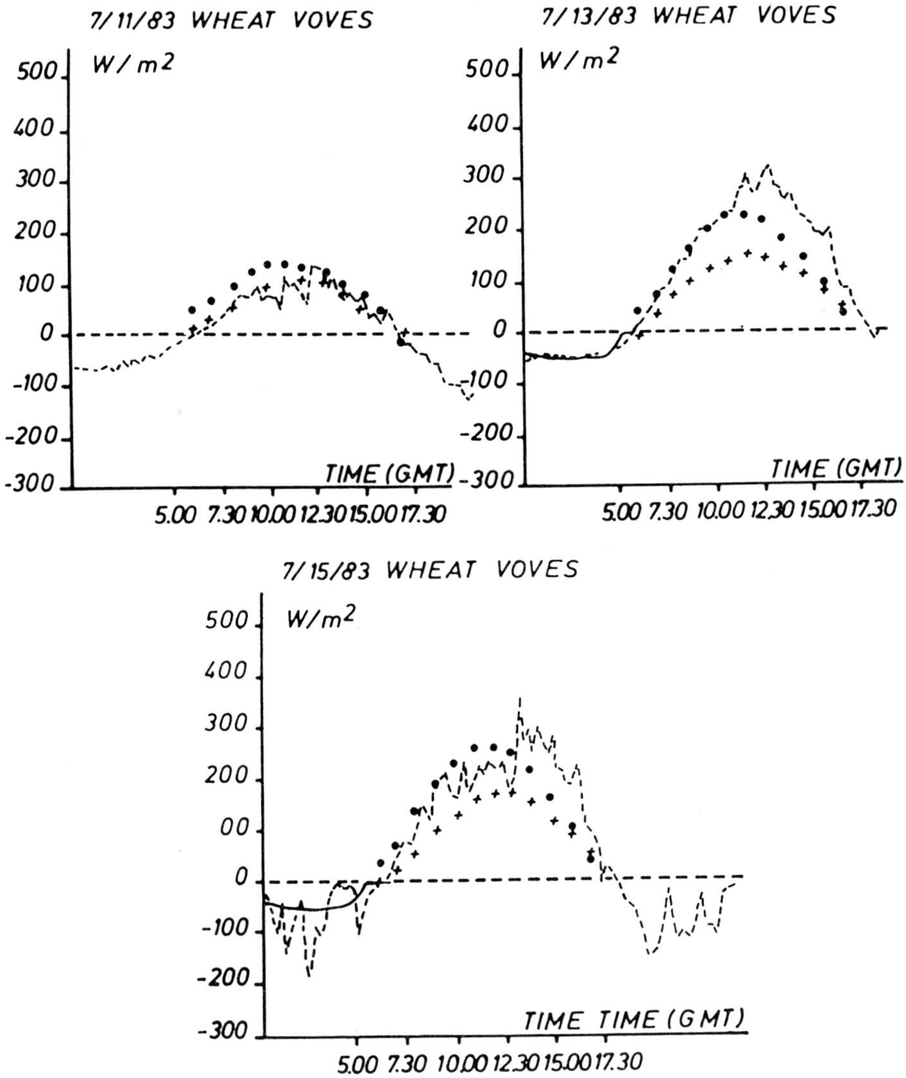

Figure 8. Comparison of sensible heat fluxes obtained from satellite and measured data over an agricultural region in France. The satellite derived fluxes are plotted with points (vegetation model) and crosses (bare soil model) (from Bernard et al., 1986).

These models may then be used more as methodological tools to test the ability of the equation set used to quantify the diurnal variation of simulated fluxes and T_s and to assess the sensitivity of the various parameters involved, rather than to compute fluxes from satellite data routinely.

4.3 Methods for Computing ET from Satellite DATA

Basically, these methods use the same approach as those previously described, but they are oriented towards a direct use of satellite data, thus involving various degrees of simplification as compared to the global basic equation set. They may be classified according to their requirements in terms of time frequency for satellite data.

a) Methods based on several data per day. The Fourier analysis of thermal diurnal cycles, developed in GSTS Strasbourg (Hechinger et al., 1982) has been applied to large areas in Sahelian Mali within the GAMP project. ET computations were derived using hourly values of albedo and surface temperature from METEOSAT (Abdellaoui et al., 1984), and appear to give realistic values. Further development of this approach is presently limited by the low precision of surface temperatures derived from operational satellites, which are the only possible source of such frequent T_s data.

b) Methods based on the day-night T_s difference. These methods are based on the concept of thermal inertia, basically defined by Watson (1975). The TELL-US model, developed by Rosema et al. (1978), which uses HCMM data, allows us to compute ET values for bare soils from the day-night temperature difference, knowing the soil thermal inertia (Figure 9).

Price (1980) proposed some simplifications for the adaptation of the TELL-US model, while Klaassen and Rosema (1979) proposed an extended version for the case of vegetation. Carlson et al. (1981) and Camillo et al. (1983) also developed models along the same lines. Applications were performed using HCMM data, but technical problems encountered in obtaining reliable day-night differences with this satellite and its short lifetime have limited the number of operational examples. NOAA-AVHRR could be used in the same manner, but recent applications have not appeared in very recent years.

c) Method based on midday surface temperature. It was primarily proposed by Jackson et al. (1977) and was based upon measurements over wheat canopies in Phoenix, from which a simplified linear relationship appeared

$$ET_d - R_{n\ d} = A + B(T_s - T_a) \qquad (10)$$

thus allowing the computation of daily ET_d from daily net radiation $R_{n\ d}$ and the instantaneous deviation $T_s - T_a$ near midday.

This relationship evidently appears as a first-order approximation which supposes that clear conditions persist throughout the day (in

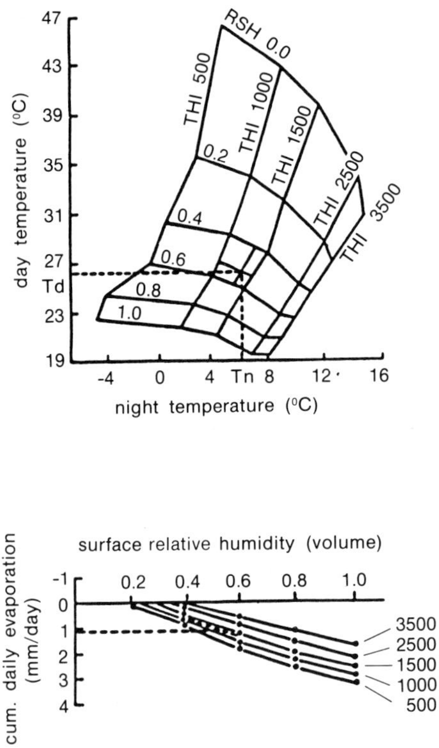

Figure 9. Look-up tables for the computation of ET (from Galli de Paratesi, 1981).

order that T_s-T_a near midday can be representative of daily integrated values) and assumes a nearly constant value for the heat exchange coefficient (which, for a given surface, defines a mean value which is only valid for a period of several days).

Our experiments in Crau gave a satisfactory assessment of ET, both in dry and irrigated zones ($\pm$ 1.0 mm/day) by the following expression (Figure 10)

$$ET_d - R_{n\ d} = 1.07 - 0.25\ (T_s-T_a)$$

The coefficient 0.25 was lower than the value 0.66 obtained by Jackson et al. (1977), which may be explained by the much lower value of the roughness parameter of the short grass considered here.

This result was interesting enough to lead us to look at the theoretical significance of this simplified relationship, in relation with the basic equations previously presented. From this analysis (Seguin and Itier, 1983) it turns out to be well founded, even if it was primarily derived from an empirical approach (Figure 11). Apart from its practical limitations (stable clear weather), the main problem lies in

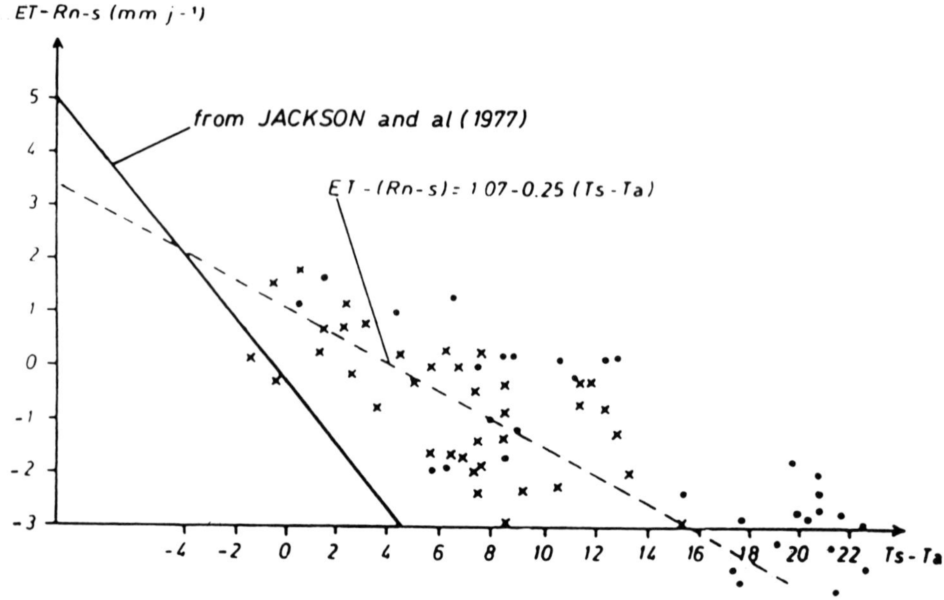

Figure 10. Relationship between ET-R_n and T_s-T_a for 32 days in the summer 1978 on the Crau site (x irrigated, . dry) (from Seguin et al., 1982).

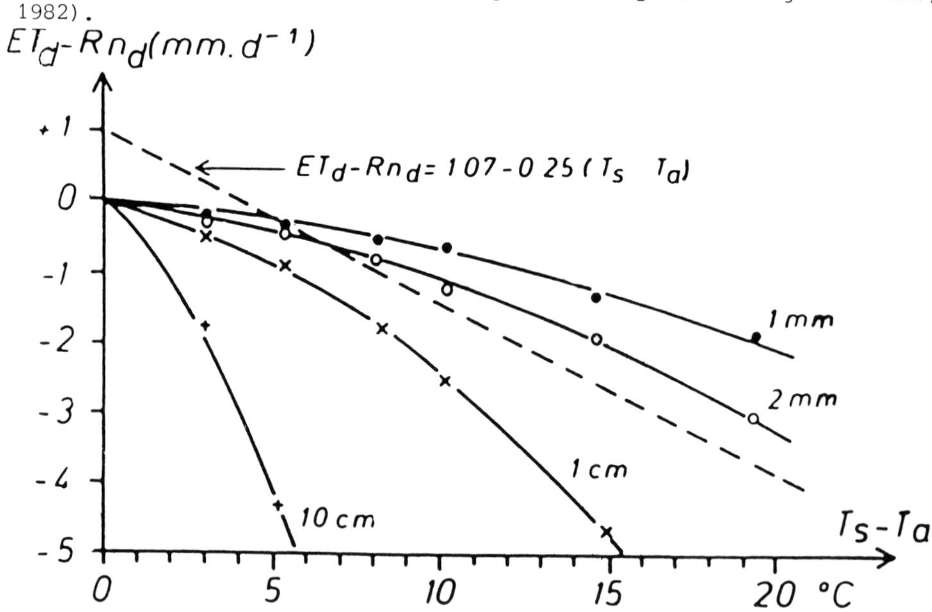

Figure 11. Comparison of the theoretically derived relationship between ET_d-R_{nd} and T_s-T_a for various values of z_o and the experimentally derived equation on the Crau site (from Seguin and Itier, 1983).

adapting the coefficient B to the roughness of the corresponding surface. Values may be proposed for surfaces ranging from bare soil to wheat (for instance), but it is difficult to set something for taller canopies, because of insufficient knowledge about heat exchange for these surfaces.

For a given surface, the simplified relationship may be used to compute, not individual daily values (for which a large dispersion exists due to the variability of the heat exchange coefficient from one day to the next), but cumulative values for periods of some days

$$\sum_{o}^{n} ET_d = \sum_{o}^{n} R_{n\,d} + nA - B \sum (T_s - T_a) \tag{11}$$

this expression allowing us to incorporate the "stress degree day" $\sum(T_s-T_a)$ which was also defined and experimented on by Jackson et al. (1977).

An application of these concepts was recently achieved in the Sahelian region (see Seguin et al., 1986), where acceptable linear relationships were established between derived $\sum T_s$ from METEOSAT data and $\sum P$, the cumulative rainfall during the rainy season (see Figure 12).

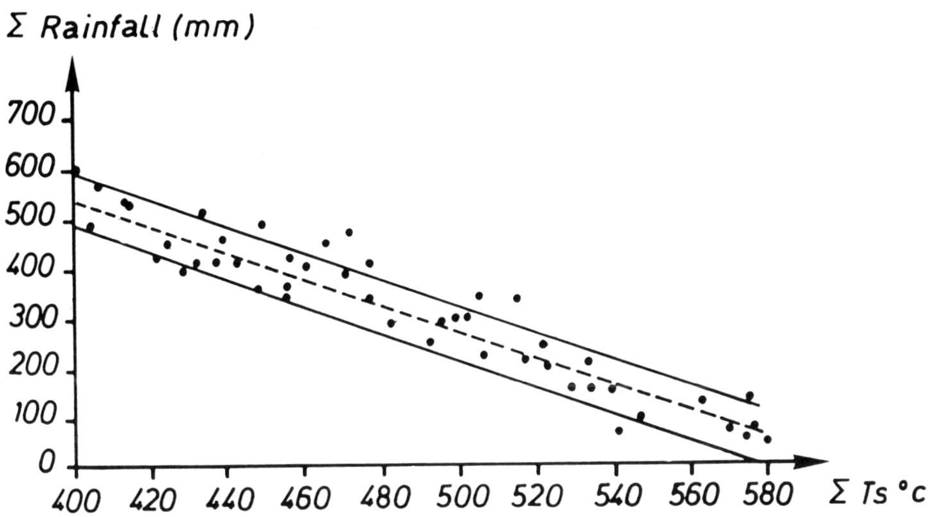

Figure 12. Relationship between $\sum T_s$ and $\sum P$ at the end of the rainy seasons (1984 and 1985 in Senegal) (from Seguin et al., 1986).

This result is a consequence of the simplified relationship, because ET is a more or less constant fraction of P in these regions and the low resolution of METEOSAT (5 km) incorporates a wide variety of grass and crops which are almost homogeneous at the large scale.

As was also shown by Nieuwhenhuis et al. (1985) and Rambal et al. (1985) this simplified approach then appears as a possible tool for deriving first-order estimates of ET spatial variations for periods of several days with clear stable weather. In order to overcome this limitation, we have developed a combination with a meteorological water balance model (see Vidal et al., 1987) which allows us to interpolate ET computations during cloudy days, by the adjustment of T_s model outputs to satellite data.

Moreover, in order for it to be adapted to NOAA-AVHRR (which appears at present as the most suitable satellite for these applications), the simplified relationship may be directly computed from this model, using the local meteorological data, with T_s corresponding to NOAA local hour (near 2.30 pm) and T_a being the maximum air temperature (easily accessible from ground data) (Figure 13).

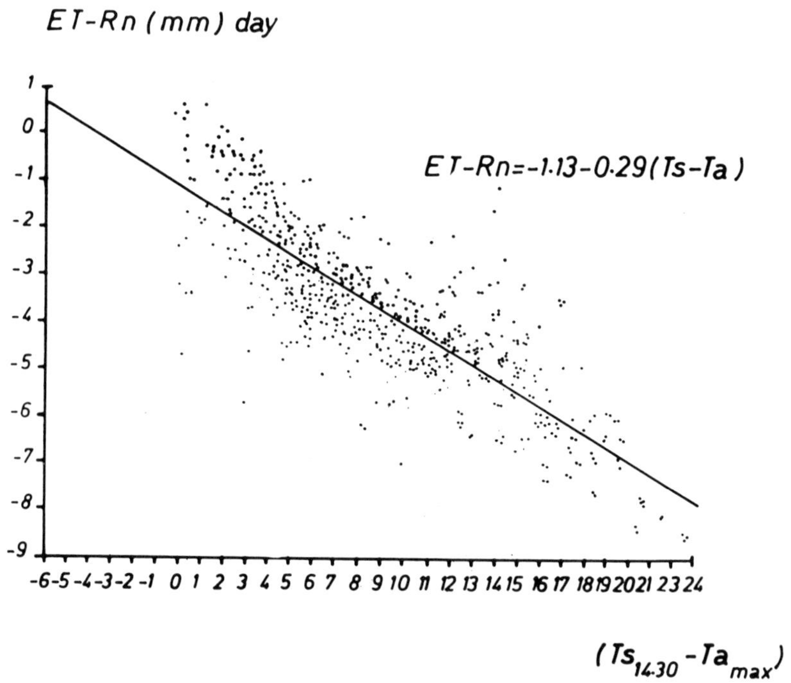

Figure 13. The simplified relationship computed from the earth water balance model for meteorological data of Avignon and St-Gilles (from Vidal et al., 1987).

We are testing this relationship at present and trying to adapt the coefficients to the roughness of various natural surfaces. One of the most severe limitations to progress is the lack of ground independent measurements of ET on the required scale (say 1 m^2), which would allow the validation of our computations. We hope to derive useful information from the large HAPEX-MOBILHY experiment, which took place in Southwestern France in summer 1986.

5. CONCLUSION

The potential of satellite thermal IR for agrometeorology is significant, especially in two fields:

- the mapping of thermal characteristics of agricultural regions on a scale (1 to 5 km, depending upon the satellite used) not accessible to classical ground networks;
- the monitoring of temperature and water balance for the same regions, allowing us to detect, quantify and localize in space the extension of climatic hazards such as frost, drought, etc.

Because of its somewhat complex significance and also of the lack of operational systems for processing thermal IR data, this potential has been only approached in recent years. One can predict, however, a significant development in the near future, together with the use of vegetation indices, to which it constitutes a necessary complement for the assessment of crop environmental conditions.

REFERENCES

Abdellaoui, A., Becker, F., Olory-Hechinger, E. and Raffy, M., 1984: 'Influence des conditions de mesure sur l'estimation de l'inertie thermique et de l'évapotranspiration à partir des données Météosat', IIe Colloque International Signatures Spectrales d'Objets en Télédétection, Bordeaux (France), September 1983, Ed. by INRA, Les Colloques de l'INRA No.23, p.475-485.

Bernard, R., Taconet, D. and Vidal-Madjar, D., 1986: 'Towards a satellite system to monitor the spatial and temporal behaviour of the soil water extent', Proc. IGARSS 86 Symp., Zürich, September 1986, Ed. by ESA, Publ. ESA SP-254, p.751-753.

Brown, K.W., 1974: 'Calculations of evapotranspiration from crop surface temperature', Agric. Met., 14, 199-209.

Brown and Rosenberg, 1973: 'A resistance model to predict evapotranspiration and its application to a sugar beet field', Agron. J., 65, 341-347.

Brunet, Y., 1984: 'Modélisation des échanges sol nu-atmosphère: essai de validation locale et influence de la variabilité spatiale du sol', Thèse de Docteur-Ingénieur, Univ. de Grenoble, p.207.

Brutsaert, W., 1982: Evaporation into the Atmosphere, D. Reidel (Ed.), 299.

Byrne, G.F., Kalma, J.D. and Streten, N.A., 1984: 'On the relation between HCMM satellite data and temperatures from standard meteorological sites in complex terrain', Int. J. Rem. Sens., 5, 65-79.

Camillo, P.J., Gurney, R.S. and Schmugge, J.J., 1983: 'A soil and atmospheric boundary layer model for evapotranspiration and soil moisture studies', Water Resour. Res., 19, 2, 371-380.

Carlson, T.N. and Boland, F.E., 1978: 'Analysis of urban/rural canopy using a surface heat flux/temperature model', J. Appl. Meteor., 17, 998-1013.

Carlson, T.N., Dodd, J.K., Benjamin, S.G. and Cooper, J.N., 1981: 'Satellite estimation of the surface energy balance, moisture availability and thermal inertia', J. Appl. Meteor., **20**, 67-87.

Caselles, V., Gandia, V. and Melia, J., 1983: 'Significance of apparent temperature measurements carried out by the HCMM satellite over area of vegetation', Agric. Meteor., **30**, 77-83.

Chen, E., Allan, L.H., Bartholic, L.F. and Gerber, J.F., 1982: 'Delineation of cold-prone areas. Using night-time SMS GOES thermal data: effects of soil and water', J. Appl. Meteor., **21**, 1528-1537.

Deardorff, J.W., 1978: 'Efficient prediction of ground surface temperature and poisture with inclusion of a layer of vegetation', J. Geoph. Res., **83**, C4, 1889-1905.

Fuchs, M., Tanner, C.B., Thurtell, G.W. and Black, T.A., 1969: 'Evaporation from drying surfaces by the combination method', Agric. J., **61**, 22-26.

Galli de Paratesi, S., 1981: 'Heat capacity mapping mission. Investi gation No. 25 (Tellus project)', Final report to NASA, 239.

Garratt J.H., 1978: 'Transfer characteristics for a heterogeneous surface of large aerodynamic roughness', Quart. J. Roy. Met. Soc., **104**, 491-502.

Hechinger, E., Raffy, M. and Becker, F., 1982: 'Comparison between the accuracies of a new discretization method and an improved Fournier method to evaluate heat transfers between soil and atmosphere', J. Geoph. Res., **87**, 7325-7339.

Heilman, J.L. and Kanemasu, E.T., 1976: 'An evaluation of a resistance form of the energy balance to estimate evapotranspiration', Agron. J., **68**, 607-611.

Huband, N.D.S. and Monteith, J.L., 1986: 'Radiative surface temperature and energy balance of a wheat canopy', Bound. Layer Meteor., **35** (1/2), 1-18, 107-116.

Jackson, R.D., Reginato, R.J. and Idso, S.B., 1977: 'Wheat canopy temperature: a practical tool for evaluating water requirements', Wat. Resour. Res., **13** (3), 651-656.

Jackson, R.D., 1982: 'Canopy temperature and crop water stress', in Hillel Ed. Advances in Irrigation, Academic Press Network, 2, 43-85.

Kalma, J.D., Byrne, G.F., Johnson, M.E. and MacLaughlin, G.P., 1983: 'Frost mapping in southern Victoria: an assessment of HCMM thermal imagery', J. Climat., **3**, 1-19.

Kerr, Y.H., Fisher, M. and Seguin, B., 1983: 'Frost risks mapping by use of AVHRR data', Comm. presented at the LARS Symp., Purdue, USA, June 1983.

Klaasen, W. and Rosema, A., 1979: 'Generalization of the TELL-US model to vegetated surfaces', EARS., Delft (Pays-Bas). Rapport au JRC Ispra, 18.

Lagouarde, J.P., Valery, P., Belluomo, P. and Soulier, A.M., 1983: 'Cartographie des topoclimats forestiers. Mise au point d'une méthodologie d'analyse de l'effet du relief sur les thermographies: application aux données HCMM sur le nord-est du Massif Central', Agronomie, **3**, 1011-1018.

Nieuwenhuis, G.J.A., Smidi, E.H. and Thunissen, H.A.M., 1985: 'Estimation of regional evapotranspiration of arable crops from thermal infrared images', Int. J. Rem. Sens., 6, 8, 1319-1334.
Price, J.C., 1982a: 'On the use of satellite data to infer surface fluxes at meteorological scales', J. App. Meteor., 21, 8, 1111-1122.
Price, J.C., 1982b: 'Estimation of regional scale evapotranspiration through analysis of satellite thermal-infrared data', IEEE Trans. Geos. Rem. Sens., GE-20, 3, 286-292.
Rambal, S., Lacaze, B., Mazurek, K. and Debussche, G., 1985: 'Comparaison of hydrologically simulated and remotely sensed actual evapotranspiration from some Mediterranean vegetation formations', Int. J. Rem. Sens., 6, 8, 1475-1481.
Recan, M., 1982: 'Simulation numérique du comportement thermique et hydrique d'un sol nu. Application à l'étude de l'évaporation par télédétection', Thèse Docteur-Ingénieur, Inst. Nat. Polytech., Toulouse, 98.
Rosema, A., 1975: 'Simulation of the thermal behaviour of bare soils for remote sensing purposes', in Heat and Mass Transfer in the Biosphere, Proc. Symp. Dubrovnik, September 1974. Ed. D.A. de Vries and N.H. Afgan, Scripta Book Company, Washington, 109-125.
Rosema, A., Bijleveld, J.H., Reiniger, P., Tassone, G., Blythe, K. and Gurney, R.J., 1978: 'Tell-us, a combined surface temperature soil moisture and evaporation mapping model', 12th Int. Symp. Rem. Sens., Manilla (Philippines).
Seguin, B., 1981: 'Bioclimatological aspects of crop production', in Applications of Remote Sensing to Agricultural Production Forecasting, A. Berg (Ed.), Balkema, 33-47.
Seguin, B., Baelz, S., Monget, J.M. and Petit, V., 1982a: 'Utilisation de la thermographie IR pour l'estimation de l'évaporation régionale', I. Mise au point méthodologique sur le site de la Crau. Agronomie, 2(1), 7-16. II. Résultats obtenus à partir des données de satellite, Agronomie, 2, 113-118.
Seguin, B., Brunet, Y. and Perrier, A., 1982b: 'Estimation of evaporation: a review of existing methods and recent developments', EGS meeting Symp. on Evaporation, Leeds (Great Britain), August 1982, 21.
Seguin, B. and Itier, B., 1983: 'Using midday surface temperature to estimate daily evaporation from satellite thermal IR data', Int. J. Rem. Sens., 4, 2, 371-383.
Seguin, B., 1984: 'Estimation de l'évapotranspiration à partir de l'infrarouge thermique', IIe Colloque Int. Signatures Spectrales d'Objets en Télédétection, Bordeaux, 12-16 September 1983, Ed. by INRA (Les Colloques de l'INRA No.23), 427-446.
Seguin, B., Lagouarde, J.P. and Kerr, Y., 1985: 'Estimation of regional evaporation using midday surface temperature from satellite thermal IR data', Proc. ISLSCP Conf., Rome, December 1985, Ed. by ESASP 248, 339-345.
Seguin, B., Assad, E., Freteaud, J.P., Kerr, Y. and Lagouarde, J.P., 1986: 'Water balance monitoring in sehelian regions with thermal IR and vegetation index data from meteorological satellites', 20th Symp. on Remote Sensing of Environment, Nairobi (Kenya), December 1986, to be published in Proc. Ed. by ERIM, Ann Arbor, Michigan.

Soer, G.J.R., 1980: 'Estimates of regional evapotranspiration and soil moisture conditions using remotely sensed crop surface temperatures', Rem. Sens. Env., 4, 2, 371-383.

Stone, L.R. and Horton, M.L., 1974:'Estimating evapotranspiration using canopy temperatures: field evaluation', Agron. J., 66, 450-454.

Sutherland, R.A., Langford, J.L., Bartholic, J.F. and Bill, R.G., 1979: 'A real-time satellite data acquisition analysis and display system. A practical application of the GOES network', J. App. Meteor., 18, 355-360.

Thom, A.S., 1976: 'Momentum, mass and heat exchange of plant communities', in Vegetation and the Atmosphere, Ed. by J.L. Monteith, 57-109, Academic Press, London, Vol.1.

Vauclin, M., 1978: 'Modèle local des transferts de masse et de chaleur entre le sol et l'atmosphère. Problèmes posés par son extension spatiale', in Actes du Colloque National de l'ASP Evolution des Climats, ed. by F. Becker and C. Pastre, II.76 - II.105.

Vidal, A., Kerr, Y., Lagouarde, J.P. and Seguin, B., 1987: 'Télédétection et bilan hydrique: utilisation combinée d'un modèle agrométéorologique et des données de l'IR thermique du satellite NOAA-AVHRR', Agric. For. Meteor., 39, 155-175.

RADIATION BUDGET AT THE GROUND FROM SATELLITE DATA

E. Raschke
Institute of Geophysics and Meteorology
University of Cologne
D-5000 Köln 41, Fed. Rep. of Germany

ABSTRACT. More than ever before the modelling of climate and circulation processes requires accurate data on radiative energy exchanges between the earth's surface and the atmosphere. Required accuracies are estimated to range between 2-20 Wm^{-2} for global and monthly averages. The four components of the radiation budget at the ground can still not be derived with sufficient accuracy from satellite measurements and from correlative data of conventional surface-based origin. The general possibilities are discussed in this paper. Basic research is still required to establish a reliable error budget, and ground truth is essential.

1. INTRODUCTION

The heat exchange between the ground, oceans and sea ice surfaces and the atmosphere has always attracted scientists of a variety of relevant disciplines. Recent progress in modelling of the climate and of circulation processes has enhanced the demand for accurate parameterization of this heat exchange over the entire earth and with a spatial resolution which fits into such models. A large variety of numerical studies (e.g. Mintz, 1982) has already shown that the model results are very sensitive to the changes of the heat fluxes at the surface. Therefore, major efforts are now being undertaken to develop methods to measure them on a worldwide basis and to parameterize them in terms of those field quantities which can be provided in numerical models. These efforts are part of the work performed within the scope of large international projects, such as the ISCCP (International Satellite Cloud Climatology Project) and the ISLSCP (International Satellite Land Surface Climatology Project). Surface-based measurements provide only limited data sets since even an excellent network will not cover the wide gaps over the oceans. Thus, satellite-based measurements are one of the primary data sources.

We will concentrate here on the possibilities of extracting information on the radiation budget components at the earth's surface from satellite-based measurements, which are primarily made for

F. Toselli (ed.), Applications of Remote Sensing to Agrometeorology, 241–255.

operational use in meteorological analyses and forecasts. Special observations are not yet available nor will they be made in the near future. The measurements of the Earth Radiation Budget Experiment (ERBE) are only useful for computations of the radiation budget at the top of the atmosphere.

The radiative fluxes of the surface are the fluxes of up- and downward solar ($M_s\uparrow$ and $M_s\downarrow$) and terrestrial heat radiation ($M_t\uparrow$ and $M_t\downarrow$). The net radiation is the sum of them, where fluxes directed towards the surface are counted as positive:

$$M_n = M_s\downarrow - M_s\uparrow + M_t\downarrow - M_t\uparrow \quad \text{(e.g. in Wm}^{-2}\text{)} \tag{1}$$

The spectral regions range from about 0.3 to 3.5 μm for M_s and from 4 to 50 μm for M_t. In all further considerations we shall also assume that values of M_n and of its components are to be determined for finite areas of the size ranges used in numerical circulation models varying from 10 to 250 km.

In this paper it was not possible to present a complete survey of all relevant references.

2. BUDGET OF SOLAR RADIATION AT THE GROUND

2.1 Introduction

The upward solar flux $M_s\uparrow$ is determined by the surface albedo, ρ_s, according to:

$$M_s\uparrow = \rho_s \cdot M_s\downarrow \tag{2}$$

Two quantities need to be determined independently: $M_s\downarrow$, the global radiation, and ρ_s. The latter depends over many natural surfaces, in particular over oceans, on the angle of illumination. Thus, usually two values of the albedo over a given area might be required, a clear sky value depending on the Sun's zenith angle and an overcast value for nearly isotropic illumination. Thus, for a fractional cloud coverage, c, one may obtain for the mean upward flux:

$$M_s\uparrow(\rho_s,c) = (1-c)\rho_s' M_s'\downarrow + c\cdot\rho_s'' M_s\downarrow \tag{3}$$

ρ_s', ρ_s'' = surface albedo for clear and overcast sky, respectively;
$M_s'\downarrow, M_s\downarrow$ = global radiation for clear and overcast sky, respectively;
c = fractional cloud coverage ($0 < c < 1$).

Over most surfaces the clear sky albedo increases with increasing zenith-angle, over oceans from values of about 0.04 to 0.20. The overcast sea surface albedo ρ_s'' assumes here values of around 0.06 to 0.08 depending on the roughness of water surfaces (e.g. Gube et al., 1980).

Since the global radiation of a clear sky and below clouds has different spectral compositions, the values of ρ_s' and ρ_s'' also assume different values over the same surface.

2.2 Global Radiation

The imaging data of operational meteorological and other earth observations can, in general, be used as indicators for the presence of clouds. For the purposes of the ISCCP these data are used to estimate the fractional coverage, possibly at three altitudes, and also a measure for the transmittance of cloud fields over the entire Earth (e.g. Schiffer and Rossow, 1985).

In recent years several attempts have been made to determine the global radiation from satellite data. As Ellingson (1981) pointed out, two different methods had been used. One is entirely based on regressions between satellite-based radiance data and ground-based measurements with pyranometers. The other uses satellite data only as indicators of various cloud parameters and relates them either to measured or calculated values of the global radiation of clear and overcast skies. They were thoroughly discussed during an ISCCP workshop in 1985 (WCP-115, 1986).

Both methods make use of the fact that the solar radiation transmitted to the ground is highly but inversely correlated to the reflection from the atmosphere to space. This is caused by clouds. One method therefore makes use of rather simple correlations between simultaneous measurements from satellites and at the ground. The second takes into account more realistic properties of the atmosphere. In both cases, however, the reflection properties of the ground, which contribute significantly to the upward radiation fluxes, as measured by the satellites, must be taken into account.

This latter method should be explained here briefly, since it is now already in operational use to map the global radiation over Central Europe (Diekmann et al., 1986). In a first step the global radiation of clear skies (Go) is calculated for each location and time interval, using in principle look-up tables. Its magnitude will be subject to variations of the atmospheric turbidity and of altitude ranges of the ground. Some careful adjustments are, therefore, necessary.

In a second step the variation of the transmittance of the atmosphere within each pixel area is determined from satellite data, where first a mean effective cloud-cover, C_{eff}, is determined from

$$C_{eff} = \frac{L - L_{min}}{L_{max} - L_{min}} \qquad (0 \leq C_{eff} \leq 1) \tag{4}$$

where:

L = actual radiance measurement of a solar channel (e.g. first channel of the METEOSAT images);

L_{min} = same measurement during the same local time, but for clear skies over the same area;

L_{max} = radiance to be measured over complete overcast areas (= equivalent to cloud albedo or reflectance).

Maps of L_{min} are, in principle, obtained for each month and each local time on the basis of a simple statistical selection procedure, which determines the lowest radiances or albedo. Thorough corrections are required for cases of persistent cloudiness - but often also for errors in the data themselves.

On the basis of radiative transfer calculations this effective cloud cover, which is a complex variable of the actual cover and cloud optical thickness, can be related to the transmittance, τ_c, of clould fields. Thus, the real value of global radiation is obtained then from Eq.(5)

$$G_n = \tau_c * G_o \quad Wm^{-2} \tag{5}$$

The integration over a daylight period makes use of the fact that geostationary satellites can provide information every 30 min! Relations of τ_c (C_{eff}) are shown in Figure 1. They are calculated for forest and snow covered areas. Their nearly linear relationships are due to the fact that one assumes a linear relation between the upward radiance and the geometrical cloud coverage.

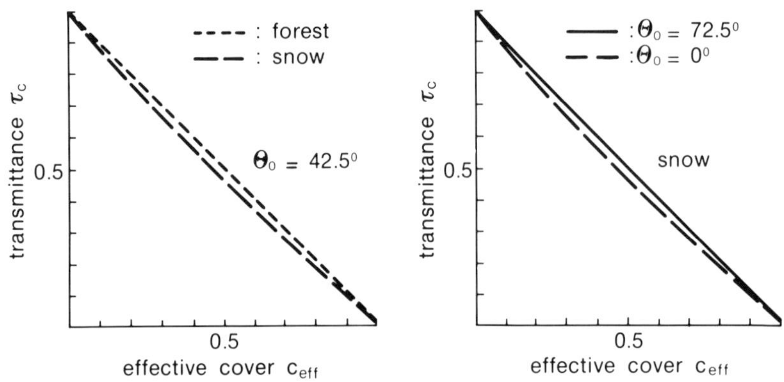

Figure 1. Relation between the transmittance τ_c of the cloudy atmosphere for global radiation and the effective cloud cover C_{eff}.

We do not intend here to judge between methods from different authors but, instead, will point to several principal difficulties. These are related to effects of the ground, to imaging capabilities of satellite radiometers, and to temporal changes of cloud fields. Further errors or uncertainties may occur in the comparisons with ground-based data of the global radiation, as measured at isolated stations with pyranometers. Here one should also take into account the different fields of view of the instruments on the ground and in space, and errors in navigating the satellite data. All these effects are in some way interrelated.

(a) Effects of ground albedo are especially important over areas with rapidly changing snow cover (e.g. many continental surfaces during the transitional seasons). Thus, complementary or correlative data should be used to account for these effects. These are information on the vegetative cover, surface structure and snow cover.

The identification of cloudfree areas over snow and icefields is not easy and requires particular care. In the visible and near infrared wavelength ranges up to about a wavelength of 1.2 μm the clould fields show little contrast to snow and ice fields underneath. In the presently available imagery their identification over snow-covered regions remains an open problem for the ISCCP. Larger contrasts occur in the signals of the 3-7 μm band of the AVHRR in NOAA-satellites, when the signals of thermal origin are removed. Similarly, the differences in the indices of refraction for ice and water particles should also enhance contrasts at 1.6 μm, where snow reflects less than low-level water clouds. Such additional spectral measurements are planned for future satellite missions (GOES-Next, Meteosat-Next). A workshop has recently been held to find solutions for these problems (WCP-131, 1987).

(b) Temporal variations of the morphology of cloud fields cause variations of the downward solar radiation. Thus, to obtain reliably accurate data of daily or monthly sums of the global radiation, one requires several measurements during the day. The exact number will vary from area to area according to the behaviour of cloud fields. In several studies (e.g. Harrison et al., 1983; Möser and Raschke, 1984; Preuß et al., 1984) it could be shown that at least 3-6 measurements per daylight period are sufficient for this purpose. These are easily obtainable from geostationary satellite. Also over the polar regions the two polar orbiters of the NOAA series provide a similarly high temporal density.

(c) Further errors are due to the narrow spectral response of most instruments and to the angular dependent reflection properties of the earth-atmosphere system to space. The magnitude of their effect onto the global radiation has not yet been systematically investigated.

(d) The spatial resolution of satellite radiometers differs considerably from that of local ground-based pyranometers, in particular when low clouds are present. These effects must be considered carefully in comparisons between values of global radiation measured directly at the ground and derived from satellite data. They might be cancelled when appropriate long averaging times of up to one month are used. No basic research has yet been carried out to allow for the design of ground-based networks or anchor-stations, which are required for a steady control of these inferred data. Gandin (1976) made some preliminary investigations into this direction.

Nevertheless encouraging results have already been reported in the literature. Several authors claim accuracies in the range between 5 and 15 percent, where ground-based data themselves are already subject to an uncertainty of about 5 percent. The same accuracy range is obtained when instead of the high-resolution data those of the much further reduced ISCCP-format B2 (30 km, sampled) are used and analysed (Raschke et al., 1987). This latter data set, now operationally produced with a global calibration and a low data rate, is of high potential value for a global inspection of solar radiation at ground levels.

In Figures 2-4, values of global radiation, which were determined from Meteosat data in the B2-format, and pyranometer measurements, are compared. More results were presented during the workshop.

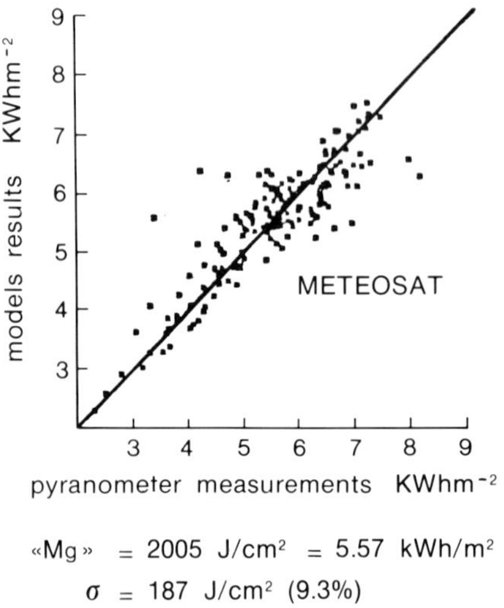

Figure 2. Monthly averages of daily sums of global radiation as calculated from METEOSAT data (June 1983) and measured simultaneously at 187 stations. <Mg> = average of all, σ = standard error of estimate.

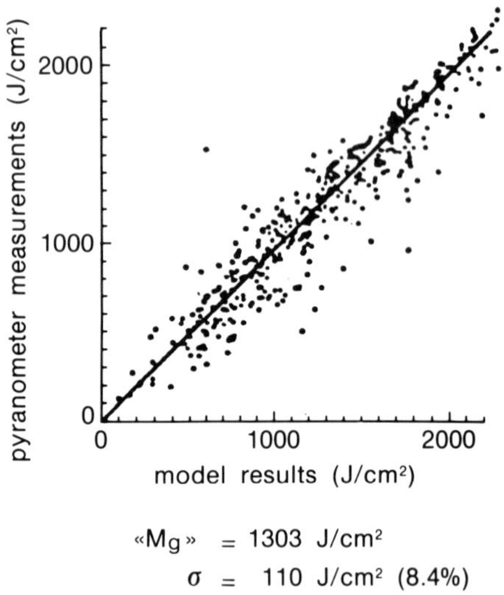

Figure 3. Daily sums of global radiation as calculated over Australia and measured simultaneously at 22 stations during July 1983. <Mg> = average, σ = standard error of estimate.

2.3 Surface Albedo

The albedo or reflectance of natural surfaces cannot be measured from space directly. It must be inferred from satellite radiance data, where corrections are necessary for

- angular dependences of upward radiances due to respective properties of surfaces and the scattering of radiation within the atmosphere;
- spectral filtering of radiation within the atmosphere;
- spectrally changing illumination of surface with different sky conditions;
- temporal changes of surface conditions, which affect their reflection properties.

In principle, the first three defects can be reduced to quite a small uncertainty with proper modelling of the atmosphere for clear skies. Since the illumination of the ground - or the global radiation - changes from clear to overcast or cloudy conditions in its spectral

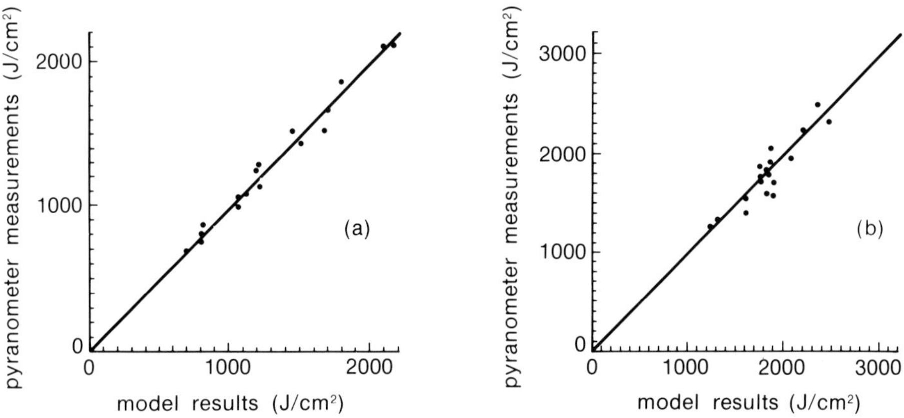

«Mg» = 1303 J/cm² σ = 52 J/cm² (4%)

«Mg» = 1780 J/cm² σ = 110 J/cm² (6.2%)

Figure 4. Monthly averages of daily sums of global radiation as calculated: (a) over Australia and measured simultaneously at 22 stations during July 1983; (b) over Asia and measured simultaneously at 18 stations during July 1983. <Mg> = average of all, σ = standard error of estimate.

composition and angular distribution, changes in the reflectance also occur as a consequence. Thus, to define a mean reflectance one may distinguish between the albedo for clear skies and that for cloudy conditions.

Here we shall demonstrate the principle. A satellite-borne radiometer measures the upward radiance $L\uparrow(\theta,\psi,\theta_o)$ which contains information on the reflection at the ground (II) and also on scattered light from the atmosphere (I). It is, in a simplified form valid for a narrow spectral interval, (e.g. $\Delta\lambda < 0.1$ µm)

$$L\uparrow(\theta,\psi,\theta_o) = \int_{z=0}^{\infty} L\uparrow(z';\theta,\psi,\theta_o) * \frac{\delta\tau_a}{\delta z}\, dz + \tau_{a,o}(\theta) * \frac{1}{\pi} * \rho_o(\theta,\psi,\theta_o) * M_G(\theta_o) \quad (6)$$

Here, in the first term the expression $\delta\tau_a/\delta z$ describes the change of the atmospheric beam transmittance with height, and $L\uparrow$ in the integral is a result of multiple scattering of light into satellite direction given by the zenith and azimuth angles θ,ψ, θ is the Sun's zenith angle. In the second term $\tau_{a,o}$ is the total atmospheric beam transmittance in the θ direction and M_G is the global radiation, parts of which are reflected into direction θ,ψ. In fact, the surface reflectance ρ_o is often strongly anisotropic, thus one should in the second term consider separately the reflection of direct and of diffuse radiation. Eq.(6) must be folded with the spectral filter function of the radiometer.

The "inversion", i.e. determination of ρ_o, needs information on atmospheric parameters (water vapour content, dust load) but also on the spectral reflection characteristics of surfaces. In the latter, very particular changes may occur between various soil types but also between bare soil and vegetation cover. One also needs some information on the angular anisotropy of the surface reflectance.

Thus, we should actually envisage obtaining data sets on the surface albedo for clear and overcast conditions, at least over those surfaces where large differences occur. These are the oceans, which cover about 70% of the earth's surface. There, the overcast albedo ranges between values of 0.05 and 0.06, where that of the cloudfree areas may assume values between 0.04 and 0.20 for high and low Sun, respectively. Albedoes of bare soil surface depend on the soil moisture, seasonal variations due to vegetation are to be expected. It must be kept in mind, however, that the product of the absorptance (= 1-albedo) with global radiation is the energy consumption at the surface. Thus, small uncertainties in our knowledge of the surface albedo over oceans may cause small errors, while drastic variations due to snow or vegetation cover over continents need careful consideration.

Several authors in recent years have analysed data of a variety of satellites, including those of the narrow-band radiometers in earth-resources satellites. These are all clear sky albedoes, obtained either by manual selection or computer generated minimum albedo maps (e.g. Geleyn and Preuß, 1983; Briegleb et al., 1986; Pinty et al., 1985). These clearly show seasonal changes (Kukla and Robinson, 1980), but no detailed error assessment has yet been made. Therefore, one must advocate here proper field experiments to measure area averages of the surface albedo from aircraft for intercomparison with results derived from satellite data. Such experiments will be performed during an ISCCP-Regional Experiment (ICE) and within several field studies of the ISLSCP.

3. BUDGET OF TERRESTRIAL RADIATION AT THE GROUND

This is described by the difference between fluxes of downward atmospheric radiation and the upward component, the latter resulting from emission of heat radiation by the surface and some of the reflection of the downward atmospheric radiation

$$M_{t,net} = M_t{\downarrow} - \rho_t * M_t{\downarrow} - M_t{\uparrow} \qquad (7)$$

where ρ_t = reflectance of natural surfaces for $M_t{\downarrow}$

and $$M_t = \varepsilon * \sigma * T_o^4 \qquad (8)$$

with:

ε = emittance for heat radiation; considered here as an average for the entire spectral range of heat radiation;

σ = Stephan-Boltzman constant;

T_o = effective surface temperature (in Kelvin).

Over most continental surfaces with low thermal capacity the terrestrial net radiation adjusts very sensitively and quickly to the heat supply by other heat fluxes, in particular to the downward heating by downward atmospheric and solar radiation. Under cloudy conditions, it assumes values close to zero. Thus, one may not need to determine its components separately, but should directly calculate $M_{t,net}$ as a weighted average of clear and cloudy conditions.

Over water surfaces, however, where the surface temperature adjusts very slowly to rapid changes of the heat supply from the atmosphere, the upward thermal heat radiation will seldom be compensated by downward heat radiation in the presence of clouds. While over bare soil the surface temperature might change rapidly over ranges of 10-20 K, the variations of near surface water temperatures are much smaller than 0.5 K. In the following we will, therefore, consider both radiation flux components separately. Values for the emittance range between 0.85 and 0.98 and the reflectance for ρ_t of downward atmospheric radiation range between 0.15 and 0.02. They depend on soil properties and negative cover. Since the small reflection of downward atmospheric radiation partly compensates for the incomplete non-black emission from most surfaces, the uncertainties due to the insufficiently known reflection and emission properties of surfaces will remain within error limits of about 4-10 Wm^{-2} (these are estimates), which are probably small compared to other error sources. Usually one assumes $\varepsilon = 1$.

3.2 Downward Atmospheric Radiation, M_t

The downward terrestrial or atmospheric radiation, $M_t{\downarrow}$, is primarily affected by the temperature and moisture of the lowest 100-200 metres of the atmosphere. The relative contribution of radiation emitted from clouds depends on cloud thickness, bottom height, and also on the water vapour concentration below the cloud layer. It may reach values of up to 30%.

A numerical error analysis recently made by Schmetz (Schmetz, 1984; Schmetz et al., 1986) indicated that with clear skies the uncertainties of the temperature profile by ± 1.5 K and of the moisture profile by ± 0.3 gm^{-3} caused uncertainties of the downward atmospheric radiation of ± 5 to 8 Wm^{-2} or of ± 1 to 2 percent. The cloud effects depend on the altitude and range between ± 5 Wm^{-2}, ± 4 Wm^{-2}, ± 2 Wm^{-2} for low (1 km), middle (5 km) and high (8 km) cloud layers, if uncertainties of 0.5 km in the bottom altitude are assumed. These are again about 1-2 percent of the expected mean fluxes. This brief discussion of errors shows that accurate knowledge of air temperature fields near the surface is important for estimates of the downward radiation. On the other hand, the information cannot be obtained with this accuracy from retrievals of satellite data only. The moisture field analyses, if available at all, are also of little use, since moisture measurements are subject to too large an error.

Schmetz et al. (1986), therefore, recommend taking reasonable climatic averages. According to the error budget analysis, which is quoted above, the variation of moisture explains only a small portion of the variation of the downward atmospheric radiation. Of greater importance is the downward emission from cloud bottoms. Since the satellite "sees" only the top level clouds, large errors may occur in cases of multilevel cloudiness, when satellite data are only used to infer cloud bottom heights and temperature. Correlative data, e.g. such as the large cloud statistics by Hahn et al. (1983), or values on the possible condensation level, might be used to predict the probability of low level cloud occurrence, when the satellite "sees" only high-level clouds.

The study by P. Schmetz (1984) enabled the extraction of maps for the downward atmospheric radiation fluxes ALPEX temperature analyses. Other studies have been carried out by Minker (1984), Darnell et al. (1983) and Meerkötter (1984 - private communication). The group of Darnell (1983,1986) uses temperature data which are determined directly from TOVS-measurements. These contain also cloud effects. Schmetz et al. (1986) use operational analyses of the near surface (1000 hPa, 850 hPa) temperature fields and climate means of the moisture. The latter is indeed a poorly known quantity. The effects of clouds have been considered so far only under daylight conditions, where the optical thickness of cloud fields is determined from their transmittance τ_c (see section 2.1) for solar radiation and the height from simultaneous measurements of the emission to space in the infrared. Cloud bottom heights, temperatures and emittance are then estimated on the basis of constant geometrical data and of the optical thickness (see above).

Some results are shown in Figures 5 to 7.

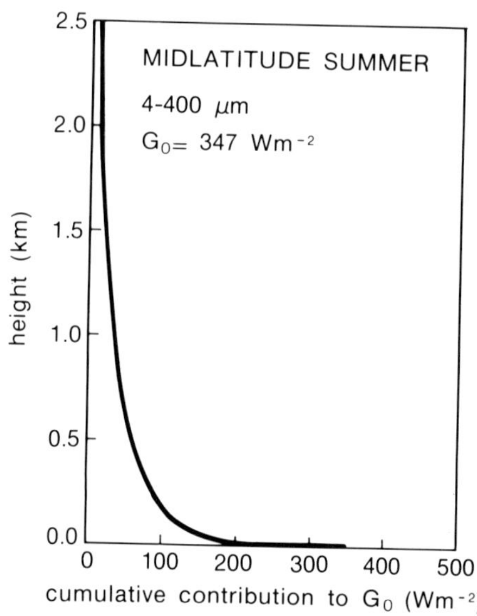

Figure 5. Cumulative contribution of the atmospheric column to the clear sky downward longwave flux G_o at the surface for a standard midlatitude summer atmosphere, i.e. the curve represents that part of G_o which stems from layers above corresponding level (Schmetz et al.,1986).

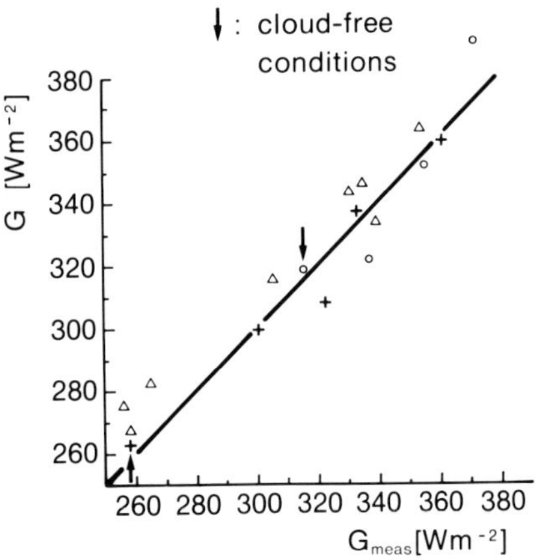

Figure 6. Comparison between ground measurements of the downward longwave flux G_{meas} and the model estimations G. Different signs denote G at three stations: O: PUKK; +: Hamburg; Δ: Braunschweig. Mean deviation is 5.5 W/m^2; RMS error is 12.2 W/m^2 (Schmetz et al., 1986).

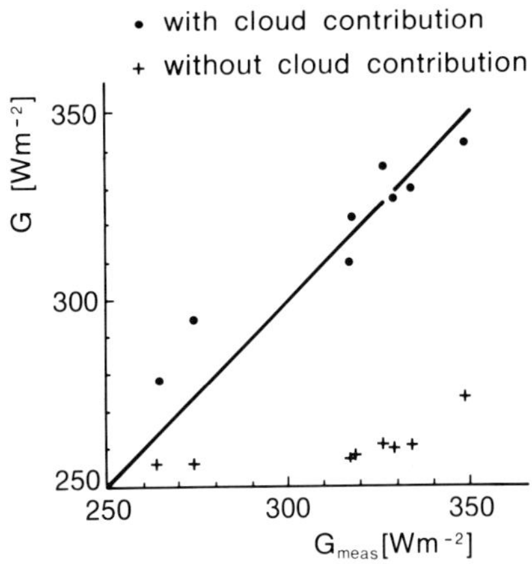

Figure 7. Comparison between the ground measured downward longwave flux at four stations in West Germany and the DLF as estimated with the model for the 29th and 30th of April at 1200 GMT. Crosses denote the model estimated clear sky DLF. Mean difference is 3.9 W/m^2; RMS difference is 11.6 W/m^2 (Schmetz et al., 1986).

So far only a few intercomparisons with ground-based measurements have been possible since the network of stations measuring the downward atmospheric radiation is too sparse. Darnell et al. (1983,1986) report on error ranges of 6.5 percent between results derived from TOVS data and with measurements from 4 stations distributed over the US.

3.2 Upward Heat Radiation from Surfaces, M_t

Its calculation requires a knowledge of the "effective" surface temperature, besides the quantities ε and ρ_t mentioned earlier. Over most continental areas this temperature is not always equal to the real surface temperature. It may rather correspond to an average of air temperature within the height range of the vegetation cover. Extremely high differences may occur for its determination over mountainous areas.

All satellite measurements for determinations of surface temperature are made in narrow spectral bands to avoid the effects of atmospheric gases. These are located near 3.7 μm. On the other hand, the spectral emittance values of natural surfaces in those spectral ranges often deviate considerably from those of the whole interval, in particular over deserts. Furthermore, in the thermal infrared region, where most measurements of the surface temperature are made, clouds are not transparent for the heat radiation from surfaces underneath. An attempt was, therefore, made by Susskind et al. (1987) to use simultaneous measurements of the heat emission in the microwave region, as available from the TOVS-package onboard the NOAA satellite. But these suffer from strong effects of surface and soil properties on the effective emittance. Susskind et al. (1982) tried to "calibrate" such data versus simultaneous measurements in the infrared over cloudfree areas, but no critical assessment of the error budget has been published so far.

A reliable mapping of the surface temperature by satellite data, therefore, appears possible only over oceans, where short fluctuations of the surface temperature, which occur over land due to variations of all heat fluxes near the surface, are not observable. The temporal resolution can be as short as 3-5 days to allow for cloud removal over each area observed.

With these assumptions, global maps of the sea surface temperature are now produced operationally with an accuracy ranging between 0.5 and 1.5 K in most areas of the world's oceans. This causes an uncertainty of the calculated values for the upward flux by about 2-3 percent, which corresponds to 5-10 Wm^{-2}. Over continental areas and snow, however, the surface temperature is very sensitive to changes of the other heat fluxes to the surface. Thus, its modelling for all cases of overcast or fractional cloud cover would require a more or less complex model of all heat fluxes or at least a parameterization with respect to global radiation, downward atmospheric radiation, cloud cover, wind and also the heat transfer properties of each particular area. Such a complete modelling has not yet been tried. If one accepts a reduction in time to monthly averages, then over most areas the surface temperature might be inferred from averages of the shelter temperature (2 m), if such quantities are known over all regions of the world!

In summary, no proven method has yet been established with reliable accuracy to map fluxes of upward heat radiation under clear sky and also cloudy conditions over continental surfaces derived from satellite data. Some basic research is still required.

4. CONCLUDING REMARKS

On the present status of mapping, over the entire globe, all four components of the radiation budget at ground with primary use of observations from operational satellites, we can state the following:

a) Reliably accurate methods are now available to map the downward solar or global radiation at the ground over most continents and the oceans. They still need careful checks when applied to data measured over snow and ice fields.
b) Changes of the surface reflectance for solar radiation can be observed by means of the operational imaging data, as far as these changes occur in the spectral ranges of these measurements. Detailed error analyses are still required, however, when such spectral albedoes are converted into albedoes for the total global radiation.
c) The downward atmospheric radiation depends primarily on the air temperature fields near the ground and to a smaller extent on cloud bottom temperature. The temperature field near the ground can be

inferred with the required accuracy from direct observations only. Climatological values of the air moisture near the ground can be used to calculate it. Satellite data provide in particular information on the positions and height of cloud fields, but not on the height of cloud bottoms.

d) The largest uncertainties occur, when the surface temperatures are mapped over continental areas and snow fields under cloudy conditions due to their quick response to changes of all heat fluxes at the surfaces.

For the radiation budget components, mentioned under b, c and d, detailed error analyses are still required on their sensitivity to uncertainties to input parameters.

The verification of each of the four components mentioned above, if desired from satellite data, requires ground-truth. For the two downward radiation fluxes this can already be provided over several experimental sites, which are already established or will be built up as anchor stations within the ISLCSP. Furthermore, during the next decade, member nations of WMO are encouraged to update the radiation network, recording to recommendations spelled out by WCP (1984) and (1986). In particular, measurements of the downward atmospheric radiation are missing. Validations of upward flux values require aerial survey measurements from aircraft.

The polar ice and snow fields are still open questions, where the high surface albedo on the one hand, and also thin haze or cloud layers on the other hand, introduce large uncertainties. Thus, additional space-borne measurements of other quantities might be required. The thin cirrus layers are also effective error sources over all other regions of the globe. They are difficult to observe in the operational imaging data.

On the other hand, there is now an urgent need to map over the entire globe all the components mentioned above and in a very unique manner. The ISCCP-data sets in the B-format, which are data reduced in their spatial and temporal resolution, will provide a basis for this purpose, since they are relatively compact but have been carefully checked with respect to malfunctions of the different instruments. Some research is already underway on the use of these data sets.

REFERENCES

Briegleb, B.P., P. Minnis, V. Ramanathan and E. Harrison, 1986: 'Comparison of Regional Clear-Sky Albedoes Inferred from Satellite Observations and Model Computations', J. Climate Appl. Meteor., **25**, 214-226.

Chou, Ming-Dah, 1985: 'Surface Radiation in the Tropical Pacific', J. Climate Appl. Meteor., **24**, 83-92.

Darnell, W.L., S.K. Gupta, W.F. Staylor, 1983: 'Downward Longwave Radiation at the Surface from Satellite Measurements', J. Climate Appl. Meteor., **22**, 1956-1960.

Darnell, W.L., S.K. Gupta, W.F. Staylor, 1986: 'Downward Longwave Surface Radiation from Sun-Synchronous Satellite Data: Validation of Methodology', J. Climate Appl. Meteor., 25, 1012-1020.

Diekmann, F., E. Raschke, H. Woick, 1986: 'Operational Mapping of Global Radiation over Central Europe', Meteor. Rundschau, 39, 59-63.

Ellingson, R.G., 1981: 'On Estimating the Radiation Budget at the Surface Using Satellite Observed Radiance', University of Wisconsin-Madison.

Gandin, L.S. and Kagan, 1976: 'Statistical Methods of Interpreting Meteorological Data', Gidrometeoizdat, Leningrad, 359.

Geleyn, J.F. and H.-J. Preuß, 1983: 'A Near Data Set of Satellite-derived Surface Albedo Values of Operational Use at ECMWF', Arch. Meteor. Geophys. Bioklim., Ser.A, 32, 353-359.

Gube, M., J. Schmetz and E. Raschke, 1980: 'Comments on "The Albedo of Water as Function of Latitude"', Monthly Weather Review, 108, 830-831.

Hahn, C.J., S.G. Warren, J. London and R. Chervin, 1983: Atlas of Simultaneous Occurrence of Different Cloud Types over the Ocean, NCAR/TN-201-STR, National Center of Atmospheric Research, Boulder, Colorado, USA.

Harrison, E.F., P. Minnis and G.G. Gibson, 1983: 'Orbital and Cloud Cover Sampling Analyses for Multisatellite Earth Radiation Budget Experiments', J. of Spacecraft and Rockets, 20, 491-495.

Idso, S.B. and R.D. Jackson, 1969: 'Thermal Radiation from the Atmosphere', J. Geophys. Res., 74, 5397-5403.

Kondratiev, K. Ya, J.F. Mizonova and A.N. Otto, 1964: 'Spectral Albedo of National Surfaces', P.A. Geoph., 59, 207-286.

Kukla, K. and D. Robinson, 1980: 'Annual Cycle of Surface Albedo', Monthly Weather Review, 108, 56-68.

Martin, M. and P. Berdahl, 1984: 'Characteristics of Infrared Sky Radiation in the United States', Solar Energy, 33, 321-336.

Mintz, Y., 1982: 'The Sensitivity of Numerically Simulated Climates to Land Surface Conditions', in Eagleson (ed.): Land Surface Processes Atmospheric General Circulation Models, Cambridge University Press.

Möser, W. and E. Raschke, 1984: 'Incident Solar Radiation over Europe Estimated from METEOSAT-Data', J. Climate Appl. Meteor., 23, 166-170.

Monteith, J.L., 1960: 'An Empirical Method for Estimating Longwave Radiation Exchanges in the British Isles', Quart. J. R. Meteorol. Soc., 87, 171-179.

Pinker, R.T. and L.A. Corio, 1984: 'Surface Radiation Budget from Satellites', Monthly Weather Review, 102, 209-215.

Pinty, B. and G. Szejwack, 1985: 'A New Technique for Inferring Surface Albedo from Satellite Observations', J. Climate Appl. Meteor., 24, 74-75.

Pinty, B. and G. Szejwack, 1985: 'Surface Albedo over the Sahel from METEOSAT Radiances', J. Climate Appl. Meteor., 24, 108-113.

Preuß, H.-J., E. Raschke and M. Daniel, 1984: 'Studies of Spaceborne Radiation Budget Measurements', Meteor. Rundschau, 37, 52-62.

Preuß, H. and J.F. Geleyn, 1980: 'Surface Albedoes Derived from Satellite Data and Their Impact on Forecast Models', Arch. Meteor. Geophys. Bioklim., Ser.B, 29, 345-356.

Raschke, E., A. Gratzki and M. Rieland, 1987: 'Estimates of Global Radiation at Ground from the Reduced Data Sets of the International Satellite Cloud Climatology Project', J. of Climatology, 7, 205-214.
Raschke, E., 1986: 'Mapping the Downwelling Atmospheric Radiation at the Earth's Surface - A Research Strategy', NASA-CR 178504, Langley Res. Center, Hampton, 21.
Rockwood, A.A. and S.K. Cox, 1976: 'Satellite Inferred Surface Albedo over Northwestern Africa', Dep. Atm. Sci., Colorado State Univ., Fort Collins, Atmospheric Science, Paper 265, 63.
Schiffer, R.A. and W.B. Rossow, 1985: 'ISCCP Global Radiance Data Set: A New Resource for Climate Research', Bull. Am. Meteor. Soc., 66, No.12.
Schmetz, P., 1984: 'Großräumige Bestimmung der Gegenstrahlung aus Satelliten- und Analysedaten', Diplomarbeit, Inst. Geoph. und Meteorol. Univ. zu Köln.
Schmetz, P., J. Schmetz and E. Raschke, 1986: 'Estimation of Daytime Downward Longwave Radiation at the Surface from Satellite and Grid Point Data', Theor. Appl. Climatol., 37, 136-149.
Shine, K.P., 1984: 'Parametrisation of the Shortwave Flux over High Albedo Surfaces as a Function of Cloud Thickness and Surface Albedo', Quart. J. R. Met. Soc., 110, 747-764.
Smith, W.L. and H.M. Woolf, 1983: 'Geostationary Satellite Sounder (VAS) Observations of Longwave Radiation Flux Report on "Satellite Systems to Measure Rad. Budget Parameters and Climate Change Signals"', Igls., Austria, WCP.
Stuhlmann, R., M. Wiegner, H. Knottenberg, J. Wirth and D. Hennings, 1985: 'Calculations of the Planetary Albedo from Satellite Radiance Measurements", Adv. Space Res., I, 133-139.
Susskind, J., J. Rosenfield, D. Reuter and M.T. Chahine, 1982: "The GLAS Physical Inversion Method for Analysis of HIRS2/MSU Sounding Data', NASA Techn. Memorandum 84936, Goddard Space Flight Center, Greebelt, Md.
Susskind, J., J. Rosenfield, D. Reuter and M.T. Chahine, 1982: 'The GLAS Physical Inversion Method for Analysis of HIRS2/MSU Sounding Data', NASA Techn. Memorandum 84936, 101, Goddard Space Flight Center, Greenbelt, Md.
Susskind, J. J. Rosenfield and D. Reuter, 1983: 'An Accurate Radiation Transfer Model for Use in the Direct Physical Inversion of HIRS2 and MSU Temperature Sounding Data', J. Geoph. Res., 88, 8550-8568 (C13).
WCP-48, 1984: 'Report of the Meeting of Experts on the Future Activities of the World Radiation Centre'(WRC-Leningrad)', WMO-Geneva.
WCP-115, 1986: 'Report of the "Workshop on Surface Radiation Budget for Climate Application"', WMO/TD, 109, 140.
WCP-131, 1987: 'Report of the "ISCCP" Workshop on Cloud Algorithms in the Polar Regions', WMO/TD, 170, 42.

MICROWAVE REMOTE SENSING OF SOIL MOISTURE

Thomas Schmugge
USDA-ARS Hydrology Laboratory
Beltsville, Maryland 20705 - USA

ABSTRACT. Because of the large contrast between the dielectric constant of liquid water and that of dry soil at microwave wavelengths, there is a strong dependence of the thermal emission and radar backscatter from the soil on its moisture content. This dependence provides a means for the remote sensing of the moisture content in a surface layer approximately 5 cm thick. The feasibility of these techniques has been demonstrated from field, aircraft and spacecraft platforms. The soil texture, surface roughness, and vegetative cover affect the sensitivity of the microwave response to moisture variations with vegetation being the most important. It serves as an attenuating layer which can totally obscure the surface. Research has indicated that it is possible to obtain 5 or more levels of moisture discrimination and that a mature corn crop is the limiting vegetation situation.

1. INTRODUCTION

The microwave remote sensing of soil moisture relies on the large contrast between the dielectric constant of water ($\sim$ 80) and that of dry soil (3.5). This arises from the ability of the electric dipole of the water molecule to align itself in response to the electric field at microwave frequencies. Figure 1 is a plot of the dielectric constant for water and ice from low frequencies to beyond the microwave. Note that for ice the dielectric constant is large at low frequencies but because of the binding of the water molecule in the solid the motion is inhibited at about 10^4 Hertz. For liquid water this reduction in the molecule's ability to rotate does not occur until about 10^{10} Hertz, i.e. in the microwave range, which we take to be from 1 GHz to 100 GHz (1 GHz = 10^9 Hz, wavelengths between 30 and 0.3 cm). The dielectric constant is of importance here because it describes the propagation characteristics of an electromagnetic wave in the medium. These characteristics include the velocity of propagation, the wavelength in the medium and the absorption of energy in the medium. The square root of the dielectric constant is the index of

F. Toselli (ed.), Applications of Remote Sensing to Agrometeorology, 257–284.

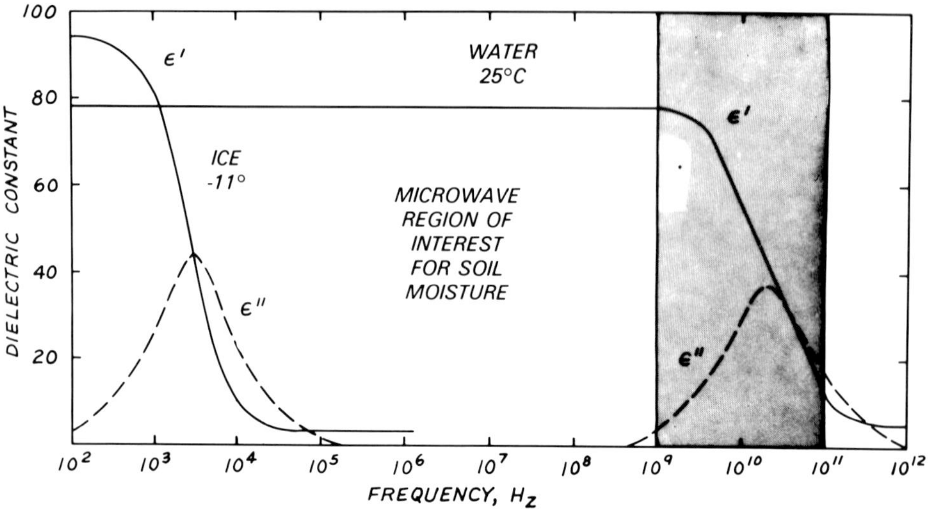

Figure 1. Dielectric constant for liquid and solid (ice) water from 10^2 Hz to 10^{12} Hz.

refraction for the material. It is the dielectric contrast at the boundary between two media which determines the reflection and transmission coefficients of electromagnetic wave at such a boundary.

In this paper we will describe how this variation of the dielectric constant for soil can be used to remotely sense its moisture content and that frequencies below 5 or 6 GHz are most effective. This capability is limited to a surface layer about 5 cm thick and is affected by surface properties such as roughness and vegetation cover.

2. DIELECTRIC PROPERTIES OF SOILS

When you mix water with soil the dielectric constant of the mixture will change dramatically, increasing from about 3 or 4 for the dry soil to almost 30 for wet soils. Figure 2 presents measurements for three soils ranging from a sandy loam to a heavy clay. These measurements were at a frequency of 1.42 GHz or a wavelength of 21 cm. Qualitatively the behaviour for the three soils is similar in that the real part of the dielectric varies from about 3 to greater than 20. However, it is clear that there are differences among the soils, namely with the rate of increase at the lower soil moistures. The heavier soils have a region of slow increase and then above a transition value there is a region of sharp increase. It was noted that this flat region was a function of soil type being wider for the heavier soils (Wang and Schmugge, 1980). The simple explanation is that the first water added to a soil is tightly bound to the soil surface and has dielectric properties somewhere between those of bound molecules in ice and those of the freely rotating molecules in the liquid. It is

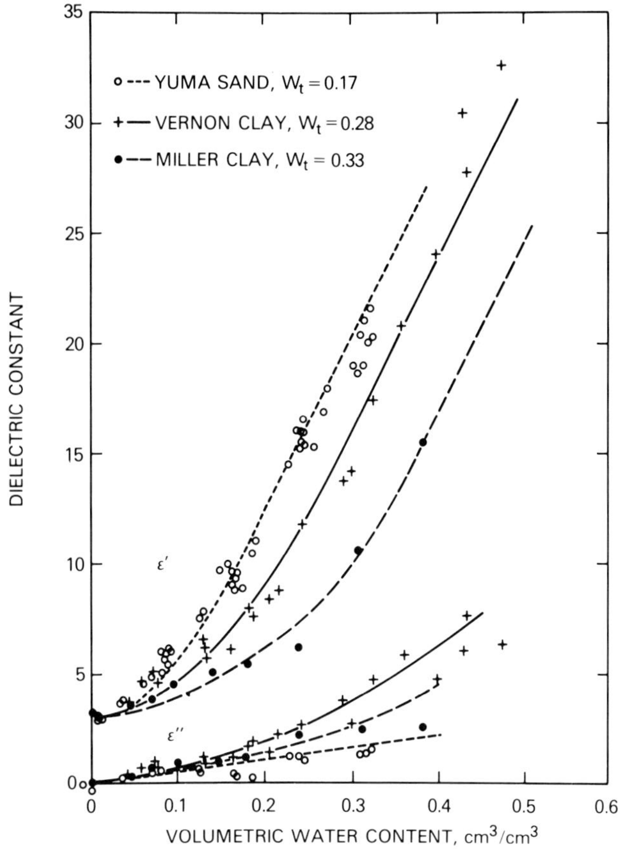

Figure 2. Laboratory measurements of the dielectric constant for three soils ranging from a light sand to a heavy clay at a frequency of 1.4 GHz.

only after there are several layers of water on particle surface that the water begins to behave like a liquid in terms of its dielectric properties.

This variation of the soil's dielectric constant with moisture can produce a variation in the soil's emissivity from 0.95 for dry soils to 0.6 or less for wet soils with changes of a corresponding magnitude for soil's reflectivity. These variations have been observed by both passive and active microwave sensors. The former are radiometers which observe the variations in the thermal emission from the soil due to emissivity changes. The latter are radars which transmit a pulse of electromagnetic energy and then measure the backscattered return which will be a function of the soil's reflectivity.

In an analysis of a considerable amount of data of the type shown in Figure 2 for a wide range of soils, Wang and Schmugge (1980) have found that the transition moisture (W_t) is linearly correlated to the estimated wilting point (WP) for the soils. Here WP is taken to be the soil moisture at a tension of 15 bars. The relationship is:

$$W_t = 0.165 + 0.49\ WP \tag{1}$$

Using this relationship an empirical model was developed to describe the dielectric behaviour of moist soils based on a simple dielectric mixing formula. The equations are:

$$k = W_c k_x + (P-W_c)k_a + (1-P)k_r, \quad W_c < W_t \tag{2}$$

with $$k_x = k_i + (k_w-k_i)(W_c/W_t)k\beta \tag{3}$$

and $$k = W_c k_x + (W_c-W_t)k_w + (P-W_c)k_a + (1-P)k_r, \quad W_c \geq W_t \tag{4}$$

with $$k_x = k_i + (k_w-k_i)\beta \tag{5}$$

Here P is the porosity of the dry soil; k_a, k_w, k_r and k_i are the dielectric constants of air, water, rock and ice, respectively. k_x is the dielectric constant of the initially absorbed water and varies from the value for ice at $W_c = 0$ and approaching the value for water at W_t. The parameter β was chosen to yield the best fit of the equations for k to the data and is given by $\beta = 0.48 - 0.57\ WP$. These equations were used to obtain the curves in Figure 2. The values of the parameters used were: $p = 0.5$; $k_i = (3.2, 0.1)$; $k_r = (5.5, 0.2)$ and $k_a = (1, 0)$ where the two values represent the real and imaginary parts of the dielectric constants for the materials.

Dobson et al. (1985) have proposed a four component dielectric mixing model which describes the soil water system as a host medium of dry soil solids containing randomly distributed and oriented disc shaped inclusions of bound water, bulk water and air. The separation of the soil water into bound and bulk components depends on the soil texture or more directly on the specific surface area of the soil. With models such as these it is possible to study the sensitivity of a soil's dielectric properties on such things as density, texture, salinity, etc. Figure 3 is an example; it shows the dependence of the dielectric constant on density for a sandy loam soil. At zero soil moisture the difference between the three curves is due to the different amount of parent material present. The range of densities shown produces a 15 to 20 percent spread of values in the real part. The spread in the imaginary part is almost negligible. Figure 4 shows the sensitivity to specific surface or texture at a density of 1.3 g/cc. The soils range from a loamy sand to a clay and exhibit the type of behaviour shown in Figure 2, i.e. the flat region at low soil moistures for the clay soil.

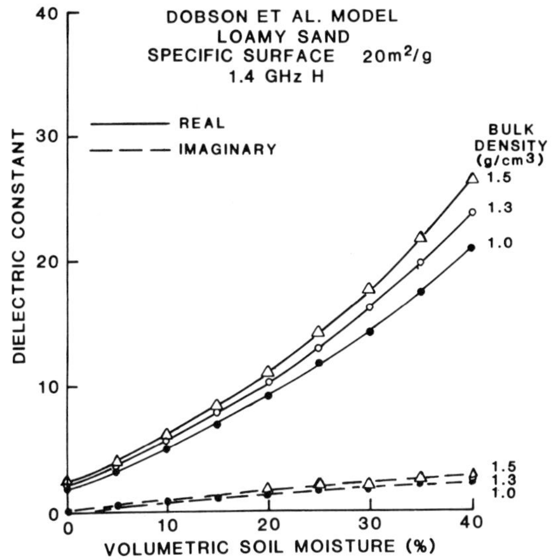

Figure 3. Dependence of the dielectric constant for a loamy sand soil on its bulk density. Symbols in Figures 3 and 4 are points at which the calculations were done.

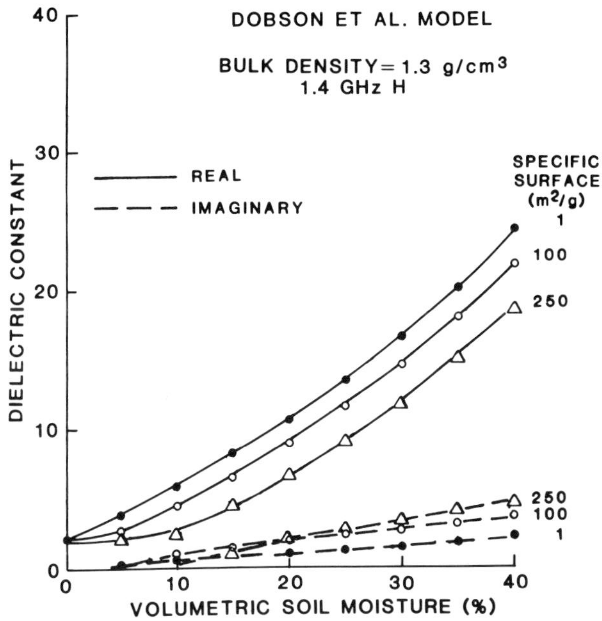

Figure 4. Dependence of the dielectric constant on specific surface or texture for soils ranging from sand (1) to clay loam (250).

3. EMISSIVITY AND REFLECTIVITY

Knowledge of the dielectric properties can be used in the Fresnel equations to calculate the reflectivity and emissivity for a soil with a smooth surface (Jackson, 1962). In their simplest form they are:

$$R_V = \left| [\cos - \sqrt{(k-\sin^2\theta)} \;/\; \cos\theta + \sqrt{(k-\sin^2\theta)}] \right|^2 \tag{6}$$

where θ is the incidence angle and k is the dielectric constant for the soil. For the horizontal polarization the equation is

$$R_h = \left| [k\cos - \sqrt{(k-\sin^2\theta)} \;/\; k\cos\theta + \sqrt{(k-\sin^2\theta)}] \right|^2 \tag{7}$$

Recall that the horizontal polarization is the one with the electric field parallel to the soil surface and the vertical polarization is where the electric field has a component perpendicular to the soil surface. Figure 5 is a plot of these equations for wet and dry soil,

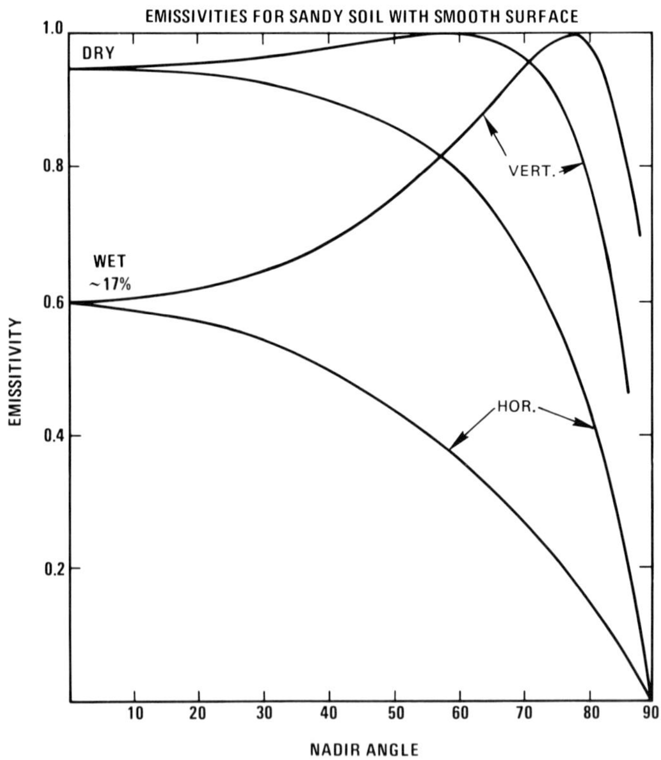

Figure 5. Calculations from Fresnel equations of the emissivity for wet and dry sandy soils with smooth surfaces. Brewster's angle for dry soil is about 58° and for wet soil 78°.

here, we have assumed that the emissivity is given by

$$e_{h,v} = 1 - R_{h,v} \tag{8}$$

The plot shows that at nadir there is a decrease in the emissivity from 0.95 when dry to 0.6 when wet. This sensitivity continues out to about an angle of 30°, at which point the polarizations begin to diverge. As the angle increases beyond this point the e_v approaches 1 at what is called the Brewster's angle given by

$$\mathrm{Tan}_B = \sqrt{k} \tag{9}$$

4. ATMOSPHERIC EFFECTS

One of the other major advantages for microwave remote sensing is the relative transparency of the atmosphere at these wavelengths. There are two interactions of microwave radiation with the atmosphere: absorption arising from the transition between the molecular energy states of the atmospheric gases and the scattering and absorption by hydrometeors in clouds and rain. The atmospheric gases which absorb microwave radiation are water vapour and oxygen. The H_2O absorption lines are at frequencies of 22.2 GHz (1.35 cm) and 183 GHz (0.164 cm) while the O_2 absorption is due to a band of lines centered about 60 GHz (0.5 cm) and a single line at 118 GHz (0.25 cm). The effect of these two gases can be calculated using their known absorption coefficients and their densities in the atmosphere. The results (Figure 6) show that for frequencies below 10 GHz (wavelengths > 3 cm) the effect of these gases is negligible with there being greater than 99 percent transmission through the entire atmosphere. Figure 6 shows that even for light rain or cloud situations there is greater than 95 percent transmission.

The effect of clouds will depend on the phase and size of the hydrometeors. An electromagnetic wave incident on a particle produces a motion of the charges contained within the particle. These moving charges will absorb some of the incident energy and reradiate some, producing a scattered field. Thus, some of the incident energy is scattered out of the beam and some is absorbed by the particle. As we saw in Figure 1, the dielectric properties of the liquid state are much greater than those of the solid and, therefore, the effects of the liquid hydrometeors will be greater than those of ice. The major effects of clouds and rain are:

1) Clouds of ice particles, i.e. cirrus clouds, have negligible effect on the microwave radiation because of the small particle sizes and the low dielectric constant of the ice.
2) Because of the small particle sizes in non-raining clouds, e.g. stratus and fair weather cumulus, there is negligible effect for frequencies below about 15 GHz. The absorptivity for clouds of these types is proportional to the square to the frequency.
3) For rain, the scattering by the larger hydrometeors is significant for frequencies above 3 GHz, as seen by the use of weather radars at these frequencies.

ATMOSPHERIC TRANSMISSION FOR THREE SKY CONDITIONS

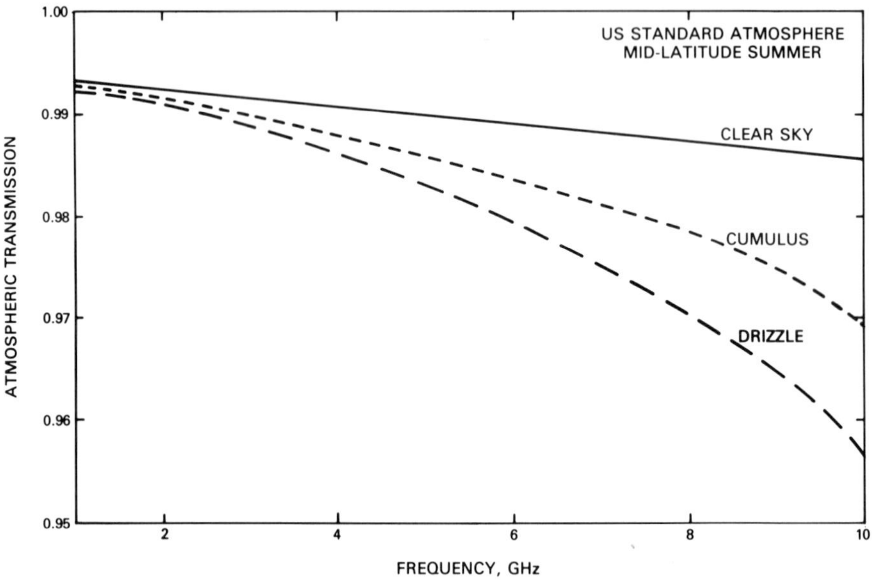

Figure 6. The atmospheric transmission at the lower microwave frequencies from 1 to 10 GHz. These were calculated with a radiative transfer model for the atmosphere.

As we will see later the most effective frequencies for the microwave remote sensing of soil are between 1 and 5 GHz and so the effects of the atmosphere can, in general, be neglected.

A more complete discussion of atmospheric effects can be found in Chapter 5 of the Manual of Remote Sensing (Colwell, 1983) and in Volume I of the book Microwave Remote Sensing by Ulaby et al. (1982).

5. MICROWAVE RADIOMETRY

A microwave radiometer measures the thermal emission from the surface and at these wavelengths the intensity of the observed radiation is proportional to the product of the thermodynamic temperature of the soil and the surface emissivity (Rayleigh-Jeans approximation to the Planck radiation law). This product is commonly called the brightness temperature (T_B) and is given by

$$T_B = \tau(RT_{sky} + (1-R)T_{soil}) + T_{atm} \tag{10}$$

where R is the surface reflectivity and τ is the atmospheric transmission. This equation is represented schematically in Figure 7. The first term is the reflected sky brightness which depends on the atmospheric

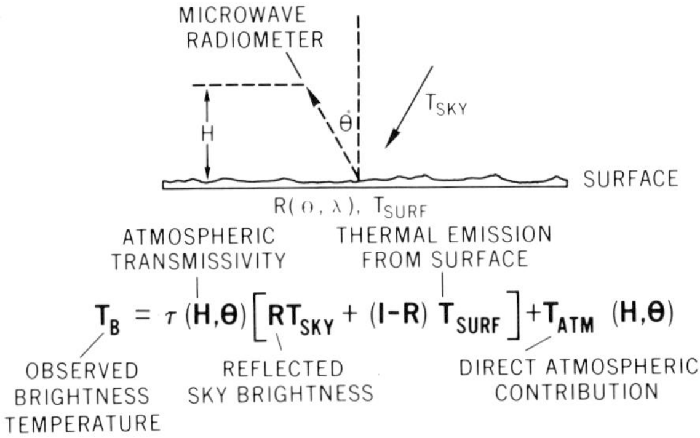

Figure 7. Schematic of the sources of microwave radiation observed by a radiometer at a height h and at angle θ.

conditions and frequency. For the frequencies of interest to us, i.e. < 10 GHz, T_{sky} = 5 to 10 K for the normal range of atmospheric conditions with 3 K of it being the constant cosmic background radiation. The third term is the direct atmospheric contribution and as noted will be about 5 K. The atmospheric transmission will typically be about 99 percent so we are left with the emission from the soil, i.e. the second term as the main contributor to T_B. Thus, Eq.(10) reduces to

$$T_B = (1-R)T_{soil} = eT_{soil} \tag{11}$$

This is true for an isothermal soil when there are variations of soil temperature the factor eT_{soil} is replaced by the integral:

$$T_B = e \int_{\infty}^{0} T(z)\alpha(z)\exp\left[-\int_{\infty}^{0} \alpha(z')dz'\right]dz \tag{12}$$

where T(z) and α(z) are the temperature and absorptivity at the depth z in the soil. The integral essentially gives the intensity of the upwelling thermal radiation from the soil and is a weighed average of the soil temperature over the electromagnetic skin depth of the soil. The factor e, the emissivity, gives the fraction of this upwelling radiation which is transmitted into the air and it is this factor which shows the primary sensitivity to the soil moisture content. The critical factor is: what is the thickness of the layer at the surface whose dielectric properties determine e. It is this thickness which determines the soil moisture sampling depth for microwave sensors. An early theoretical study of the radiative transfer in a soil by Wilheit (1975) has shown that this depth is only on the order of a few tenths of a wavelength thick or about 2 to 5 cm at the 21-cm wavelength. This result has been confirmed by experimental work of Newton et al. (1982) and Wang (1987). In Newton's work the dry down of a saturated field was observed with radiometers at 3 frequencies (1.4, 5 and 10.7 GHz) and the temporal variation of the responses was compared with that observed the soil moisture measured in 0-9, 0-5 and 0-2 cm layers. It was found that the highest two frequencies indicate drying at a rate faster than that seen in the 0-2 cm layer, while the 1.4 GHz results indicated drying at rate between the 0-2 and 0-5 cm layers. The conclusion was that the higher two frequencies were sampling a layer thinner than 2 cm while the 1.4 GHz was sampling a layer whose thickness was between 2 and 5 cm.

Wang's approach was to use a large number of field measurements of both the microwave emission and the soil moisture profile. Using Wilheit's model with the measured profiles he calculated the expected emissivities and compared them with the soil moistures in several layers. By comparing these curves with measured emissivity variation with soil moisture he concluded that the 1.4 GHz radiometer was responding to the moisture in a layer between 2 and 5 cm thick.

5.1 Field Verification

As part of a study of salinity effects on the soil's emissivity, Jackson and O'Neill (1987) made a careful series of field measurements over loamy sand soil having a smooth surface and compared the results with those expected from the two dielectric constant models described above. The results are shown in Figure 8. The emissivity ranges from

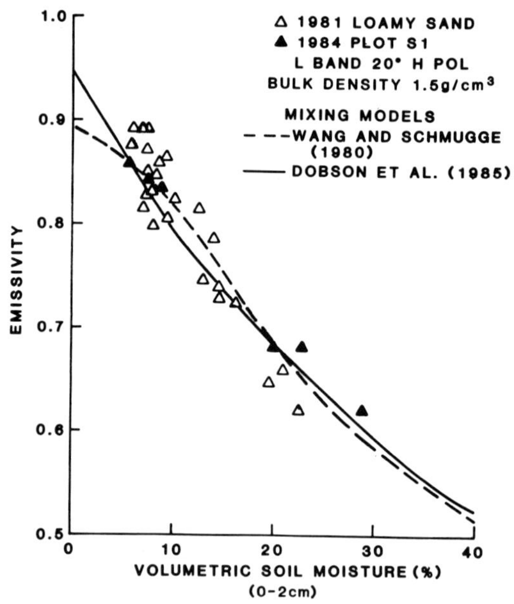

Figure 8. Observed and predicted relationships between emissivity and volumetric soil moisture for a bare, smooth, no salinity loamy sand soil at 1.4 GHz (L-band) (from Jackson and O'Neill (1987)).

0.6 for the wet soil ($\approx$ 30 percent volumetric soil moisture) to 0.9 for the dry soil ($\approx$ 8 percent). The calculations from the two models agree reasonably well with each other and with the data which indicates that the basic sensitivity of microwave emissivity to soil moisture variations is well understood and the basic theory is verified. The complications arise when such real factors as surface roughness and vegetative cover are added to the problem.

5.2 Surface Roughness

As we have seen, the emissivity of a smooth surface can be calculated using the Fresnel equations (Eqs.(6) and (7)). When the surface is rough the situation is more complicated; simply stated, the reflectivity will decrease and thus the emissivity will increase. This results from the increased surface area that can emit radiation. There have been numerous attempts to calculate quantitatively the effects of roughness, but the efforts are hampered by the difficulty of experimentally quantifying the surface roughness.

The magnitude of the effect is shown in Figure 9 which presents

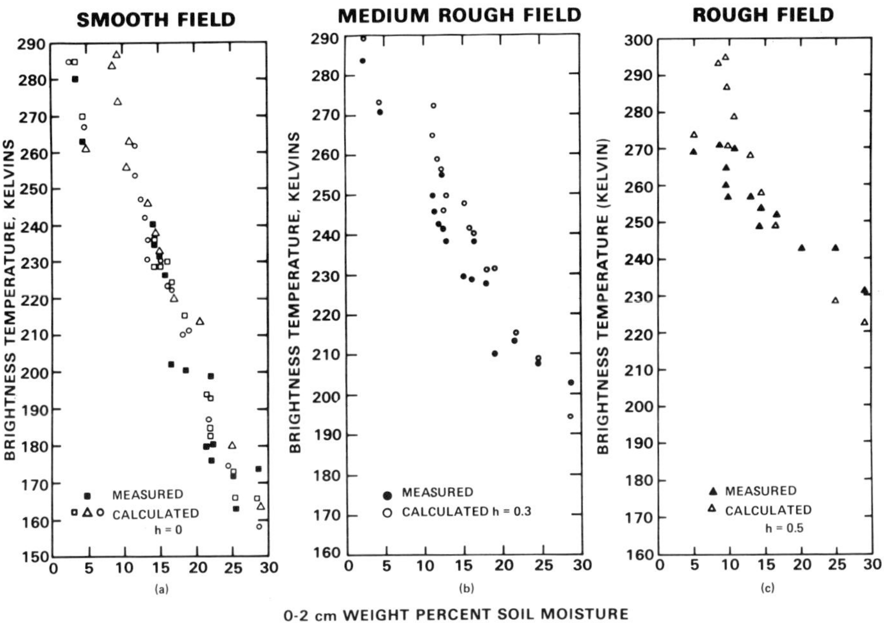

Figure 9. Roughness effects on the microwave emission from soils at a wavelength of 21 cm. The solid symbols are the measured values and the open symbols are calculations from Eq.(13) using h = 0 in (a), h = 0.3 in (b), and h = 0.5 in (c) (from Choudhury et al. (1979)).

the results of field measurements at the 21-cm wavelength for three surface roughnesses classified as smooth, medium rough, and rough. These fields had root mean square (RMS) height variations of 0.92, 2.6 and 4.3 cm, respectively. The range of T_B between wet and dry conditions decreased from about 120 K for the smooth field to 40 K for the rough field. Using a simple model, Choudhury et al. (1979) have shown the increase in emissivity given by:

$$\Delta e = r_o [1 - \exp(-h)] \tag{13}$$

where r_o is the smooth surface reflectivity and h is an empirically derived roughness parameter that is proportional to the RMS height variations with $h = 0$ for a smooth surface. For dry fields $r_o < 0.1$, the effect of roughness on the emissivity will be small, whereas for wet fields $r_o = 0.4$, the effect is larger. In Figure 9 there is little difference in T_B for the dry field conditions while there is about a 60 K effect for the wet fields. These results were for fields which were specially prepared for roughness studies and represent the extremes of surface conditions and, for example, it is expected that the roughest condition would only be encountered immediately after deep plowing. Figure 10 is an example of results for more normal conditions.

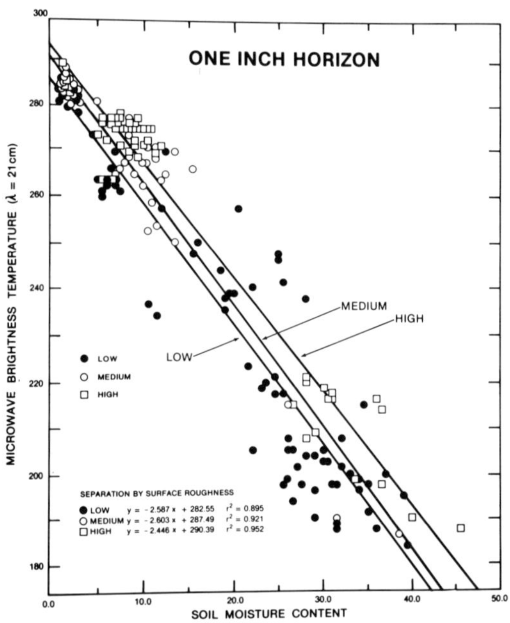

Figure 10. Scatter plot of T_B versus soil moisture for aircraft observations at the 21-cm wavelength over a test site in Hand County, South Dakota. The data were segregated into 3 roughness categories based on ground photographs (from Owe and Schmugge (1983)).

These were data acquired from an aircraft platform at an altitude of 300 m over a typical agricultural region in the Great Plains of the United States. The data were segregated according to a qualitative assessment of roughness from photographs and it is seen that the range of variation due to roughness is small for these more typical conditions. The regression slopes were not significantly altered by the roughness and were, on the average, 8 K higher than the smoothest. These fields covered the range of roughnesses that normally occur in an area which is mostly planted in small grains or in pastures. These results point out while roughness can have a very significant effects, its naturally occurring range may not be as great as the extremes observed in the prepared plots for truck experiments.

Another possible approach would be to use supporting data from other sensors to correct for roughness effects. In analyzing a data set containing both 1.6 GHz scatterometer and 1.4 GHz radiometer data, Theis et al. (1986) observed that the influence of surface roughness on the radar backscatter (σ^o) was substantial at large incidence angles while the sensitivity remained about the same at 40° incidence angle. They were able to use this fact to quantify the degree of roughness which was then used to correct for roughness in the radiometer data. Using this approach they improved the correlation (R^2) of T_B with soil moisture from 0.69 to 0.95. Although conducted on only a limited data set, these encouraging results indicate the potential of multi-sensor techniques for improving microwave estimates of soil moisture.

5.3 Vegetation Effects

The simplest way to think of the effects of vegetation on the microwave emission from the soil is in terms of a cloud which can absorb and re-emit radiation, i.e. similar to the atmospheric situation. This assumes that scattering is negligible, which is true at the longer wavelengths. For a sufficiently thick layer of vegetation, only the radiation from the vegetation itself will be observed. The brightness temperature above the canopy is given by

$$T_B = T_V[\ell + (e_S-1)\exp(-2\tau_p) + e_S(T_S/T_V-1)\exp(-\tau_p)] \tag{14}$$

where $\tau_p = \ell\alpha\sec\theta$ is the one way canopy absorption factor or optical depth, ℓ and α are thickness and absorptivity of the vegetation, and T_S and T_V are the temperatures of the soil and vegetation. If we assume that the vegetation and soil temperatures are approximately equal this equation reduces to

$$e = T_B/T_S = \ell + (e_S-1)\exp(-2\tau_p) \tag{15}$$

Since the temperatures are expressed in Kelvins, the ratio T_S/T_V typically will be within 1 or 2 percent of one. Kirdiashev et al. (1979) has expressed the optical depth as

$$\tau_p = \frac{\mu \sec(\theta)}{3\lambda} \; 10^{-5} \; Qmk'' \tag{16}$$

where μ is a plant shape parameter and Q the dry biomass in units of 100 kg/ha, k'' is imaginary part of the dielectric constant for water, and m is the gravimetric moisture content of the vegetation. Grouping the constants together we get the following expression for τ_p

$$\tau_p = b_2 \, W \tag{17}$$

where $W = Qm$ is the weight of water per unit area (Jackson et al., 1982). Figure 11 is a plot of τ versus W from a number of field experiments.

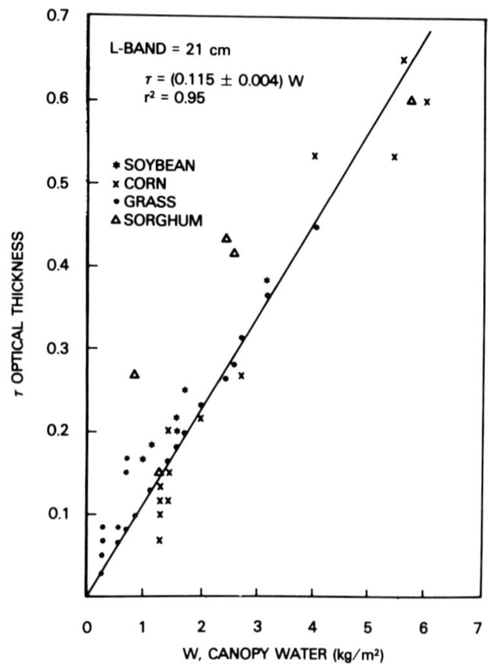

Figure 11. Variation of optical depth (τ) determinations with vegetation water content for fields covered with 4 different crops.

Here τ was determined from the microwave response and W measured (Mo et al., 1982). Thus, to quantify the effect of vegetation on the microwave emission from soils it is necessary to have some estimate of the plant above ground biomass.

A possibility for doing this lies with the use of visible and near infrared reflectance data which is strongly dependent on vegetation conditions. In one such analysis, Theis et al. (1984) used the perpendicular vegetation index calculated from visible and near IR as an estimator of plant biomass. The PVI, which is directly related to plant biomass, was used as a vegetation corrector in algorithm to estimate soil moisture from 21-cm brightness temperature data. With this correction the correlation between the microwave data and soil moisture improved from 0.09 to 0.75. This is an indication of the promise that the use of multispectral data can greatly improve our ability to estimate soil moisture remotely.

6. ACTIVE MICROWAVE APPROACHES

An active microwave sensor or radar sends out a pulse of microwave radiation and then measures the intensity of the radiation reflected back to it as represented schematically in Figure 12. The intensity of

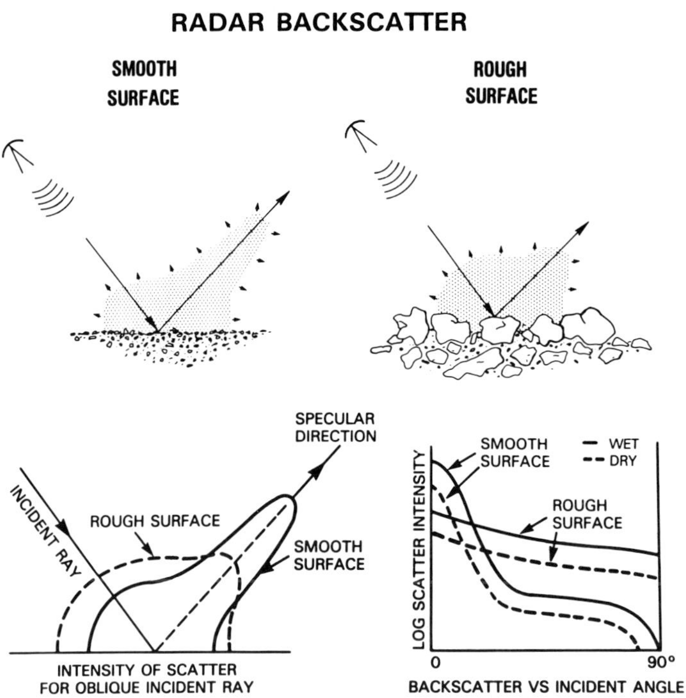

Figure 12. Schematic diagram of radar intensity patterns showing the different behaviours of smooth and rough surfaces. The backscatter coefficient (σ^{o}) is the intensity in the direction of incidence.

the return for a smooth surface is mostly reflected or scattered into the specular direction and that some scattering source, such as surface roughness, is required to obtain significant intensity in the backscatter direction. This is represented in the lower right of the figure showing that the backscattered intensity is at a maximum for the

0° angle for the smooth surface while for the rough surface the return is more isotropic.

An advantage of the radar is that the energy in the received pulses can be angularly separated into the returns from different locations on the ground. Thus, if the radar is on board an aircraft, it is possible to produce a backscatter image of the ground as the aircraft flies along. If the coherent nature of radiation is utilized it is possible to obtain very high resolution images from what are called synthetic aperture radars (SAR). The details of this technique are beyond the scope of this paper and the reader is referred to the book by Ulaby et al. (1982; 1986) on Microwave Remote Sensing for more information.

Analogous to the optical reflectivity of terrain, σ^o describes the scattering properties of the surface in the direction of the illuminating source. The scattering behaviour of terrain is governed by the geometrical and dielectric properties of the surface (or volume) relative to the wave properties (wavelength, polarization, and angle of incidence) of the incident radiation. Since the dielectric properties of the soil are strong functions of its moisture content, σ^o will be dependent on the soil moisture of a surface layer whose thickness is determined by the properties of the soil at the wavelength of observation. This thickness will be approximately the same for both active and passive microwave approaches, i.e. on the order of a few tenths of a wavelength as discussed earlier. As with the passive microwave case the dependence of σ^o on soil moisture will be a function of surface roughness and vegetation. In general it can be said that the active microwave response is more dependent on roughness and less on vegetation than the passive microwave response.

6.1 Look Angle and Roughness Effects

The dependence of σ^o on incidence angle and surface roughness has been studied extensively by scientists at the University of Kansas (Ulaby et al., 1978; 1979; Dobson and Ulaby, 1981) using a truck mounted 1-18 GHz radar. A sample of the results for wet field conditions is shown in Figure 13 where plots of σ^o versus angle are presented for 5 surface roughnesses and 3 frequencies.

At the longest wavelength (1.1 GHz) the smoothest field appears to be almost specular, in that there is a very strong response at nadir which decreases rapidly as the angle increases. There is a decrease in σ^o of almost 40 dB as the angle increases from 0 to 30°. For the roughest field the response is almost isotropic with less than a 5 dB change with angle. It is only at the longest wavelength that any of the fields displayed this specular behaviour with all of the fields appearing rough at the shortest wavelength (7.25 GHz). At all 3 frequencies there is an angle where the curves for the different roughnesses intersect, thus observations at this angle would be independent of surface roughness. At 4.25 GHz this intersection occurs at an angle of 10° and it was this combination of incidence angle and frequency that yielded the best sensitivity to soil moisture independent of surface roughness.

ACTIVE MICROWAVE DEPENDENCE ON ROUGHNESS

DATA FROM UNIV. OF KANSAS

WET SOILS: 0.34 — 0.4 g/cc IN TOP cm

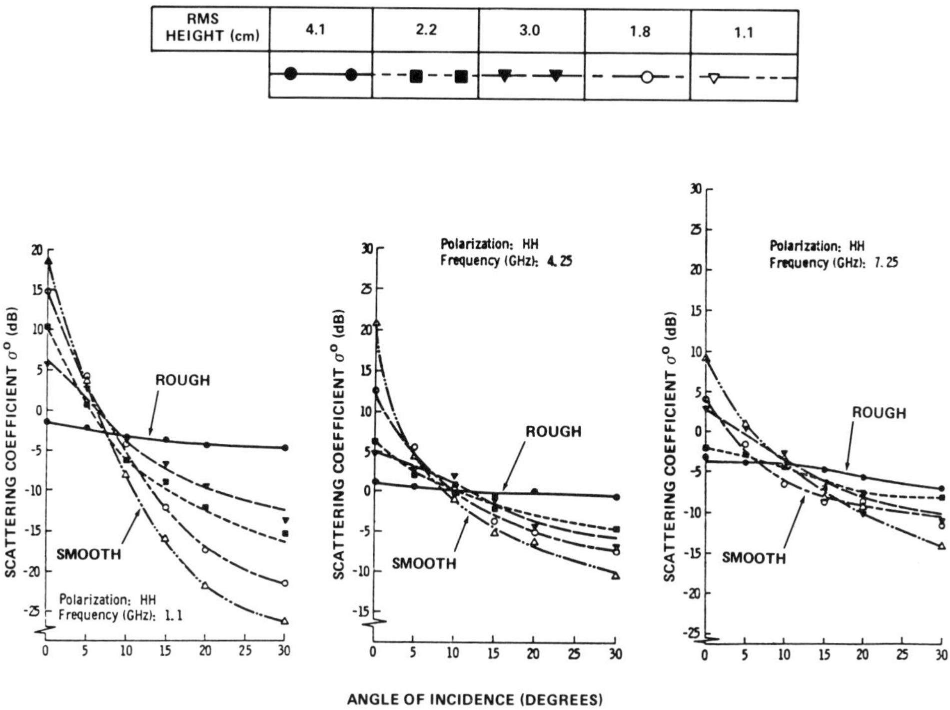

Figure 13. Angular variation of σ^o at 3 frequencies in dB for 5 bare fields having the RMS surface height variations as indicated. All the fields were in a wet condition; measurements were made from a tower at the University of Kansas.

These experiments were repeated for dry fields with similar results except that the curves are down by about 10 dB. Figure 14 presents the difference at the 4.25 GHz frequency for the two extremes of roughness; it is seen that the wet and dry curves are generally parallel and that the smooth and rough curves would intersect at about the angle of 10°.

Based on these results, Ulaby et al. (1978) recommended the following set of optimum parameters for radar observations of soil moisture: frequency = 4-5 GHz, θ = $7\text{-}17^o$, and horizontal transmit and horizontal receive (HH) polarization. The basic result has been confirmed by a number of experiments performed by the French group at the Centre de Recherches Physiques de l'Environnement (Bernard et al., 1982).

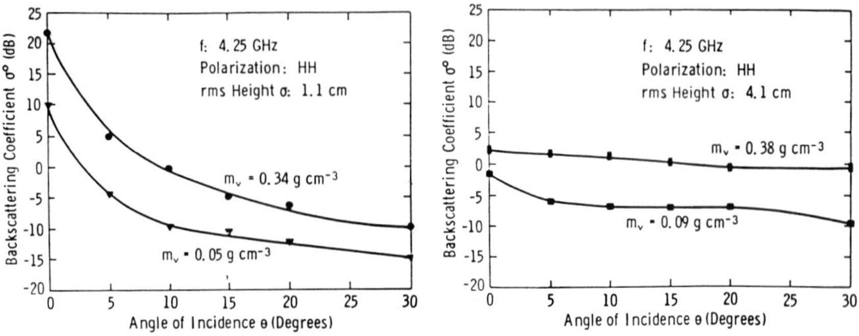

Figure 14. Angular variation of σ^{O} at 4.15 GHz for the two extreme roughness conditions at both the wet and dry moisture levels.

6.2 Vegetation Effects

A vegetation layer covering the soil will absorb and scatter some of the microwave radiation incident on it. As we discussed in the passive microwave section the absorption is primarily due to the water content of the vegetation. The precise mechanism or the centres for the scattering are not understood very well at present and are the subject of much current research both experimental and theoretical.

The effect of vegetation on σ^{O} expressed in natural units (m^2/m^2) is schematically represented in Figure 15 in terms of the backscatter

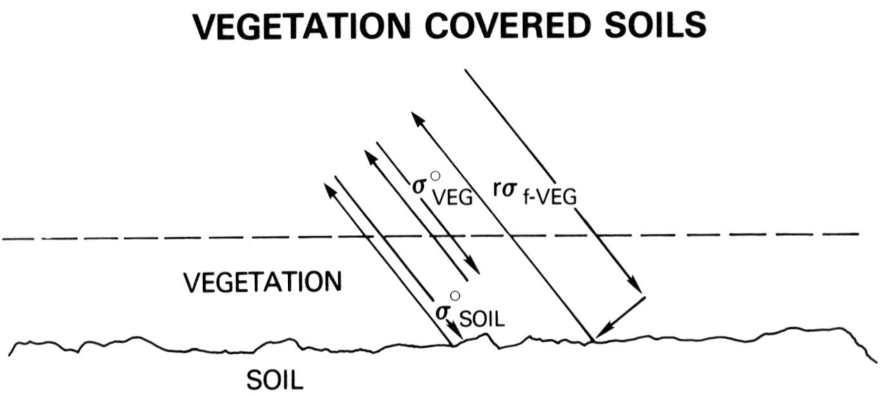

$$\sigma^{\circ}_{CANOPY} = \sigma^{\circ}_{VEG} + \sigma^{\circ}_{SOIL}/L^2 + r\sigma_{f\text{-}VEG}/L^2$$

I II III

Figure 15. Schematic representation of the sources of radar backscatter from vegetated soil, where L is the canopy loss factor.

coefficient $\sigma^o{}_v$ of the vegetation and the loss factor L $(L = \exp[-\tau])$ for the vegetation. The 3 terms contributing to the radar backscatter are: (1) the ground surface scatter term; (2) the volume scatter term representing scattering from the vegetation layer; and (3) the surface/volume scatter term accounting for the scattering from both the surface and the volume, i.e. radiation forward scattered by the vegetation and then reflected by the soil or vice versa. Note that the third term tends to reinforce the increase of backscatter with increasing soil moisture.

From the analysis of extensive field data, Ulaby et al. (1982;1983) have determined values for both $\sigma^o{}_v$ and the loss factor L of a number of different types of vegetation. These results are presented in Figure 16 where curves representing the data for several types of vegetation are shown along with the values for $\sigma^o{}_v$ and L. As expected, because of its structure, corn has the largest backscatter. They found that the scattering from the vegetation is dominant at low soil moisture values, below about 50 percent of field capacity, but above that level the observed σ^o is dominated by the soil contribution with the absorption by the vegetation being compensated by its backscatter.

Fung and Eom (1985) performed a theoretical study on the effect of vegetation on the sensitivity of active and passive microwave sensors for soil moisture sensing using a scattering model. They studied the reduction in sensitivity as a function of the scattering albedo (ω)

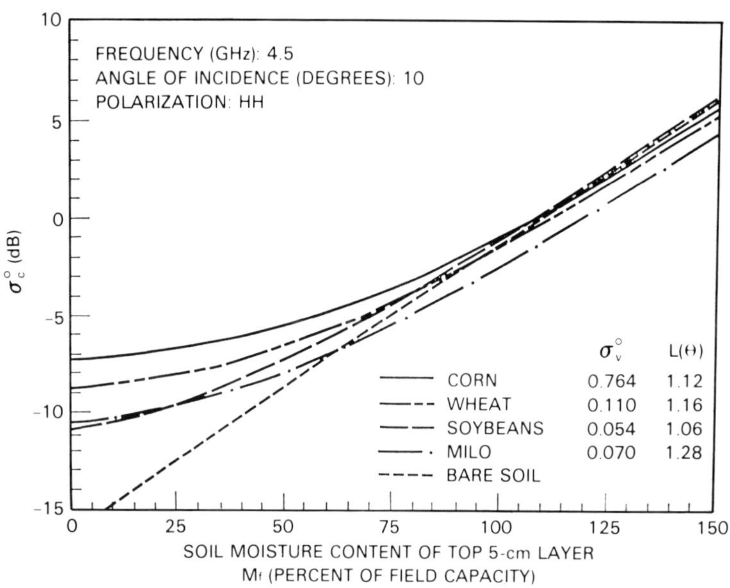

Figure 16. Regression results of trower measurements for L and $\sigma^o{}_v$ for 4 crops. The soil moisture values are expressed for a percent of field capacity (FC) as a means of normalizing for the different soils (Schmugge, 1980).

for the vegetation and the optical depth (τ). The results for the active case show that at angles of 10° or less the canopy backscatter is dominated by that from the soil, but as the angle increases the backscatter from the vegetation becomes dominant. Data indicate that at C-band wavelengths (5 to 6 cm) an assumed scattering albedo of 0.3 appears reasonable for several types of vegetation. For this value of ω the results show the least reduction is sensitivity as a function of optical depth for angles around 10°. But at angles of 20° or more there is considerably stronger dependence on both ω and τ due to the increased path length in the vegetation. The analysis for the passive case shows the expected exponential fall-off in sensitivity with the optical depth τ, but little dependence on angle or scattering albedo. Thus, if knowledge of the vegetation biomass is available, it may be easier to compensate for the effect of vegetation in the passive case. They found that for angles $> 20^{\circ}$ the sensitivity reduction in the active case is about the same as that for the passive and that at the near nadir incidence angles the active sensor has a smaller loss in sensitivity than for passive sensors of the same wavelength.

6.3 Spacecraft Results

The flights of microwave sensors on recent satellites, e.g. Skylab, Seasat, Nimbus 5, 6 and 7, and SIR A and B flights of the shuttle, have provided opportunities to do case studies on the remote sensing of soil moisture. The S-194 instrument on Skylab was a non-scanning 1.42 GHz radiometer with a 110 km field of view. With such coarse spatial resolution it is difficult to compare the sensor response directly with ground measurements. However, there have been several indirect comparisons. Figures 17 and 18 show comparisons of the Skylab and Nimbus 7 SMMR (Scanning Multichannel Microwave Radiometer) brightness temperatures observed for several passes over the state of Texas in the United States with Antecedent Precipitation Index (API) which is about the only moisture parameter available for comparison when retrospective analyses are performed (Wang, 1985). The error bars represent the standard deviation of the API values obtained for each resolution element. In Figure 17, S-194 data are presented for two regions of Texas: A in the semi-arid high plains area and B in the humid eastern part of the state which has considerable forest cover. This increase in vegetation cover causes the reduced slopes observed in the data for region B. Figure 18 contains the similar analyses for the two lower frequency channels of SMMR, 6.6 and 10.7 GHz, which have comparable resolution to the S-194 sensor. For the low vegetation region (A) the sensitivities are about the same as those for the 1.4 GHz data but for region B, with the higher vegetation, the sensitivity to soil moisture variations as represented by API values is almost gone. This illustrates the greater effectiveness of the lower frequencies for soil moisture sensing in the presence of vegetation.

The higher resolution of the SAR's on Seasat and the shuttle afforded the opportunity to compare satellite observations directly with ground measurements. Blanchard and Chang (1983) did this for Seasat

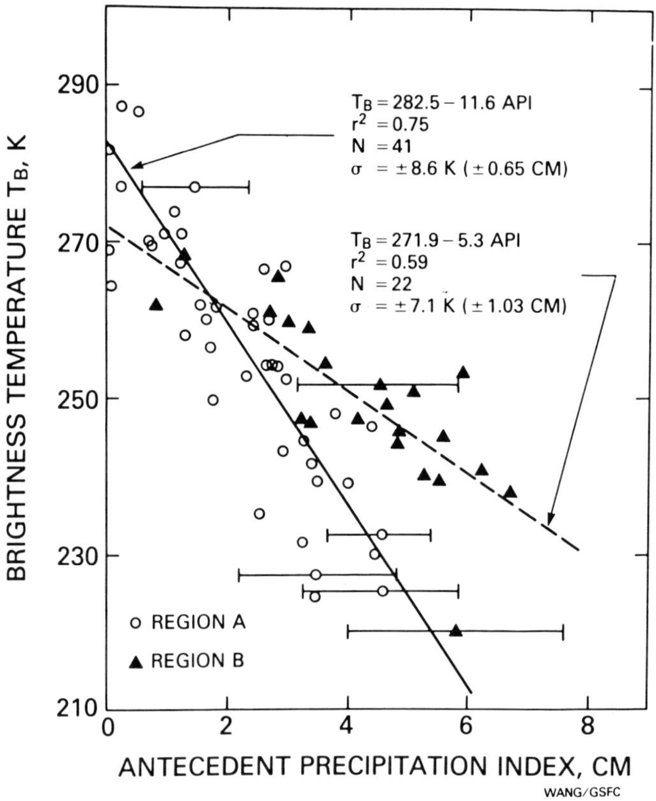

Figure 17. Satellite observations of the T_B at 1.4 GHz versus API over two regions in the state of Texas.

data over a test site in the Oklahoma panhandle. They compared the digitally processed Seasat σ^o data with ground measurements of soil moisture for bare, alfalfa and milo fields. They found a strong correlation of σ^o and soil moisture with an $r^2 = 0.7$. Working with the SIR-B data from October 1984 investigators working in California (Wang et al., 1986) and in Illinois (Dobson and Ulaby, 1986) found stronger correlations ($r^2 > 0.8$) for a variety of field conditions. However, all of these investigators found that there were such strong dependences of σ^o on surface features such as roughness and vegetation cover that it is difficult to extract soil moisture information from a single SAR observation. Most of these difficulties can be overcome by acquiring data at multiple dates, angles and/or frequencies to aid in the extraction of the soil moisture information.

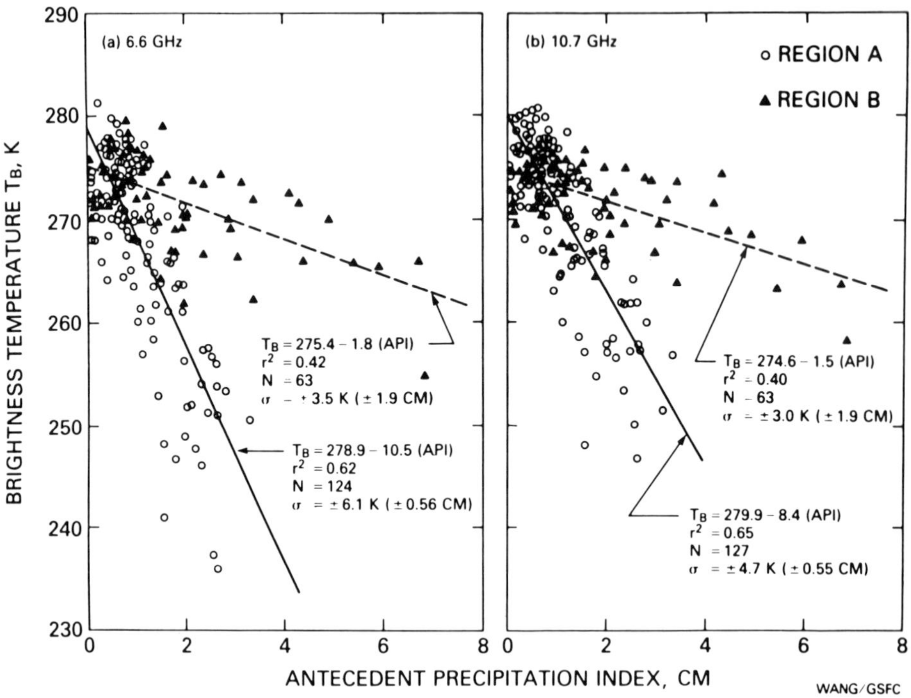

Figure 18. Satellite observations of T_B versus API over two regions in the state of Texas; a) SMMR results at 6.6 GHz and b) at 10.7 GHz.

7. DISCUSSION AND CONCLUSIONS

The results presented here are examples of the progress that has been made in improving our fundamental understanding of the use of microwave sensors for the remote sensing of soil moisture. The next step in the process should be a demonstration of the capabilities of these sensors for determining surface soil moisture. An example of this type of result is given in Figure 19. Here 1.4 GHz emissivities as measured from an aircraft platform were used to estimate the 0 to 2.5 cm soil moisture and compared with ground measurements for this layer. The data are from a series of 9 flights over an agricultural area in Hand County, South Dakota, in the United States. The algorithm used to extract the soil moisture values was developed using calculated emissivities for actual soil moisture profiles measured at the US Water Conservation Laboratory in Phoenix, Arizona. The resulting calculated emissivity versus soil moisture was adjusted for estimated surface roughness and vegetation effects and then applied to the observed 1.4 GHz emissivities to determine the soil moisture. The RMS differences between the observed and calculated values was 4 to 5 percent. In an analysis of the data from these flights, Owe et al. (1982) found the average value

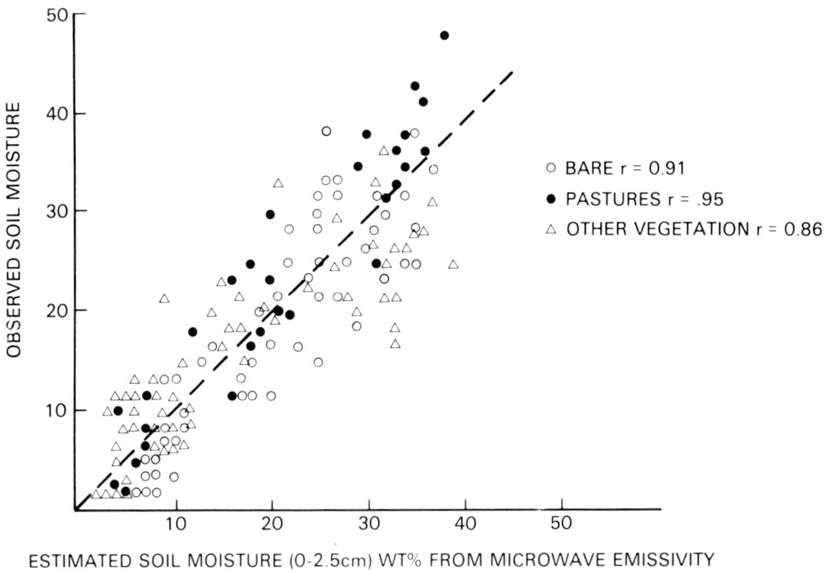

Figure 19. Comparison of measured and estimated values of soil moisture. The latter were derived from aircraft observations of microwave emissivity at 1.4 GHz.

of the coefficient variation (CV = standard deviation/mean) for the 0 - 2.5 cm layer was 0.25 with it being greater at the lower soil moisture values. Thus, the RMS difference observed in Figure 19 is comparable to the standard deviation observed in the ground measurements of soil moisture, particularly for the wetter cases. In another study, 1.4 GHz radiometer data were obtained over 2700 km^2 area in the Texas panhandle in an attempt to assess preplanting soil moisture conditions (Jackson et al., 1987). In this case land use and soil texture maps were used with the microwave emissivity to make a map of the estimated soil moisture for the area.

As described here, microwave remote sensing techniques can provide estimates of the soil moisture content for a surface layer about 5 cm thick. This depth is shallow compared to the 1 to 2 metre rooting depth of many crops. Estimating the root zone soil moisture from surface measurements has been studied using correlation techniques (Blanchard et al., 1981) and modelling studies (Jackson, 1980) which assume a moisture profile in hydraulic equilibrium. The conclusion from both approaches was that, if the water content of the surface 10 cm is known, the moisture content in the top metre could be calculated within acceptable limits, and that the lowest errors were obtained when the surface water contents were measured just before dawn.

Using a similar technique, Kondratyev et al. (1977) in the USSR obtained large area estimates of the pre-planting moisture stored in the top metre of the soil. In their study they required knowledge of the moisture gradient for each soil type in the test area. This gradient describes the moisture profile assuming hydraulic equilibrium conditions. Figure 20 is an example of their results which shows a high degree of variation in the soil moisture. In comparing the ground measurements with the aircraft results it should be recalled that the latter are estimates for approximately 10 km^2 while the former are point values.

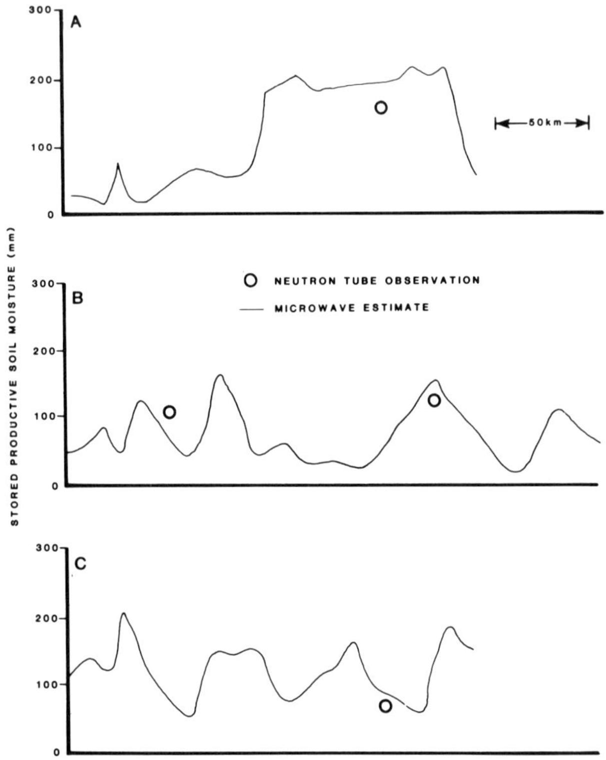

Figure 20. Soil moisture estimated from aircraft microwave observation: (a), (b), (c) three different flightlines. The curve is the estimate from the microwave approach and the open circles are from neutron soundings.

This approach has been implemented on an operational basis for irrigation scheduling, drainage, and structural surveys in at least one large district of the Moldavian SSR. Armand et al. (1981) report that aircraft radiometers are being used for water management in a 150,000 km^2 area. Data are collected once a week over each location and then processed and distributed within 24 hours. As an indication of the cost effectiveness of the remote sensing approach the authors claim that the cost is approximately \$0.40/km^2 using the remote sensing approach versus \$3/km^2 using conventional techniques.

Repetitive observations of the surface soil moisture can also be used to estimate soil properties. Camillo et al. (1986) coupled moisture and energy balance models with remotely sensed data of surface soil moisture to estimate the soil's hydraulic conductivity, matrix potential, and a soil texture parameter. The derived conductivities agreed with values measured with an infiltration ring. Potentially, this is a very useful technique for estimating the soil's average hydraulic properties over large areas, as has been demonstrated using a helicopter borne C-band scatterometer (Bernard et al., 1986).

Alternatively, knowledge of the surface layer moisture can be used to estimate moisture fluxes at the soil surface. These are of interest in themselves and can be used in moisture balance models to estimate the moisture in the profile. Barton (1978) in Australia used soil moistures as estimated from an airborne radiometer in a model to estimate the evapotranspiration from a grassland. Bernard et al. (1981) in France used a similar approach using simulated radar backscatter data. In follow-up papers they have verified the technique using field and airborne measurements with a C-band radar system (Prevot et al., 1984; Soares et al., 1987). Both groups reported considerable success in estimating ET rates.

The microwave remote sensing of soil moisture is at a threshold at the present time. Theoretical and experimental research over the past 10 to 15 years has pretty much defined the capabilities of both the active and passive microwave approaches. Briefly summarized they are: the ability to measure the moisture content of a surface layer about 5 cm thick to a relative accuracy of 10 and 15 percent; the measurement can be made under all weather conditions and through light to moderate vegetation conditions, the limiting case appears to be a mature corn crop; and the factors of surface roughness and soil texture will introduce uncertainties into the soil moisture determinations which can be mostly accounted for with ancillary data. Before these systems are flown on spacecraft it will be necessary to demonstrate the utility of such a remotely sensed soil moisture measurement with further experiments of the type described above. It is always important to realize that remote sensing measurements will not provide as accurate or as deep a measurement of soil moisture as can be obtained by conventional in-situ measurements, but they do provide a means for getting repetitive measurements over a large area of the surface soil moisture condition and thus provide a unique opportunity to obtain previously unattainable information about the land surface.

REFERENCES

Armand, N.A., Oleksich, V.N., Shinkaryuk, V.G. and Shutko, A.M., 1981: 'Remote determination of the moisture content of soils in the irrigated lands of Moldavia', Gidromekh. Melior. (NASA Tech. Transl.) 58-60.

Barton, I.J., 1978: 'A case study comparison of microwave radiometer measurements over bare and vegetated surface', J. Geophys. Res., 83, 3513-3517.

Bernard, R., Vauclin, M. and Vidal-Madjar, D., 1981: 'Possible use of active microwave remote sensing data for prediction of regional evaporation by numerical simulation of soil water movement in the unsaturated zone', Water Resources Res., 17, 1603-1610.
Bernard, R., Martin, Ph., Thony, J.L., Vauclin, M. and Vidal-Madjar, D., 1982: 'C-band radar for determining surface soil moisture', Rem. Sens. Environ., 12, 189-200.
Bernard, R., Soares, J.V. and Vidal-Madjar, D., 1986: 'Differential bare field drainage properties from airborne microwave observations', Water Resources Res., 22, 869-875.
Blanchard, B.J., McFarland, M.J., Schmugge, T.J. and Rhoades, E., 1981: 'Estimation of soil moisture with API algorithms and microwave emission', Water Resources Bull., 17, 767-774.
Caillo, P.J. and O'Neill, P.E., 1986: 'Estimating soil hydraulic parameters using passive microwave data', IEEE Trans. Geosci. Remote Sensing, GE-24, 930-936.
Choudhury, B.J., Schmugge, T.J. and Mo, T., 1982: 'A parameterization of effective soil temperature for microwave emission', J. Geophys. Res., 87, 1301-1304.
Colwell, R.N. (Ed.), 1983: Manual of Remote Sensing, 2nd ed., Am. Soc. of Photogrammetry, Falls Church, Virginia.
Dobson, M.C. and Ulaby, F.T., 1981: 'Microwave backscatter dependence on surface roughness, soil moisture, and soil texture: Part III, soil tension', IEEE Trans. Geosci. Remote Sensing, GE-19, 51-61.
Dobson, M.C., Ulaby, F.T., Hallikainen, M.T. and Reyes, M., 1985: 'Microwave dielectric behavior of wet soil: Part II, dielectric mixing models', IEEE Trans. Geosci. Remote Sensing, GE-23, 35-46.
Dobson, M.C. and Ulaby, F.T., 1986: 'Active microwave soil moisture research', IEEE Trans. Geosci. Remote Sensing, GE-24, 23-36.
Dobson, M.C. and Ulaby, F.T., 1986: 'Preliminary evaluation of the SIR-B response to soil moisture, surface roughness, and crop canopy cover', IEEE Trans. Geosci. Remote Sensing, GE-24, 517-526.
Fung, A.K. and Eom, H.J., 1985: 'A comparison between active and passive sensing of soil moisture from vegetated terrains', IEEE Trans. Geosci. Remote Sensing, GE-23, 768-775.
Jackson, T.J., 1980: 'Profile soil moisture from surface measurements', J. Irrigation Drainage Div., IR-2, ASCE, 81-92.
Jackson, T.J., Schmugge, T.J. and Wang, J.R., 1982: 'Passive microwave sensing of soil moisture under vegetation canopies', Water Resources Res., 18(4), 1137-1142.
Jackson, T.J. and O'Neill, P.E., 1986: 'Microwave dielectric model for aggregated soils', IEEE Trans. Geosci. Remote Sensing, GE-24, 920-929.
Jackson, T.J., Hawley, M.E. and O'Neill, P.E., 1987: 'Preplanting soil moisture using passive microwave sensors', Water Resources Bull., 23, 11-19.
Jackson, T.J. and O'Neill, P.E., 1987: 'Salinity effects on the microwave emission of soil', IEEE Trans. Geosci. Remote Sensing, GE-25, 214-220.
Kondratyev, K.Y., Malentyev, V.V., Rabinovich, Y. and Shulgina, E.M., 1977: 'Passive microwave remote sensing of soil moisture', in

Proc. 11th Int. Symp. Remote Sensing Environ., 2, 164-1661.

Mo, T., Coudhury, B.J., Schmugge, T.J., Wang, J.R. and Jackson, T.J., 1982: 'A model for microwave emission from vegetation covered fields', J. Geophys. Res., 87, 11229-11237.

Mo, T., Schmugge, T.J. and Wang, J.R., 19 : 'Calculation of the microwave brightness temperature of rough soil surfaces: bare field', IEEE Trans. Geosci. Remote Sensing, GE-25, 47-54.

Newton, R.W., Black, Q.R., Makanvand, S., Blanchard, A.J. and Jean, B.R., 1982: 'Soil moisture information and thermal microwave emission', IEEE Trans. Geosci. Remote Sensing, GE-20, 275-281.

Owe, M., Jones, E.B. and Schmugge, T.J., 1982: 'Soil moisture patterns observed in Hand County, South Dakota', Water Resources Bull., 18, 949-954.

Owe, M. and Schmugge, T.J., 1983: 'Microwave radiometer response to soil moisture at the 21-cm wavelength', presented at the Amer. Soc. Agronomy, Annual meeting (Washington, D.C.) 1983.

Prevot, L., Bernard, R., Taconet, O. and Vidal-Madjar, D., 1984: 'Evaporation from a bare soil evaluated from a soil water transfer model using remotely sensed surface soil moisture data', Water Resources Res., 20, 311-316.

Schmugge, T.J., 1980: 'Effect of soil texture on the microwave emission from soils', IEEE Trans. Geosci. Remote Sensing, GE-18, 353-361.

Schmugge, T.J., O'Neill, P.E. and Wang, J.R., 1986: 'Passive microwave soil moisture research', IEEE Trans. Geosci. Remote Sensing, GE-24, 12-22.

Soares, J.V., Bernard, R., Taconet, O., Vidal-Madjar, D. and Weill, A., 1987: 'Estimation of bare soil evaporation from airborne measurements', to be published in J. of Hydrology.

Theis, S.W., Blanchard, B.J. and Blanchard, A.J., 1986: 'Utilization of active microwave roughness measurements to improve passive microwave soil moisture estimates over bare fields', IEEE Trans. Geosci. Remote Sensing, GE-24, 334-339.

Theis, S.W., Blanchard, B.J. and Newton, R.W., 1984: 'Utilization of vegetation indices to improve microwave soil moisture estimates over agricultural lands', IEEE Trans. Geosci. Remote Sensing, GE-22, 490-496.

Ulaby, F.T., Batlivala, P.P. and Dobson, M.C., 1978: 'Microwave backscatter dependence on surface roughness, soil moisture, and soil texture: Part I, bare soil', IEEE Trans. Geosci. Electron., GE-16, 286-295.

Ulaby, F.T., Bradley, G.A. and Dobson, M.C., 1979: 'Microwave backscatter dependence on surface roughness, soil moisture, and soil texture: Part II, vegetation covered soil', IEEE Trans. Geosci. Electron., GE-17, 33-40.

Ulaby, F.T., Moore, R.K. and Fung, A.K., 1982-86: 'Microwave remote sensing: Volumes I, II and III', Artech House Inc., Dedham, Massachusetts.

Ulaby, F.T., Aslam, A. and Dobson, M.C., 1982: 'Effects of vegetation on the radar sensitivity to soil moisture', IEEE Trans. Geosci. Remote Sensing, GE-20, 476-481.

Wang, J.R. and Schmugge, T.J., 1980: 'An empirical model for the complex dielectric permittivity of soils as a function of water content', IEEE Trans. Geosci. Remote Sensing, **GE-18**, 288-295.

Wang, J.R., 1985: 'The effect of vegetation on soil moisture sensing from orbiting microwave radiometers', Remote Sensing Environ., **11**, 141-151.

Wang, J.R., Engman, E.T., Shuie, J.C., Rusek, M. and Steinmeier, C., 1986: 'The SIR-B observations of microwave backscatter dependence on soil moisture, surface roughness, and vegetation covers', IEEE Trans. Geosci. Remote Sensing, **GE-24**, 510-516.

Wang, J.R., 1987: 'Passive microwave sensing of soil moisture: the frequency dependence of microwave penetration depth', to be published in IEEE Trans. Geosci. Remote Sensing, **GE-25**.

Wilheit, T.T., 1978: 'Radiative transfer in a plane stratified dielectric', IEEE Trans. Geosci. Electronics, **GE-16**, 138-143.

THE VEGETATION INDEX AND THE STUDY OF VEGETATION DYNAMICS

J.P. Malingreau
Commission of the European Communities
Joint Research Centre - Ispra Establishment
21020 Ispra (Va) - Italy

ABSTRACT. The presentation reviews the basic principles supporting the derivation and analysis of vegetation indices from spectral data. The relationships between the normalized difference vegetation index and the vegetation characteristics is discussed, with emphasis upon dynamic processes important in biomass accumulation. Since time-series of vegetation index data are mainly available at low resolutions, problems related to the time and space dimensions of satellite data are given special attention. Time-series of NOAA AVHRR data are used for illustrating the characteristics of the vegetation index for major ecosystems. Temporal development curves are interpreted in terms of seasonality, agricultural practices and plant growth. Major research needs are identified.

1. THE PHYSICAL BASIS FOR DERIVING VEGETATION INDICES, DEFINITION AND OBJECTIVES

1.1 The development and use of vegetation index analysis has received much attention in the last few years and there is to-day little in vegetation monitoring which does not involve the use of one or more of such spectral measurements. What are vegetation indices, how are they derived, how do they relate to the plant and canopy characteristics, how are they used? are a few questions which will be answered here. There is a growing body of research literature devoted to the subject which can be usefully consulted for additional information on this very relevant subject in the field of remote sensing.

1.2 Vegetation index analysis must be based on a good knowledge of the spectral properties of vegetation as well as of the relationships between such spectral properties and plant growth and development. The very mechanism of plant production - the reaction process of photosynthesis - is driven by light. It is the leaf structure which regulates the contact between the atmosphere (the source of CO_2) and the mesophyll cells which contain the water required for the reaction process to take place. These plant organs thus maintain an optical environment

F. Toselli (ed.), Applications of Remote Sensing to Agrometeorology, 285–303.

in which the photosynthetically active radiation (PAR: 0.4 - 0.7 μm) can interact with the chloroplast-containing cells which are the site of photosynthesis.

When incident solar radiation interacts with green leaves, a high level of absorption takes place in the 0.4 - 0.7 μm region, while in the 0.7 - 1.3 μm there is high reflectance and transmittance. The first phenomenon is due to the presence of active chloroplasts while the second is explained by the scattering of radiation caused by the leaf structure (Allen et al., 1973). At the level of plant canopies, spectral absorption and reflection determine the spectral radiance of the soil-plant complex and this parameter provides the basis for passive remote sensing. The objective is thus to derive information on the nature and state of the vegetation by using specific wavelengths selected to provide a strong signal from the vegetation and a good spectral contrast with the background. Figure 1 shows several regions of the spectrum where such parameters can be evaluated (Tucker, 1978).

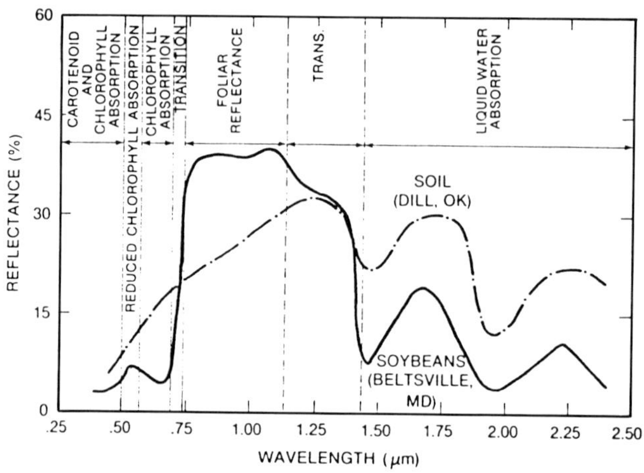

Figure 1. Delineation of the 0.4 - 2.5 μm region into spectral intervals where different biophysical properties of green vegetation control the reflectance of incident solar irradiance from the vegetation in question. Sample spectral reflectance curves for green vegetation and soil are also included to illustrate why some wavelengths have greater spectral contrasts than others (from Tucker, 1978).

One must keep in mind that the usefulness of a particular wavelength region can be slightly modified by the type of soil, the density of the canopy and the geometry of the sun viewing angle configurations, all conditions not always easy to control in a "real situation" (see Table I).

TABLE I

Parameters and relationships involved in the determination of crop directional reflectance. References: 1-NAS, 1970; 2-Rao et al., 1979; 3-Colwell, 1974; 4-Colwell, 1974; 5-Sults, 1972a; 6-Sults, 1972b; 7-Sults et al., 1972; 8-Tucker, 1977; 9-Tucker et al., 1977; 10-Tucker et al., 1979; 11-Kauth et al., 1976; 12-Gausman et al., 1973a; 13-Gausman et al., 1973b; 14-Gausman et al.,1976; 15-LeMaster, 1979; 16-Egbert et al., 1980; 17-Richardson et al., 1977; 18-Jackson et al., 1980.

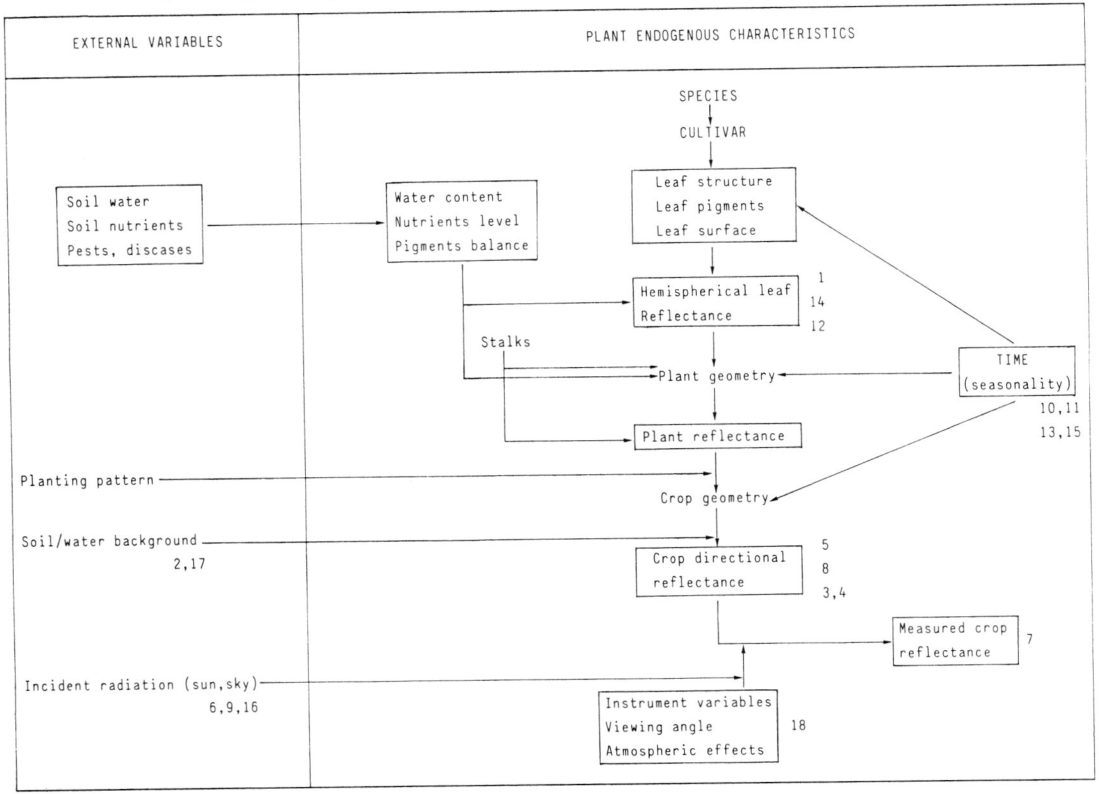

1.3 Vegetation indices are particular combinations of spectral responses in different wavelength bands which emphasize a particular feature of the vegetation.

The development and use of vegetation indices is guided by three general objectives:

- To enhance, through an appropriate combination of spectral bands, the relevant vegetation features. Ideally, the indices should have a better defined relationship with physiological properties of the crop than individual spectral measurements.
- To standardize the representation of crop spectral response (useful in region to region or year to year comparisons).
- To reduce the dimensions of the data sets (i.e. an index reduces to one data set the n sets which contribute to its calculation).

It is also worth noting that ratio indices have the added advantage of being dimensionless; this renders the calibration of the radiance values to reference standards unnecessary (calibration of the index with respect to ground features remains, of course, necessary).

Table II presents some of the most commonly analyzed vegetation indices found in the literature. Only the normalized difference vegetation index (NDVI) (Deering et al., 1975) and the Perpendicular Vegetation Index (PVI) will be discussed here. These two mathematical combinations of spectral band radiances have been the focus of most research on the topic.

2. THE NORMALIZED DIFFERENCE VEGETATION INDEX

2.1 The NDVI is defined as the ratio of the difference between the near infrared and red reflectances to their sum or

$$\text{NDVI} = (\text{near infrared} - \text{red})/(\text{near infrared} + \text{red})$$

in LANDSAT MSS terms: NDVI = (BD7-BD5)/(BD7+BD5)
in LANDSAT TM terms: NDVI = (BD4-BD3)/(BD4+BD3)
in AVHRR terms: NDVI = (Ch2-Ch1)/(Ch2+Ch1).

The index is called normalized because it is divided by the sum of radiances and thus normalizes somewhat for differences in solar spectral irradiances. The NDVI is dimensionless and can take values from 0 to 1; negative values are sometimes found when the red reflectance is higher than the near infrared as for certain types of dry soils. The value of the NDVI is mainly determined by the difference between the near infrared response which increases with increasing vegetation in the scene, and the red response which decreases with decreasing vegetation.

2.2 Field measurements have established that there is a relationship between green biomass and the NDVI up to a certain value of the cover biomass above which the NDVI remains constant (saturation level).

TABLE II
Vegetation indexes used in remote sensing analysis

Definition (G = green band, R = red band IR = near-infrared band)	Name[1]
IR/R	7 to 5 ratio
R/G	5 to 4 ratio
$\frac{IR - R}{IR + R}$	normalized infrared/red difference
$\left(\frac{IR - R}{IR + R} + 0.5\right)^{1/2}$	transformed normalized difference
$\frac{G}{IR * R}$	
$\frac{R * IR}{G}$	
$\sqrt{(R_v - R_s)^2 + (IR_v - IR_s)^2}$	perpendicular vegetation index index (s = soil, v = vegetation)
	Tasselled cap transforms:[2]
0.332MSS4 + 0.603MSS5 + 0.656MSS6 + 0.263MSS7	- brightness
- 0.283MSS4 - 0.660MSS5 + 0.577MSS6 + 0.388MSS7	- greenness
- 0.899MSS4 + 0.428MSS5 + 0.0759MSS6 - 0.481MSS7	- yellow stuff
- 0.016MSS4 + 0.131MSS5 - 0.452MSS6 + 0.882MSS7	- nonesuch
GREENNESS - 0.9178 BRIGHTNESS + 5.58959	GRABS
$GREENNESS_v - GREENNESS_s$	Green number (v = vegetation, s = soil)

(1) names sometimes directly refer to the LANDSAT MSS bands.
(2) coefficients are for LANDSAT II MSS data.

A relationship of the type

$$LAI = a\ e^{b.NDVI} + c$$

(LAI = Leaf Area Index, m^2/m^2) has, for example, been established for wheat and it appears that the saturation level is reached for LAI of 2-3. A modelling of the radiative transfer in canopies of increasing thickness leads to similar observation by Sellers (1986). This author concludes that the NDVI "is an insensitive measure of the LAI and/or biomass when:

1) the LAI is > 2 or 3;
2) there are patches of bare ground in the field of view;
3) there is an unknown quantity of dead material in the canopy;
4) the leaf angle distribution is unknown and the solar elevation is high."

Such restrictions seem to indicate that LAI or biomass is not exactly the primary parameter to be investigated using the vegetation index approach (Figure 2).

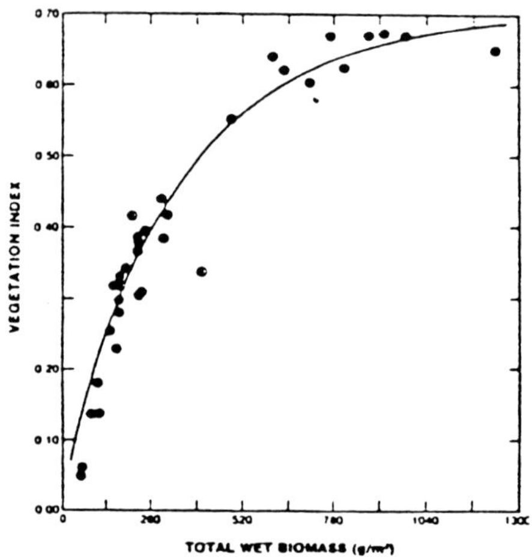

(g) $r^2 = 0.89$
$Y = 0.70 - \exp(-0.4207 - 0.0030 \cdot X)$

Figure 2. Relationship between vegetation index and total wet biomass (from Tucker, 1979).

2.3 Given the radiative transfer in the plant canopy (see introduction), relationship between the measured spectral reflectance and the energy absorbed for photosynthesis should be measurable. This has been investigated by several authors who have found a dynamic relationship between the IPAR (intercepted photosynthetically active radiation) and the vegetation index (Figure 3) (Wiegland and Richardson, 1983; Hatfield et al., 1984; Steven et al., 1983; Sellers, 1986). The relationship is dynamic in the sense that it evolves throughout the vegetation development season.

More recent work (Sellers, 1986; Tucker and Sellers, 1986) indicates that the reflectance data in the spectral band forming the NDVI provide indications on biophysical parameters (rates) such as primary productivity (via the local density of the chlorophyll and the photosynthetic capacity of the canopy) and evapotranspiration (via a variable stomatal resistance to gas and water vapour transfer). The mechanisms whereby these processes influence the reflectance properties of the canopy are detailed in the literature cited above. It can be

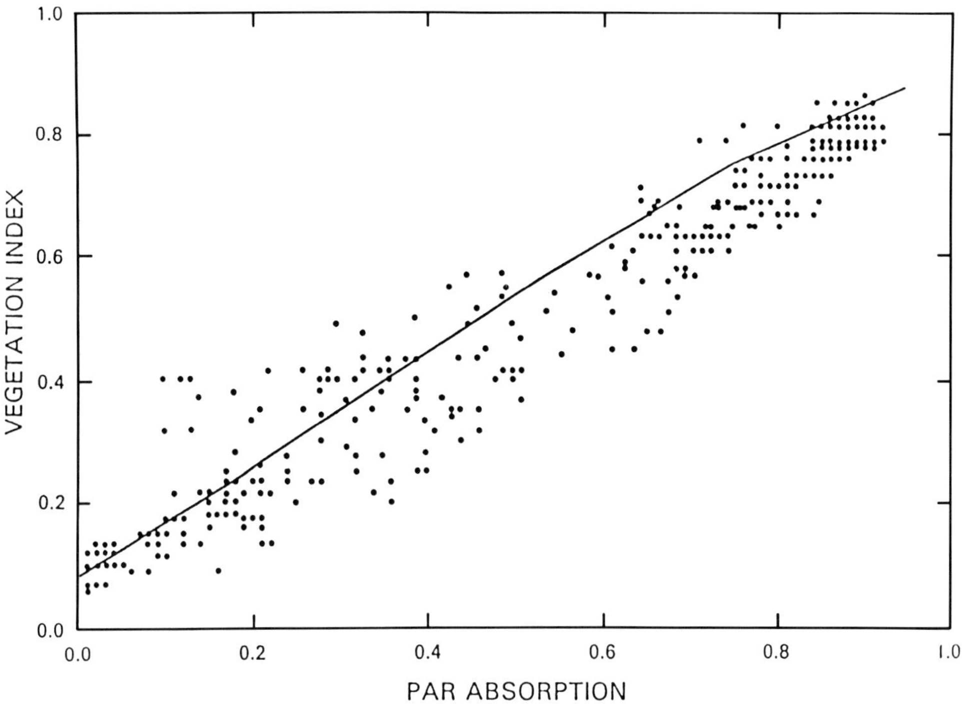

Figure 3. Relationship between vegetation index and absorbed photosynthetically active radiation (PAR) (simulation; Sellers, 1986).

derived from these investigations that the radiance properties of the vegetation provide more information about processes (i.e. photosynthesis, respiration) than about the state of the surface (i.e. leaf area index or biomass). This important conclusion leads to an examination of remote sensing data in a dynamic perspective. As they are related to biophysical processes taking place in the canopy, the vegetation indices are to be analyzed in such a dynamic framework. A time-duration dimension has to be added in the approach.

2.4 When the NDVI is plotted during the development cycle of the crop, a bell-shaped curve is usually obtained. In general terms, the growing part of the curve represents the vegetative phase, the plateau represents the transition between the vegetative and reproductive phase and the decreasing part of the curve represents the progressive senescence of the crop. The NDVI of a canopy results from the particular dynamics of the canopy in terms of greening and senescence of the leaves, as shown in Figure 4 (Bartholomé, 1987). It is important to note that such dynamics can be complex and that it is the "mixed" signal which is being analysed (! low resolution pixels).

It is useful to introduce here the concept of the integrated vegetation index. As the NDVI measurement is representative of a rate of

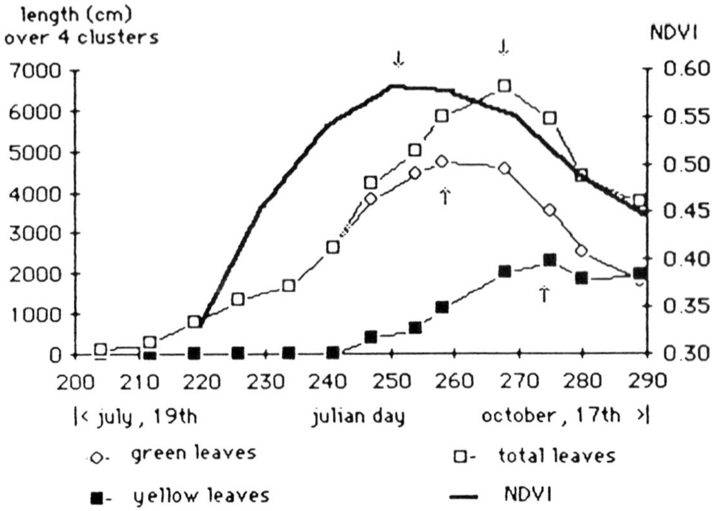

Figure 4. NDVI development for sorghum plots in Mali (Bartholomé, 1987).

photosynthetic plant activity, its integration over time should tell us more about the productive history of the plant and by inference about production, the object of our analysis. A few additional parameters are first defined (see Figure 5).

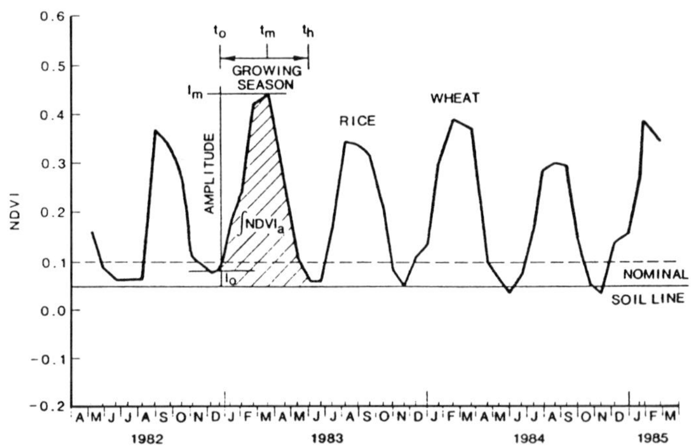

Figure 5. NDVI development curve for the wheat-rice cropping system in the Punjab (India) (GVI data) (from Malingreau, 1986).

2.5 The NDVI evolution in time can be characterized using the following:

- Base NDVI (I_o) is the "soil line" over which the vegetation grows. This value usually lies between NDVI = 0 and 0.05. As we will see in further examples, the base can vary.
- NDVI values will reach a peak which corresponds to peak photosynthetic activity and, as a rule, the end of the vegetative period (booting in cereals). This is the NDVI max (I_m).
- $I_m - I_o$ is the amplitude of the variation for a given vegetation type and a given season (green wave ..!).
- The temporal resolution (x axis) depends on the availability of data and the dynamics of crop growth. As a rule, the temporal resolution should be such that no significant vegetation-related event occurs in the non-measurement period. From a practical point of view, the temporal resolution should be such that no undue burden is placed upon the data processing system. Weather conditions often prevent the choice of an ideal temporal resolution and one should be pragmatic in this respect (see discussion on the preparation of temporal series).
- The growth cycle "apparent" starting time t_o (this does not necessarily represent the emergence of the crop as ground resolution limitations may prevent its early detection).
- t_m: time of maximum NDVI.
- t_h: time of return to soil line.
- The integrated vegetation index is the area between the base line and the NDVI dev. curve. Often, quantitative expressions of the curve are not available and one must approximate this integral

$$\int \text{NDVI} = \int_{t_1}^{t_2} \text{NDVI} \cdot dt$$

or

$$\int \text{NDVI} \cong \text{NDVI} \cdot \text{days} = \sum_{t_1}^{t_2} (\text{NDVI} \cdot n)_i$$

where n = the number of days in each period covered by a NDVI measurement (= temporal resolution).

This last relationship can be measured for the area above 0 or above the soil line (adjusted NDVI). Integration can also be effected over specific crop stages or specific values of the NDVI. For example, the grain formation period is vital for the final yield of cereals and should thus be examined carefully.

- Speed of NDVI development = dNDVI/dt.

2.6 Various studies have attempted to relate the performance of a crop to the vegetation index. Two approaches can be distinguished: one uses the $NDVI_i$ at a specific time of the crop development, the other one relates the temporal evolution of the NDVI (NDVI) to the crop performance.

Barnett and Thomson (1982) obtained a good correlation between the NDVI measured at heading and the final yield for grain in the Great Plains of the US. With other authors, they confirm that a relationship of such kind is better when the $NDVI_i$ is measured at the time of maximum vegetation index (around heading). However, this rule holds better for varieties which show reduced vegetation development because of saturation problems in high above-ground biomass conditions (Hatfield, 1981).

2.7 The integration of the vegetation index is more in tune with a dynamic perspective on vegetation development and production. At a very general level, first - as shown in Figure 6 - one can immediately see the effect of a climatic event such as a drought on the

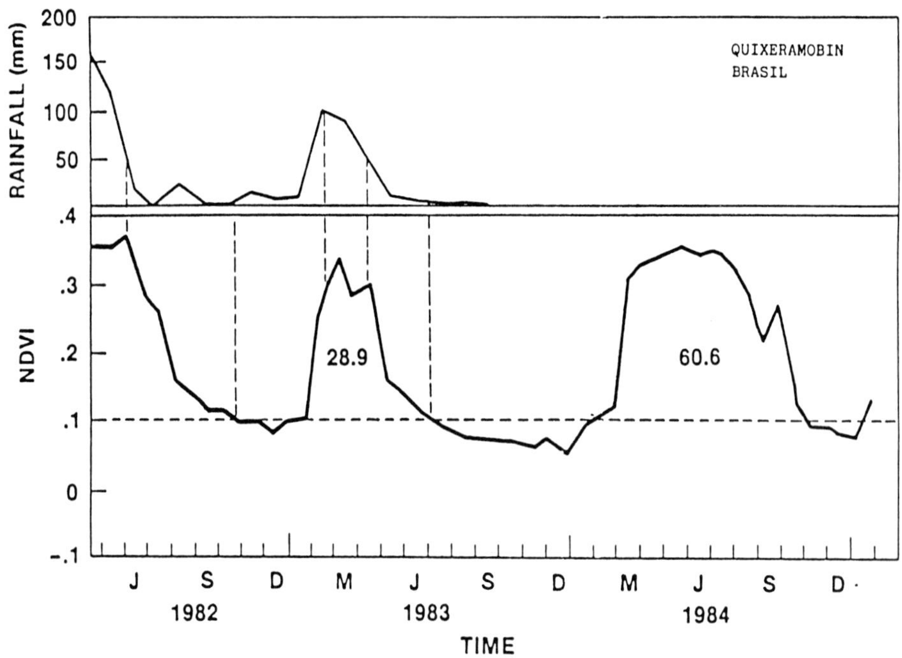

Figure 6. NDVI temporal development curve for the caatinga of Brazilian Nordeste. Data are derived from the Global Vegetation Index data set (15 km resolution); temporal resolution is 3 weeks. The drought of 1983 is well marked by the integrated NDVI values for the growing seasons (from Malingreau and Setzer, 1988).

vegetation index development of natural vegetation (Malingreau and Setzer, 1988). The graph depicts the impact of the famous 1983 drought on the caatinga vegetation of the Brasilian nordeste. The integrated index is reduced by half in 1983 as compared to 1984. Note that the maximum vegetation index for the 1983 growing season is comparable to that of the other years.

In detailed studies over a whole range of cereals, various authors have established the validity of the relationship between the integrated vegetation index and final grain yield (Pinter et al., 1981; Hatfield, 1984). These investigators all point to the sensitivity of the approach to the choice of integration period; early flowering (for 50% of plants in plot) to senescence appear to provide the best results. This requires, of course, the precise determination of the phenological stage of the crop. Field information is necessary but it appears that the NDVI development curve could give some clues in this respect. Hatfield (1983), for example, indicates that for sorghum, the $NDVI_{max}$ is reached at the end of elongation - early panicle appearance and that when the NDVI reaches $NDVI_{max}/2$, 90% of grain dry matter has been accumulated. Such results are crop and site specific but the approach opens many new avenues of investigation which are relevant to the determination of significant phenological stages by remote sensing techniques.

2.8 Finally, one should not forget that if for some crops, the biomass represents the production, many food crops are grain-crops; the measurement of green matter production is only half of the story. In this case, it is further necessary to establish a relationship between biomass and useable yield; such a relationship is, of course, variety specific (i.e. traditional varieties of rice with high biomass and low yield versus improved varieties with low biomass of erected leaves and high yield). Many alternatives are possible in such an investigation. One can assess the efficiency of the vegetative phase using the integrated NDVI during the first phase of the development curve and then adjust the final yield through an analysis of the evolution of the NDVI thereafter (Hatfield et al., 1983). A regional potential yield determined for a region during the vegetative development thus serves as a reference which is progressively adjusted as the crop progresses. In a more logical way, the integration of the NDVI during the reproductive phase - the true grain producing phase - can be linked with the final yield (Pinter et al., 1981). Successful attempts to relate the speed of development (dNDVI/dt) to grain production have also been reported (the period selected is the second part of the NDVI development curve - slope < 0 - in the Idso et al. (1980) and Baret and Guyot (1986) studies).

2.9 Of much interest for the development of operational approaches to remote sensing monitoring of vegetation is the modelling of spectral development curves. Various attempts have been made to develop mathematical expressions of vegetation development curves. Badwar (1985) pioneered such work and proposed a logistic curve of the type:

$$VI = VI_o + (VI_m - VI_o) A (t-t_o)^{\alpha} e^{[-B(t-t_o)^2]}$$

where VI_o and VI_m are the basal and maximum vegetation index and $(t-t_o)$ a relative time scale.

Many other functions have been developed (i.e. exponential expressions). Baret (1986) decomposes the radiometric evolution of a wheat crop into three phases to which he applies different models corresponding to specific physiological changes in the canopy. The mathematical treatment of vegetation index development curves presents several advantages:

- it yields a smoother curve than that provided by actual spectral measurements; as we will see in the next section, these latter are often affected by external factors;
- it permits interpolations and thus a reduced temporal frequency of measurements;
- it leads to a definition - for a crop and a region - of a "normal" development curve to which actual data can be compared (detection of anomalies, early warning, relative yield approaches, etc.).

2.10 A complete review of the growing body of research on this topic is beyond the scope of this presentation; however, the above discussion indicates that complete series of vegetation index data provide a unique material for analysis. Experimental as well as applied research efforts are being continued in this field; the analysis of the vegetation index in relation to other environmental/crop characteristics such as surface temperature appears especially promising (see section 4). All these approaches rely, however, upon the availability of adequate time series of data.

3. ESTABLISHING TIME SERIES OF REMOTE SENSED VEGETATION INDEX DATA

3.1 One can view the vegetation index analysis from a more pragmatic point of view than that briefly described above. The alternative approach relies upon the establishment of time series of NDVI data and the empirical determination of the relationships between their variations and changes taking place in the vegetation canopy. This is usually the first step adopted in a remote sensing operation and given the numerous difficulties associated with the operation, it is of interest to the "practitioner" to assess how such time series can be established.

For various reasons, adequate time series of high resolution satellite data are often not available. The frequency of observation provided by satellites such as LANDSAT and SPOT (16 and 23 days in normal operation conditions) usually does not allow us to overcome the cloud constraints which are a major nuisance in the temporal approach. Furthermore, even if high resolution - high frequency data were available, limitations in terms of data processing would quickly appear. In this time of satellite remote sensing commercialization, the cost is another element to be taken into consideration. From an operational point of view, an element of reliability must always be introduced in the constraint equation. Indeed, the purpose of monitoring is to ensure the delivery of adequate information, whatever the environmental or operational conditions. A satellite-based early warning system, for example, should not become paralyzed because cloud cover conditions are less than optimal or because the data load exceeds the capacity of the system. In some cases, a choice may sometimes have to be made between a sophisticated but less reliable system and a rough - even antiquated but robust one. Striking the right balance between all the factors to be taken into consideration is not an easy task.

We will now examine how the NOAA satellite sensor platform can satisfy current needs in terms of vegetation assessment.

3.2 The Advanced Very High Resolution Radiometer (AVHRR) system was primarily designed for meteorological applications and it is only recently that its usefulness as an instrument of vegetation surveillance has been demonstrated. The characteristics of that system are summarized in Table III.

TABLE III
Characteristics and status of the NOAA/AVHRR system

TIROS-N,	launched October 1978, NASA funded protofight
NOAA-6,	launched June 1979, NOAA funded
NOAA-7,	launched June 1971, NOAA funded
NOAA-8,	launched March 1983, NOAA funded
NOAA-9,	launched December 1984, NOAA funded
Coverage cycle	: 9 days
Scan angle range	: ± 55·4°
Ground coverage	: 2700 km
Orbit inclination	: 98·8°
Orbital height	: 833 km
Orbital period	: 102 min
IFOV	: 1·39-1·51 mrad
Ground resolution	: 1·1 km (nadir); 2·4 km (max. off-angle along track); 6·9 km (max. off-angle cross track)
Quantization	: 10 bit
Equatorial crossing:	Des. Asc.
	07.30 19.30 (NOAA-6 and NOAA-8)
	14.30 02.30 (NOAA-7 and NOAA-9)
Spectral channels :	1 2 3 4 5
Spectral range (μm):	0·58-0·68+ 0·725-1·1 3·55-3·93 10·3-11·3 11·5-12·5‡
	Status as of January 1985
NOAA-6	: Taken out of operational service 5 March 1983. Reinstated 22 June 1984.
NOAA-7	: Operational
NOAA-8	: Taken out of operational service 12 June 1984.
NOAA-9	: Operational status, January 1985

+ Channel 1 range on TIROS-N: 0·55-0·90; ‡ Not on NOAA-6 or NOAA-9.

A few of those characteristics are worth noting:

- Thanks to its very large scanning angle, the AVHRR can cover every spot on the earth nearly every day. The same characteristic leads, however, to large geometric distortions and angular effects which have to be corrected.

- The ground resolution of the instrument is 1:1 km at nadir. At such a coarse level, one deals with an areal integration of the reflectance associated with the wide range of features which can be present in the field of view. Each of those constituting features may have its own dynamic and spectral properties.

- Through sampling, a range of resolution formats are available:
  . +1.1 km in HRPT (High Resolution Picture Transmission) format received by local stations in sight of the satellite;
  .   in LAC (Local Area Coverage) format recorded onboard for only 4' or about 4000 km along track; available from NOAA on request only;
  . +4 km GAC (Global Area Coverage) data available daily for the whole world (archived since 1981);
  . +15 km GVI (Global Vegetation Index) data; normalized difference vegetation index produced weekly since April 1982.

- Time series of AVHRR data are available under different formats since 1979.

- The spectral bands are appropriate for the analysis of vegetation (CH1,CH2); thermal channels allow a measurement of surface temperature (CH3,4,5).

3.3 Temporal composition: atmospheric conditions and system geometry affect the measurement of the vegetation index in ways which are not always easy to cope with. Various geometric and radiometric correction procedures have been devised and intensive research is under way in this field. Corrections are fundamental to temporal vegetation index analysis; comparisons from one date to another are indeed not possible until a standardization of the signal environment is achieved. The topic is wide and falls outside the scope of this presentation.

The empirical approach adopted in NDVI analysis is called the "maximum value composite" (Holben, 1986). This approach relies on the observation that any atmospheric or angular disturbance will depress the NDVI; a procedure which retains the maximum value over a series of temporal images will thus yield the "best" or "cleanest" vegetation index for that period. A number of acquisitions is thus combined over the selected period (i.e. weekly, 10-day or three-weekly) and for each pixel location, the maximum NDVI is retained (see Figure 7). This procedure is often associated with a thermal cloud mask; this mask eliminates the NDVI data which have a temperature below a defined threshold (i.e. 2850 K).

Experience has shown that the maximization approach may yield series of data of acceptable quality. Errors can still be made if in the selected period the "cleanest" image is still affected by undetected atmospheric contamination (dust, hase).

An examination of the smoothness of the NDVI temporal development curve allows an evaluation of the effectiveness of the method.

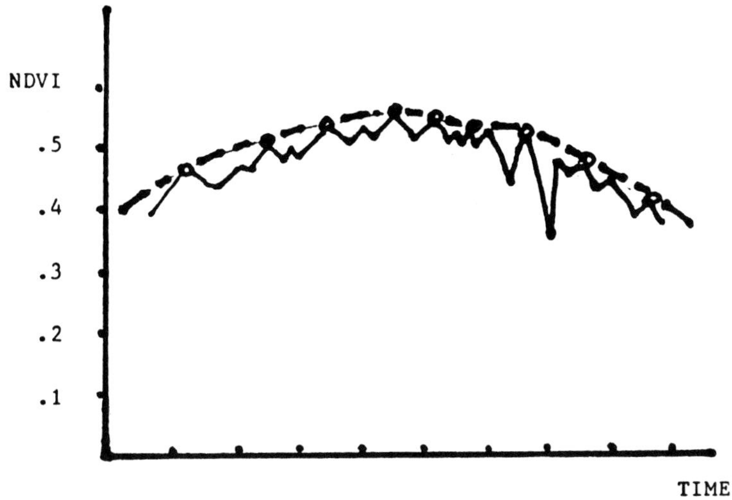

Figure 7. Graphic representation of the NDVI maximization procedure. The solid line represents the original NDVI development curve. The dashed line links all the maximum NDVI values over longer time periods. The application of this procedure yields a smoother curve whose variations are more likely to reflect the dynamics of vegetation than the influence of extraneous factors.

3.4 Vegetation index images: despite the fact that the objective of vegetation monitoring is not primarily the production of images, a pictorial representation of the distribution of NDVI pixels is useful for assessing spatial patterns and establishing relationships with environmental factors. As in any representation of remote sensing images, a colour code must be assigned to intervals of NDVI values. Illustrations of a wide range of AVHRR NDVI products shown in the course use the colour code developed by the NASA GIMMS project.

3.5 Temporal series of vegetation index: examples will be shown of NDVI development curves over selected ecosystems. Their interpretation will be made in terms of crop growth, agricultural practices and seasonality (see also Malingreau, 1986; Goward et al., 1986; Justice et al., 1986).

3.6 AVHRR data and ecosystem monitoring: one should not be under the impression that the proposed monitoring approach can stand on its own. It must be combined with other aspects of vegetation monitoring. The following flow chart represents various elements/approaches which should be considered in the framework of crop monitoring (Figure 8).

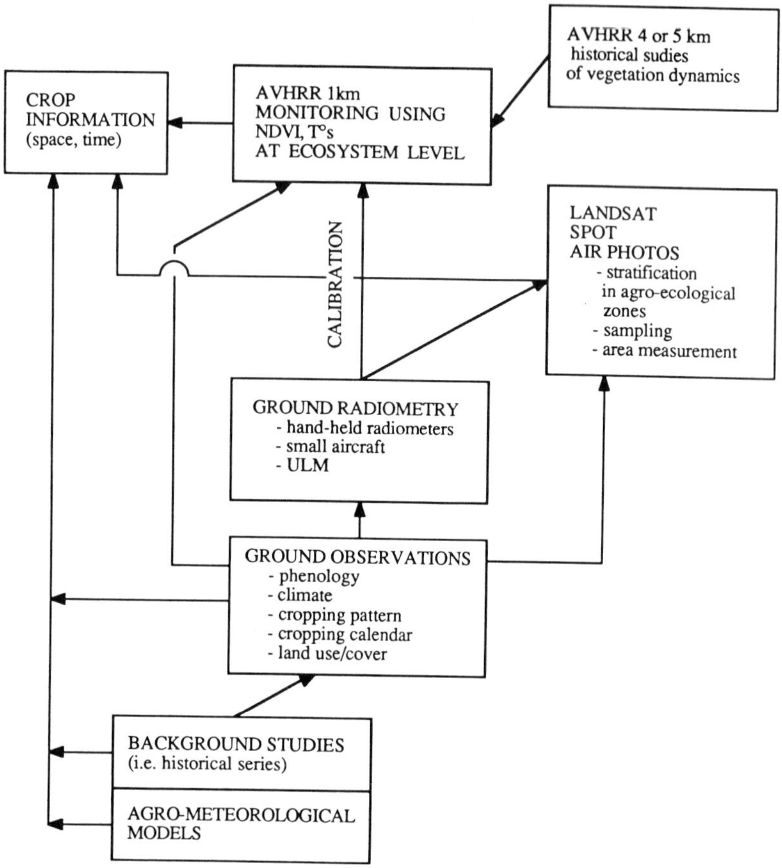

Figure 8. Remote sensing and crop monitoring.

4. THE PERPENDICULAR VEGETATION INDEX

Since the NDVI is not independent of the substrate, it may present disadvantages in situations of very low vegetation. An alternative approach has been developed which is based upon an index called the Perpendicular Vegetation Index (PVI).

A soil line is first established in the near infrared/red plane for a given soil type. The presence of vegetation causes the points to depart from the local soil line. The magnitude of the displacement with respect to the soil line has been shown to be related to the amount of growing vegetation (Richardson and Wiegand, 1977).

The PVI is calculated as follows:

$$PVI = [(IR\ soil - IR\ veg)^2 + (R\ soil - R\ veg)^2]^{1/2}$$

where IR soil, IR veg, R soil, R veg are the coordinates of the vegetation and soil points, respectively.

Trajectory PVI curves (Figure 9) represent the evolution of the IR-R reflectance throughout the growth cycle. Alternative pathways which could be taken by vegetation affected by "abnormal" circumstances can be modelled (Figure 10).

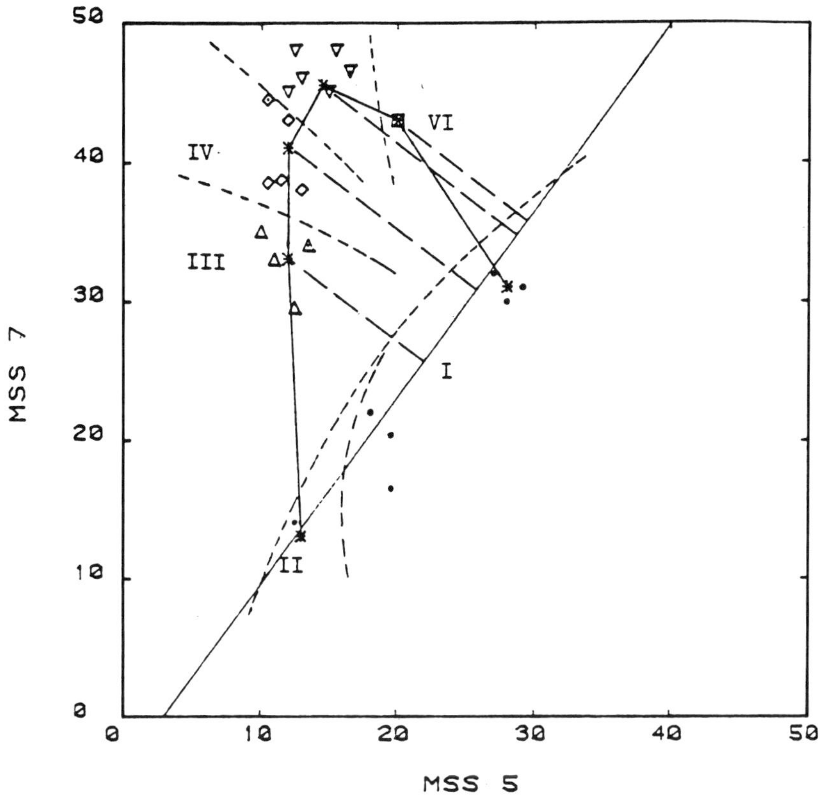

Figure 9. Trajectory curve for wetland rice development in the LANDSAT MSS infrared versus red bands. The plane is divided into six regions: I - dry soil, stubble; II - flooded paddy; III - young crop; IV - full cover; V - maturing crop; VI - ripe, harvest (Malingreau, 1980).

REFERENCES

Badhwar, G.D. and Henderson, H.K., 1985: 'Application of thematic mapper data to corn and soy-bean development stage estimation', Remote Sensing of Env., 17, 197-201.

Baret, F., 1986: 'Contribution au suivi radiométrique des cultures de

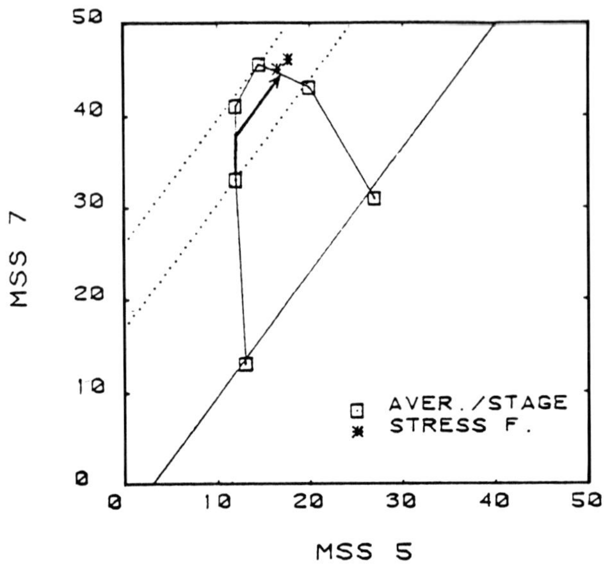

Figure 10. Wetland rice trajectory curve with hypothetical stress pathway (arrow).

céréales', Thèse, Univ. Paris Sud.

Baret, F. and Guyot, G., 1986: 'Suivi de la maturation de couverts de blé par radiométrie dans les domaines visible et proche-infrarouge', Agronomie, 6.

Barnett, T.L. and Thomson, D.R., 1983: 'Large area relation of Landsat MSS and NOAA-6 AVHRR spectral data to wheat yields', Rem. Sensing of Env., 13, 277-290.

Bartholomé, E., 1987: 'Radiometric measurements and crop yield forecasting: some observations over millet and sorghum experimental plots in Mali', Earsel Symp. Noordwijkerhout.

Deering, D., Rouse, J.W., Haase, R.H. and Schell, H.H., 1985: 'Measuring forage production of grazing units from Landsat MSS data', Proc. 10th Int. Symp. on Remote Sensing of Env., Univ. of Michigan, Ann Arbor, 1169-1198.

Goward, S.A., Tucker, C.J. and Dye, D., North American Vegetation Patterns Observed with the NOAA-7 AVHRR, Vegetatio 64, 3.

Hatfield, J.L., 1983: 'Remote sensing estimators of potential and actual crop yield', Rem. Sensing of Env., 13, 301-311.

Hatfield, J.L., Asrar, G. and Kanemasu, E.T., 1984: 'Intercepted photosynthetically active radiation in wheat canopies estimated by spectral reflectance', Rem. Sensing of Env., 14, 65-76.

Holben, B.N., 1986: 'Characteristics of maximum-value composite images from temporal AVHRR data', Int. J. of Remote Sensing, 7, 11.

Idso, S.B., Pinter, P.J., Jackson, R.D. and Reginato, R.J., 1980: 'Estimation of grain yields by remote sensing of crop senescence rates', Rem. Sensing of Env., 9, 87-91.

Justice, C.O., Townshend, J.R.C., Holben, B.N. and Tucker, C.J., 'Analysis of the phenology of global vegetation using meteorological satellite data', Int. J. of Remote Sensing, 6, 1271.

Malingreau, J.P., 'The wetland rice production system and its monitoring using remote sensing in Indonesia', Ph.D. Diss. Univ. of Cal. Davis.

Malingreau, J.P., 1980: 'Global vegetation dynamics: satellite observations over Asia', Int. J. of Remote Sensing, 7, 9.

Malingreau, J.P. and Setzer, A., 1987: 'Satellite derived vegetation index for drought monitoring in Brazil' (in preparation).

Pinter, P.J. Jackson, R.D., Idso, S.B. and Reginato, R.J., 1981: 'Multidate spectral reflectance as predictors of yield in water stressed wheat and barley', Int. J. of Remote Sensing, 21, 1.

Richardson, A.J. and Wiegand, C.H., 1977: 'Distinguishing vegetation from soil background information', Photg. Eng. and Remote Sensing, 43, 1541-1552.

Sellers, P.J., 1986: 'Canopy reflectance, photosynthesis and transpiration', Int. J. of Remote Sensing, 6, 1335.

Steven, M.D., Biscoe, P.V. and Jaggard, K.W., 1985: 'Estimation of sugar beet productivity by reflection in the red and infrared spectral bands', Int. J. of Remote Sensing, 4, 325-334.

Tucker, C.J. and Sellers, P.J., 1986: 'Satellite remote sensing of primary production', Int. J. of Remote Sensing, 7, 11.

Tucker, C.J., 1978: 'A comparison of satellite sensor bands for monitoring vegetation', Phot. Eng. and Rem. Sens., 44, 1369.

SATELLITE REMOTE SENSING OF RAINFALL

Eric C. Barrett
Remote Sensing Unit, Department of Geography,
University of Bristol
Bristol BS8 1SS - U.K.

ABSTRACT. Rainfall is of general significance to agrometeorology, not least in the monitoring of crop performance, and for crop (harvest) prediction. Unfortunately it is particularly difficult to monitor satisfactorily by conventional (rainguage) means. This paper summarises the many techniques which have been devised for satellite-assisted monitoring of rainfall, and proceeds to illustrate and exemplify them by more detailed discussion of techniques developed in the University of Bristol for operational use in many (land) areas of the world. These include manual, interactive, and objective methods, based on visible and/or infrared and/or passive microwave satellite imagery. The paper concludes with some suggestions for future strategies for operational rainfall monitoring with satellite support, plus a summary of actions required if satellites are to contribute usefully to agrometeorological and agroclimatological programmes in the near future.

1. INTRODUCTION

Rainfall is notoriously one of the noisiest meteorological parameters whether considered through space or through time, yet one of the most important in agrometeorology. At any one instant the fraction of the Earth actually receiving rain cannot be much more than one half of 1%. This falls mostly from mobile clouds or cloud systems, many of whose lifetimes are measureable in only a few hours. There is a wide range of intensities of instantaneous rainrates from practically zero to above 100 mm h^{-1}. To make matters worse, the variation of rainfall intensity with duration can be large from storm to storm as well as from region to region (Barrett and Martin, 1981).

However, it is vital for many purposes in the fields of environmental monitoring and renewable resources management to evaluate rainfall as accurately as possible. Of particular importance are evaluations of rainfall in support of hazard monitoring (especially related to flash floods, droughts, and insect pest infestations), hydrology and water management (especially ground water recharge, reservoir recharge, and basin run-off), and crop production and food security

F. Toselli (ed.), Applications of Remote Sensing to Agrometeorology, 305–326.

assessments and associated activities (especially food commodity marketing and famine relief).

Unfortunately, rainfall data from conventional sources (raingauges) are often inadequate. Salient points include the following:

1) The density of raingauges is usually much too low for many purposes;
2) The supply of data from existing raingauges is liable to a number of possible deficiencies, involving their accuracy, reliability, availability and timeliness;
3) There is widespread evidence that the supply of raingauge data is becoming generally less than previously, though the picture is very varied and improvements, while relatively rare, are not unknown;
4) The particular characteristics of rainfall ensure that, even where and when it is reasonably well observed by raingauge networks, the processes of isohyetal map construction and depth/area (volumetric) assessment of rainfall are fraught with difficulty. In short, rainfall distributions, being characteristically disjointed, cannot be constructed with complete confidence from existing raingauge networks, so that whatever the mathematical contouring procedures used, doubts inevitably surround the results obtained.

2. SATELLITE APPROACHES TO RAINFALL MONITORING

It is fortunate that the spatially continuous and highly repetitive image data provided by meteorological satellites contain information which can be used to supplement and improve the assessments of rainfall drawn from conventional sources. It is fortunate also that, in the most commonly and easily exploited image data - from the visible (VIS) and infrared (IR) regions of the electromagnetic spectrum, evidences of cloud area and many cloud types are reasonably direct. It is fortunate too that the costs of satellite reception facilities and associated systems are relatively modest and decreasing in relative terms.

It is singularly unfortunate that, in the VIS and IR wavebands, evidence of the distribution and intensity of rainfall is intrinsically indirect.

However, the demand for improved information on rainfall is so strong that a plethora of satellite rainfall monitoring techniques has grown up since the late 1960s to ensure that satellite data can be used as effectively as advances in science and technology allow. It has become commonplace to classify these techniques according to a scheme first outlined by Martin and Scherer (1973), and subsequently up-dated by Barrett (1983) (see Table I). If taken to relate to the fundamentals of any satellite technique, this classification is still appropriate today. However, several recently introduced methods and some developments of existing methods are more hybrid in nature. This has resulted both from a relatively spontaneous realisation that earlier techniques had both strengths and weaknesses and from some deliberate attempts to produce the "best possible" techniques for specific applications. An example of this second process has been the development of a set of cross-bred techniques for use by NOAA (the US National Oceanic and Atmospheric Administration) under the auspices of the AgRISTARS Project,

TABLE I
Summary of current and proposed satellite rainfall monitoring methods (after Barrett, 1982)

Method	Chief applications	Satellite	Sensor (s)	Present status
Cloud indexing	meteorology, climatology, hydrology, crop prediction, hazard monitoring, etc.	polar-orbiting geostationary	visible and/or infrared	
Climatological	crop prediction	polar-orbiting	visible and/or infrared	quasi-operational
Life history	severe storm assessment, meteorological research	geostationary	infrared	
Bispectral	meteorological research	polar-orbiting geostationary	visible and infrared	
Cloud physics	cloud research and atmospheric thermodynamics	geostationary	infrared and microwave	developmental
Passive	oceanic meteorology and climatology	polar-orbiting	microwave	
Active	cloud and rainfall research forecasting	? polar-orbiting	satellite radar (active microwave)	future

so that many of their sub-routines could be selected interchangeably by the duty operator to best suit the situations to be analysed (see Figure 1 and Moses and Barrett, 1986).

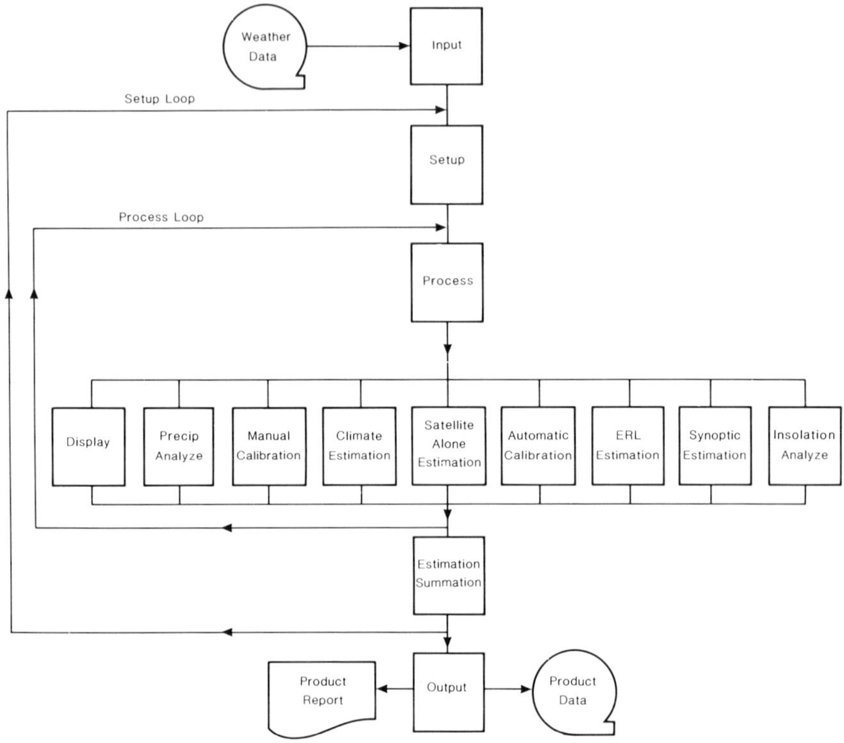

Figure 1. NOAA/NESDIS AgRISTARS precipitation estimation system (from Moses and Barrett, 1986).

To systematise the current wide range of present satellite rainfall monitoring techniques and to summarise these in a useful and convenient way, a new tabulation has been prepared recently which is shown here as Figure 2. This testifies amply to the variety and number of techniques from which the prospective user community may already choose. Two additional points may be made in this respect. First, the real choice is even wider than Figure 2 and Table II suggest, for several of the techniques listed have variants not indicated therein. Second, the possibilities for the selection of two or more complementary techniques, when one alone might not provide the range or quality of results required, are clearly very large if the entries in Figure 2 are considered permutationally.

For present purposes it will be convenient to sub-divide our detailed treatment of satellite-improved rainfall monitoring techniques into "manual interactive techniques" and "objective techniques". The manual and interactive techniques were developed first in the historical

Predictors \ Channels		ONE		TWO	
		VIS	IR	VIS	IR
One		UC-SD	GPI Michigan	HRC Montreal	
Two or more	linear		NESS Melbourne FSU NASA/Marshall NASA/GLAS Bochum	Grid history CEAS	FAO
	non-linear		CIST IIC PERMIT Reading		BIAS-3F NESDIS Cropcast ADMIT

Figure 2. Visible/infrared techniques classification matrix (after WMO, 1986).

sense and are still the most appropriate to use where stratiform, thick cirrus, or layered clouds are dominant.

In view of the large number of techniques which have been developed we will exemplify the principles and processes involved by reference to methods which have originated in, or have been at least partially developed by, the Remote Sensing Unit of the University of Bristol (see Table III. All have been used in support of national or international monitoring or management schemes in Less Developed Countries.

3. MANUAL SATELLITE RAINFALL MONITORING TECHNIQUES: e.g. THE BRISTOL METHOD

In the 1970s much attention was paid to the design, development, testing and implementation of a technique for evaluating and monitoring rainfall by satellite. This was based on expert visual interpretation of hard-copy satellite imagery, calibrated by climatology, plus meteorology whenever possible.

The central assumption in the Bristol method is that:

$$R = f(C_t, C_a, S_w, M_c) \qquad (1)$$

where R is accumulated rainfall (commonly for either 12 or 24 hr periods), C_t is cloud type according to a menu developed from a combination of theory and experimentation, C_a is fractional cloud cover area in grid squares of preselected size, S_w is synoptic weather (higher rain-rates being anticipated from larger and/or better-organised raincloud systems) and M_c is terrain influence on rainfall, here expressed in terms of a small number of altitude bands above sea-level.

TABLE II
Inventory of visible/infrared satellite rainfall monitoring techniques listed in Figure 3, with key references for each.

ADMIT	Agricultural drought monitoring integrative technique	Barrett & D'Souza (1986)
BIAS-3F	Bristol/NOAA interactive scheme	Barrett (1984); Moses (1985)
BOCHUM		Kruger et al. (1982)
CEAS	Center for Environmental Assessment Services	Lecomte (1977)
CIST	Convective infrared satellite	Griffith et al. (1981); Augustine (1984)
CROPCAST		Heitkemper et al. (1985)
FAO	Food and Agriculture Organization (United Nations)	van Dijk (1981)
FSU	Florida State University	Krishnamurti et al. (1983)
GPI	GOES precipitation index	Arkin (1979,1983)
Grid history		Martin & Howland (1985)
HRC	Highly reflective cloud	Kilonsky & Ramage (1976); Garcia (1981)
IIC	Intensity index of convection	Murakami (1983)
MELBOURNE		Del Beato & Barrell (1985)
MICHIGAN		Cutrim (1983)
MONTREAL		Lovejoy & Austin (1979A,B); Bellon et al. (1980)
NASA/MARSHALL		Robertson (1985
NASA/GLAS		Negri et al. (1984)
NESDIS	NESDIS operational convective precipitation technique	Scofield & Oliver (1977) Clark & Borneman (1984)
NESS		Whitney & Herman (1979)
PERMIT	Polar-orbiter effective rainfall monitoring integrative technique	Barrett & Power (1986)
READING		Dugdale & Milford (1985)
UC-SD	Univ. of California, San Diego	Gautier et al. (1985)

This technique has been extensively tested over tropical and extratropical regions, especially in the late 1970s, for example in support of irrigation design and Desert Locust monitoring programmes. The "Cloud Indices" formed from ($C_t \times C_a$) are translated into rainfall estimates through local region specific regressions. Figure 3 illustrates the manual mapping stages entailed by this technique.

Methods of this kind have the advantage of cheapness, being applicable to hard copy imagery, but have the disadvantage of requiring expert attention to all stages of the procedure, including transposition and mapping of synoptic weather data, analysis of the satellite imagery, and the subsequent registration of the results to other data inventories.

TABLE III

Stages in the historical development of the Bristol methods of satellite VIS and IR rainfall evaluation and monitoring.

Year(s)	Characteristics	Primary Areas of Application	Use	Time scale	Space scale	Designation
1970	Simple, singular regression relationship between cloud indices & rainfall	Tropical Far East	Climatology	1 month	$2\frac{1}{2}^{o}$sq	
1975	Multiple area-specific regressions	Northern Sumatra	Irrigation design	24 hr	18.5 km^2	The Bristol Method
1976	Estimates adjusted to current meteorological conditions	Oman	Water resource assessment	24 hr	$1/4^{o}$sq	The Bristol Method
1977/79	Product-square scale adjustments for morphoclimatic effects. Applicability to geostationary images	North-west Africa	Desert Locust control	12 hr	-1°sq	The Bristol Method
1981	Complex Global Regression and product square categorization	North-west Europe	Meteorology Hydrology Crop prediction (AgRISTARS)	6, 12 or 24 hr	Pixel resolution, degradable to any required area, regular (grid square) or irregular (e.g. drainage basin)	BIAS (the "Bristol/NOAA Inter-Active Scheme")
1982/83	Interactive image processing system methodology	North-west Europe, Central & southern Asia, Western Sahel	Meteorology Hydrology Crop prediction (AgRISTARS)	6, 12 or 24 hr	Pixel resolution, degradable to any required area, regular (grid square) or irregular (e.g. drainage basin)	BIAS (the "Bristol/NOAA Inter-Active Scheme")

TABLE III
Cont'd.

Year(s)	Characteristics	Primary Areas of Application	Use	Time scale	Space scale	Designaticn
1985	Eyeball analyses of potential rainclouds	Arabian peninsula	Climatology, meteorological (extreme event) reconnaissance	24 hr		PRC (Potential Rain-Cloud) Mapping Method
1985/86	Objective technique based on geostationary satellite visible & infrared bispectral thresholding, with or without calibration by synoptic data	Africa	Climatology, Hydrology, Crop prediction, Drought monitoring & Desertification	3 or 4 hr	Pixel resolution, degradable to any required area	ADMIT (Agricultural Drought Monitoring Integrative Technique)
1986	Objective technique based on polar-orbiter or geostationary satellite infrared brightness thresholding calibrated by synoptic data	Western Sahel	Climatology, Hydrology, Crop prediction, Drought Monitoring & Desertification	6 hr	Pixel resolution degradable to any required area	PERMIT (Polar-orbiter Effective. Rainfall Monitoring Integrative Technique)

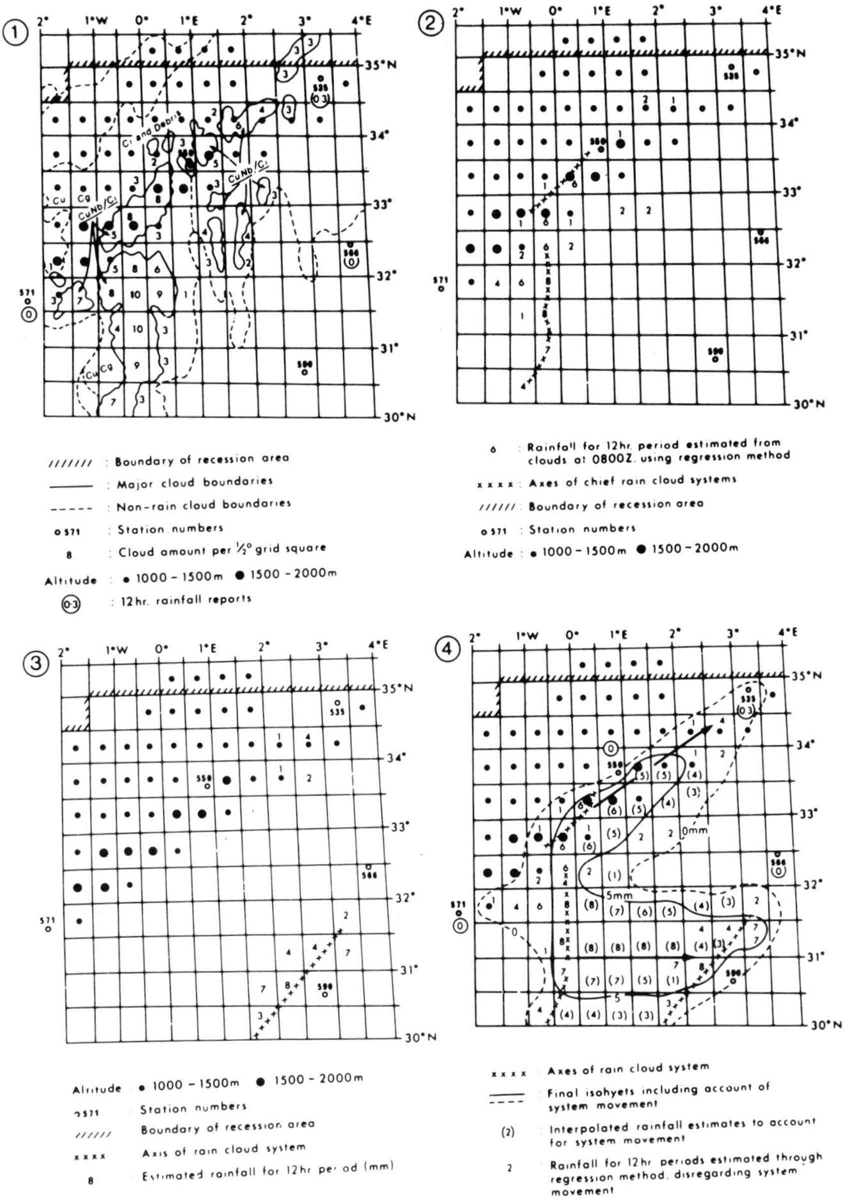

Figure 3. Stages in the Bristol method. Top left: initial cloud type/ area analysis. Top right: rainfall estimates for image at t_1. Bottom left: rainfall estimates for image at t_2. Bottom right: rainfall estimates for image t_1+t_2 (from Barrett, 1986).

4. INTERACTIVE TECHNIQUES FOR SATELLITE RAINFALL MONITORING (e.g. THE BRISTOL/NOAA INTERACTIVE SCHEME)

In the early 1980s under the auspices of the US AgRISTARS (Agriculture and Resource Inventory Surveys through Aerospace Remote Sensing) Project, a number of the more promising satellite rainfall monitoring techniques were selected for further development with NOAA/NESDIS support. The ultimate aim was to devise an integrated set of software for use in support of agricultural crop monitoring in several subcontinental-sized areas of the world. One of the techniques selected was the Bristol method. Working in collaboration, Bristol University and NOAA/NESDIS scientists developed the BIAS technique therefrom. As Figure 4 suggests, the primary intention was the development of a technique

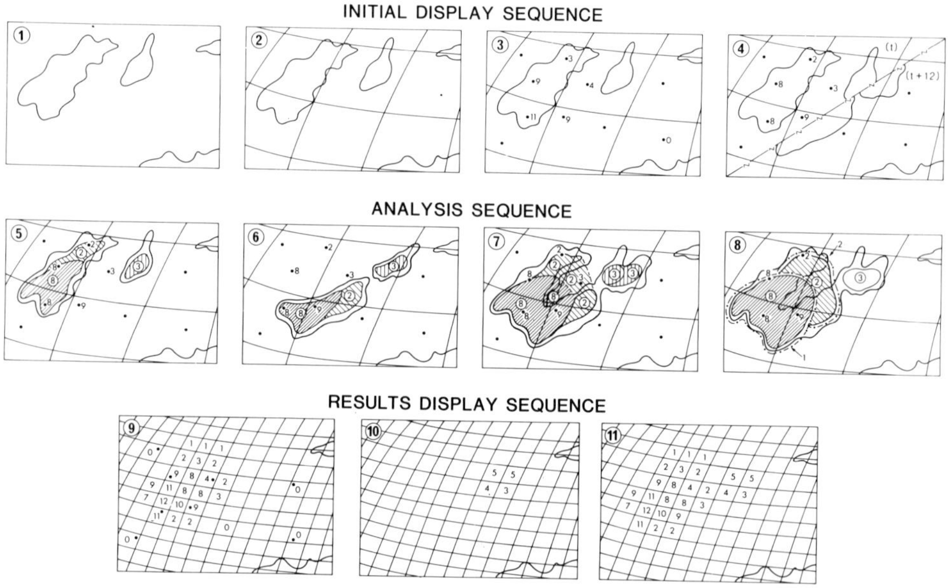

Figure 4. Stages in the Bristol Interactive Scheme (from Barrett,1986).

which closely followed the logic and structure of the Bristol Method, but which would be implemented on an interactive image processing system so that the routine aspects could be handled by a computer, leaving the interpretive aspects to be concentrated on by a skilled analyst.

At the same time, though, the opportunity was taken to effect other significant improvements in the operability of this approach. These included the following:

1) A "global regression" was constructed for immediate implementation in any new area of study (see Figure 5).
2) Different weightings were developed for application to t_1 and t_2 image analyses (for the beginning and end of each analysis period, respectively) and for t_i (the period intervening between t_1 and t_2).

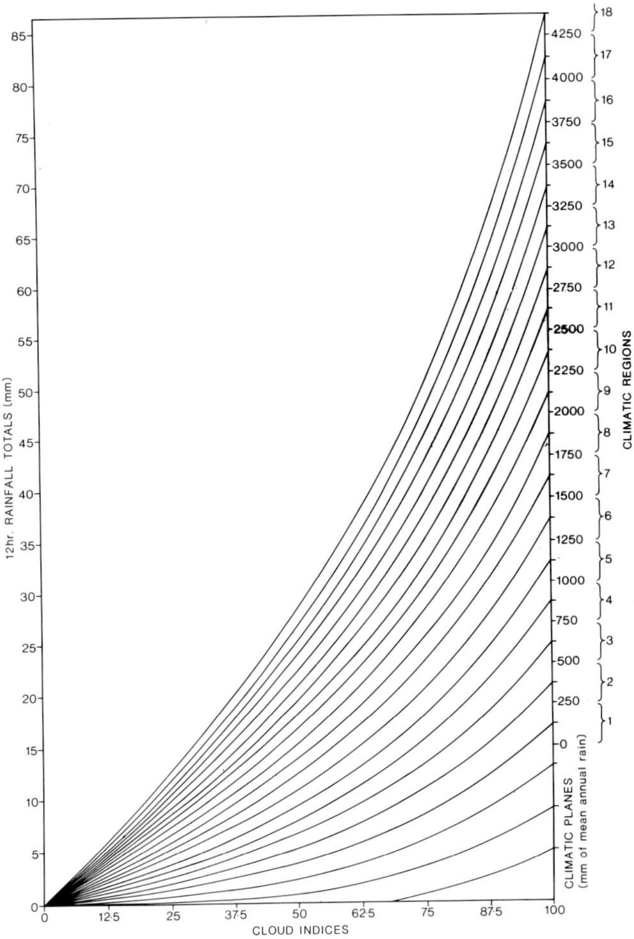

Figure 5. The BIAS Global Regression, as developed for the AgRISTARS Programme (from Moses and Barrett, 1986).

3) A full pixel-resolution terrain model was devised to input more detailed "morphoclimatological" information.
4) A "meteorological floating" scheme was devised so that, wherever and whenever possible, local station reports should be used to weight accordingly the estimates of rainfall deduced from given areas of cloud.
5) The sizes of the unit areas for which rainfall can be estimated were reduced from the rather large grid squares of the Bristol method to the full pixel resolution of the available satellite imagery.
6) The outputs could be presented at full pixel resolution, or aggre-

gated for any selected sizes of grid squares or any sizes and shapes of geographic areas (e.g. drainage basins, administrative units, etc.).

BIAS has been applied to case study situations in North and South America, Europe and Asia as well as within Africa. Verification tests over the USA carried out for AgRISTARS indicated that, even when raingauge data are quite numerous (mean station spacing of c. 300 km) and fairly regularly spaced, BIAS gives an improvement in rainfall monitoring results equivalent to between 5-25% when compared with surface isohyetal procedures alone (Moses and Barrett, 1986). Where rainfall distributions are disjointed or "spotty", and raingauge data less abundant, even larger improvements may be expected.

The principal advantage of this technique over the manual techniques is that it removes much of the drudgery from the opration so far as the manual analyst is concerned and, therefore, enables him to focus all of his attention upon the analysis of the meteorological situation involved. Its principal disadvantage is that it calls for the use of a sophisticated image-processing system to support the analyst in his activities.

5. OBJECTIVE TECHNIQUES FOR SATELLITE RAINFALL MONITORING

Whilst manual and interactive techniques are most appropriate in areas where sheet cloud is commonplace, different techniques are necessitated by convectional weather regimes. In these, there are characteristically much more rapid cycles of cloud growth, development and dissipation. Therefore, in order to identify potential rain areas it is necessary to analyse imagery more frequently than the manual or interactive techniques permit. Central to the need for geostationary satellite techniques is the demand for objective data analysis. Only in this way can the appropriate frequency of satellite images be accommodated. The families of techniques which have been devised to accommodate relatively large numbers of images each day fall into two categories, namely the "Life History" techniques and the "Bispectral" techniques. In the former, potential rainclouds are identified in infrared satellite imagery by temperature thresholdings, after which each individual element of raincloud is monitored through its individual life cycle. Unfortunately, this is demanding computationally, and has only been employed in research campaigns (see Griffiths et al., 1978,1981). As in previous sections, the techniques which will be described hereafter are under development in the Remote Sensing Unit of the University of Bristol for operational application to areas of continental or subcontinental size.

6. PERMIT (POLAR ORBITER EFFECTIVE RAINFALL MONITORING INTEGRATIVE TECHNIQUE)

As a major part of an EEC project concerned with the "Monitoring of pastureland ecosystems in the Sahel and mapping of cloud cover and rainfall" (NERC, 1986) a simple, objective method was devised for rainfall monitoring using either polar-orbiter or geostationary satellite data, based in infrared temperature thresholding, climatic information and surface station reports (as available). At its simplest this may be represented by:

$$R = f(C_t, C_n, M_c, (S_w)) \tag{2}$$

Here, C_t is cloud top temperature (areas with IR brightness temperatures below -32°C being deemed potential raincloud), C_n is the number of raincloud days per pixel, M_c is morphoclimatology (expressed as mean rainfall per rain day, recalculated on a month-by-month basis), and S_w is a synoptic weight related to raingauge data, if available (Barrett et al., 1986). Figure 6 shows a first approximation rainfall field obtained as a product of C_t and M_c for each pixel. If required, these fields may be weighted by ground truth to provide "calibrated" estimates as explained in section 7 below.

7. ADMIT (AGRICULTURAL DROUGHT MONITORING INTEGRATIVE TECHNIQUE)

This is an objective bispectral thresholding technique based on pairs of VIS and IR images during daylight hours, and on IR imagery alone at night (see Figure 7). However, the analysis of the nighttime IR-only evidence of raincloud area is statistically tied to the daytime VIS/IR evidence to prevent the areas analysed as rainclouds from "blowing up" where extensive cirrus is found. Therefore, ADMIT may be said to involve the following terms:

$$R = f(C_B, C_T, C_n, M_c, (S_w)) \tag{3}$$

where C_B is cloud brightness in the VIS, and other terms are as above. In practice, the selection of appropriate rain/no rain thresholds may be based on the evidence of bispectral scattergrams either for the whole, or some part only, of the period in question. Best results are obtained from the former. Tests over Africa as a whole (Barrett and D'Souza, 1986) suggest that some 85% of the continental area may be correctly classified in terms of rain/no rain areas, and that the related rainfall estimation fields (after multiplication through by M_c values) represent rainfall acceptably at least over periods of 10 and 30 days and upwards. For daily rainfall estimation the number of raincloud slots per day (C_n) might also be usefully invoked as an additional set of weights.

8. PASSIVE MICROWAVE POSSIBILITIES

Looking further into the future, there is hope that present research into the use of passive microwave data may lead to either improved (multispectral VIS, IR and MW) rainfall monitoring techniques now under investigation, or new (MW only) methods. Table IV summarises the three

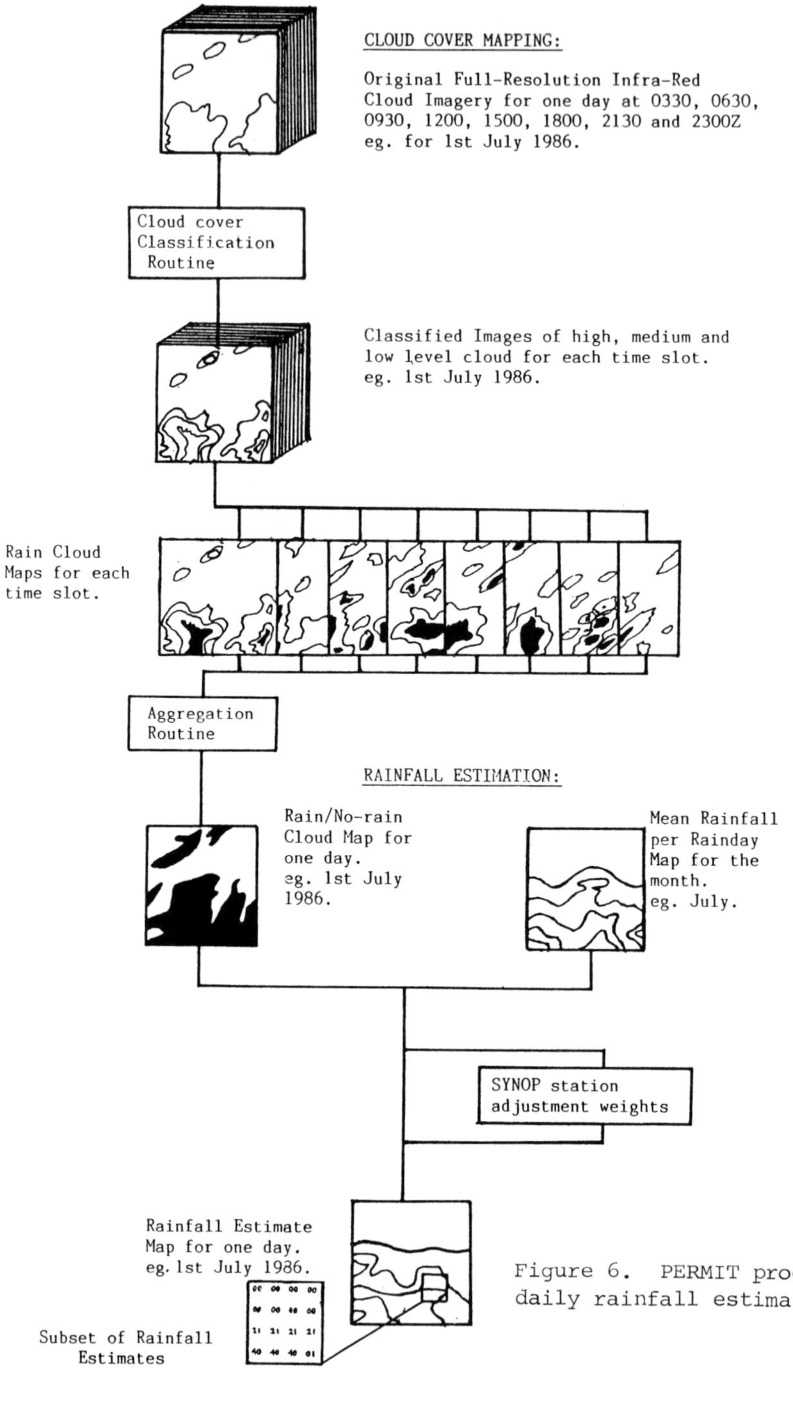

Figure 6. PERMIT production daily rainfall estimates.

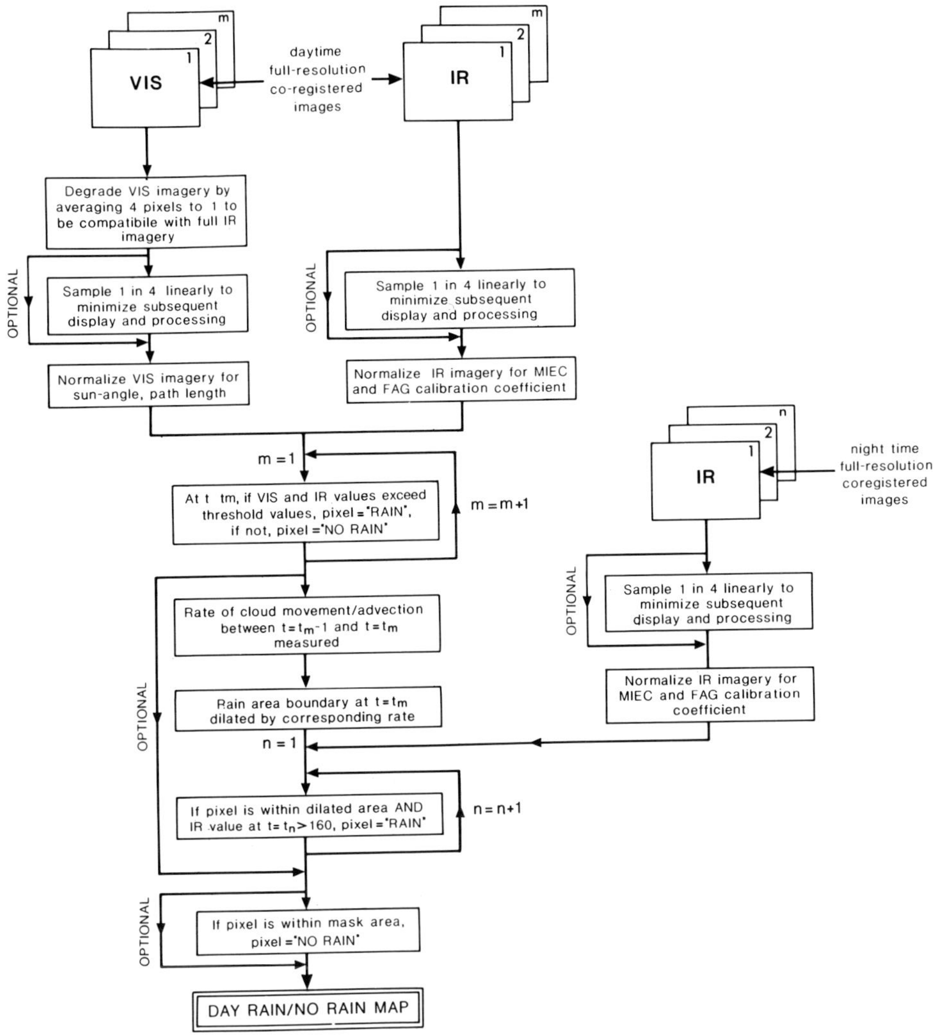

Figure 7. ADMIT production of daily rain/no rain map.

TABLE IV
Microwave imaging instruments

Instrument	ESMR-5	ESMR-6	SMMR	SSM/I
Spacecraft	Nimbus-5	Nimbus-6	Nimbus-7/Seasat	DMSP 5D-2/SX
Launch date	12-72	06-75	10-78 06-78	11-86
Expired	inoperable 08-83	partial 08-83	04-86 09-78	N/A
Frequencies/ resolutions:				
6.6 GHz			vh 121*79 km	
10.7 GHz			vh 74*49 km	
18.0 GHz			vh 44*29 km	
19.35GHz	x25*25 km			vh 70*45 km
21.0 GHz			vh 38*25 km	
22.23GHz				v 60*40 km
37.0 GHz		vh 25*45 km	vh 21*14 km	vh 38*30 km
85.5 GHz				vh 15*14 km
Primary aims of mission:	Liquid water content of clouds, sea ice and open sea coverage. Surface composition and soil type, with surface features and surface moisture.		Sea surface temperature, near sea surface winds. Water vapour, liquid water content and cloud droplet size; rainfall rate.	Continual ocean wind speed and ice coverage, age, and extent. Intensity of precipitation, cloud water content and land surface moisture.

Key: ESMR - Electrically Scanning Microwave Radiometer.
SMMR - Scanning Multispectral Microwave Radiometer.
SSM/I - Microwave Imager.

v - vertical polarization.
h - horizontal polarization.
x - non-polarized.

instruments which have proved useful to date, plus the next instrument of interest. Experience has shown that imagery at the 18.0 and 19.35 GHz frequencies provides good quantitative evidence of rainfall over oceans. Unfortunately for overland applications the physical properties and behavioural characteristics of atmospheric hydrometeors and various background services combine to present much greater problems of both rainfall area identification and rain intensity evaluation. These arise mainly because the emissivity of land is generally high, but often variable. Furthermore, these emissivities are often close to those of the rain areas themselves.

Generalised models suggest that, at 37 GHz, rain will attenuate upwelling land radiation especially if rain is heavy, but confirm that light to moderate intensity rains cannot be assessed with confidence in this way. Some hope has been engendered through either dual frequency and/or dual polarisation screening procedures, leading to the development of passive microwave rainfall algorithms currently under test. However, the best hope for the future seems to lie in multifrequency/dual polarisation approaches. For these to be tested fully we await the oft-delayed launch of the SSM/1 instrument on the next DMSP satellite, whose 85.5 GHz channel should provide assistance with both very light and very heavy rain area determination. However, since the spatial resolutions of such sensors are still relatively low, VIS and IR information will continue to be invaluable for satellite rainfall monitoring for the foreseeable future. For a general discussion of these problems and possibilities, see Bailey, Barrett and Kidd (1986).

9. OPERATIONAL STRATEGIES FOR SATELLITE RAINFALL MONITORING

Given, then, that there is a wide range of available techniques from which to choose when designing a test campaign or long-term operation, a number of considerations will have a bearing upon the selection process. These are likely to include (Barrett, 1986):

1) the size of region(s) to be monitored;
2) the physical characteristics of the region(s) (e.g. altitude, relief, aspect, etc.);
3) the dominant types of rainfall processes;
4) the nature and density of conventional observations which may be available for calibration and/or verification;
5) the availability of satellite data;
6) the skills and experience of local staff;
7) the availability of image processing facilities;
8) the purposes for which improved rainfall data are required.

Supposing that a large, varied area is involved, digital facilities are available, and a range of end users require data of different temporal and spatial resolutions, a hierarchy of techniques seems more likely to provide the sought-after outputs than any one individual method.

So far as the validity of satellite rainfall monitoring results are concerned, much research has attempted to compare satellite estimates with surface rainfall data (whether raingauge or radar), but there is still no generally accepted method for undertaking this task. One basic problem is that the satellite data are area estimates, whereas much of the ground truth is for point locations only. A further problem is that, if at least some of the available surface data are reserved for verification of the satellite or satellite improved estimates, fewer are then available for the calibration of the satellite estimates themselves, thereby probably reducing the apparent performance of the technique in question.

In a fully operational context it may be desirable to integrate all available conventional data with the satellite imagery in search of the best possible rainfall monitoring products. In such circumstances it must either be demonstrated by special ground-truth gathering campaigns, or accepted as an a priori principle, that the use of an integrated satellite-aided approach gives better results than either surface or satellite schemes can do alone.

In practice, relatively little attention has been paid (e.g. through sensitivity analyses) to the optimum way(s) of combining surface with satellite data, though it is clear that this, to some extent, may depend upon the governing purpose for which the final rainfall data are required. For example, techniques such as ADMIT and PERMIT may be refined to entail $R = f(C_T, M_C, R_O)$ and $r = f(C_T, C_B, M_C, R_O)$, respectively, where R_O is observed (e.g. raingauge) data. In such respects the following promising possibilities arise:

1) Raingauge data may be summed and averaged over an entire geographic, or administrative region, and regressed against "first guess" satellite estimates, thus providing a weighting function by which "final" estimations may be calculated.
2) Each raingauge report may be used to weight initial satellite estimates in a local "tributory" region of its own. This may be a Thiessen polygon or some other existing hydrological subdivision, or some climatically or agrometeorologically determined region, etc.

10. PRODUCTION, ORGANISATION AND PRESENTATION OF RESULTS

Any satellite rainfall monitoring technique which supplies quantitative results via a digital computer (e.g. BIAS, ADMIT and PERMIT directly, or the Bristol method indirectly if the results of manual analyses are subsequently digitized) can support a wide range of products, tailored to various end-user requirements. These include:

1) full pixel-resolution estimation fields;
2) arrays of estimates for regular areas of larger size (e.g. fitting latitude/longitude, NMC, or other existing grids);
3) aggregated values for areas of irregular shape and size (e.g. rainfall basins, other relief units, administrative areas, etc.);
4) reduced-resolution maps of rainfall, e.g. (smoothed) isohyetal maps.

However, underpinning all of this is the basic choice between centralised and disbursed monitoring, whether considered on a continental, national or some other scale. In short, it may be neither necessary nor desirable for many products to be prepared locally; centralisation may help to achieve results more economically, under conditions in which high levels of quality control may be easier to apply and maintain. On the other hand, if products can be prepared locally, it is more likely then that the local user will be able to influence the end products in a way which is best suited to his particular needs.

11. PRESENT NEEDS

If satellite rainfall monitoring techniques are to be used more widely for operational applications especially in the Developing World, a number of present needs must be satisfied. These include the following:

1) The newer (interactive and objective) techniques must be tested over longer periods than has been possible hitherto.
2) Systematic attention should be paid to the refinement of existing techniques so that general biases, special local regional factors, etc. might be recognised and accommodated.
3) Resources must be allocated to the improvement of background data sets where these are needed for calibration.
4) Attention must be paid to the most effective, economic, and efficient ways of incorporating satellite techniques into local, regional and/or continental operations.
5) Networks and methods for quality control (verification) of operational results must be designed and deployed whenever possible. Special attention must be paid to ways of comparing the point observations from surface stations with the areal estimates from satellites: there is no widely-spread procedure for this important task.
6) Suitable reception and data processing facilities must be installed, and software tailored to suit the intended operational pattern. The time and cost involved in software technology transfer are frequently under-estimated.
7) Training of satellite rainfall estimation technique operators, and education of present and potential end users is vital.
8) A continuing dialogue between scientists, technologists, operators and the user community will help ensure that the routine operations can evolve through time, and so ensure that needs and opportunities are continually as well matched as possible.

12. GENERAL CONCLUSIONS

Satellite rainfall monitoring techniques today present a wide and growing range of possibilities for improved monitoring of environmental conditions of special relevance to meteorology, climatology, hydrology and water management. This could increasingly support improved forecasting of related events and conditions of significance to human life and economy, e.g. droughts, floods, insect plagues, food shortages, and famine. Most significantly, because of the rise of interactive and objective (computer-based) techniques, end products are now extremely flexible, and can be tailored almost infinitely to suit local needs, opportunities and preferences.

REFERENCES

Arkin, P.A., 1979: 'The relationship between fractional coverage of high cloud and rainfall accumulations during GATE over the B-scale array', Mon. Wea. Rev., 107, 1382-1387.

Arkin, P.A., 1983: 'A diagnostic precipitation index from infrared satellite imagery', Tropical Ocean-Atmosphere Newsletter, 17, 5-7.

Augustine, J.A., 1984: 'The diurnal variation of large-scale infrared rainfall over the tropical Pacific Ocean during August 1979', Mon. Wea. Rev., 112, 1745-1751.

Bailey, J.O., E.C. Barrett and C. Kidd, 1986: 'Satellite passive microwave imagery and its potential for rainfall estimation over land', J. Brit. Interplan. Soc., 39, 527-534.

Barrett, E.C., 1979: 'Satellite rainfall estimation by cloud indexing methods for desert locust survey and control', Satellite Hydrology (M. Deutsch, ed.), American Water Resources Association, 92-100.

Barrett, E.C., 1983: 'Rainfall evaluation by remote sensing, problems and prospects', Hydrological Applications of Remote Sensing and Data Transmission, IAHS Publication No.145, 247-258.

Barrett, E.C., 1984: Development and Testing of BIAS-3F, The Bristol/NOAA Interactive Scheme for Satellite-improved Rainfall Monitoring (three field version) and the BIAS User's Guide (second edition), Final Report to U.S. Dept. of Commerce, Contract No. NA-83-DDC-PPP34, Remote Sensing Unit, Univ. of Bristol, Bristol, England, 90.

Barrett, E.C., 1986: Estimation of Precipitation from AVHRR and Meteosat Data Over Africa, presented at the XXth Int. Symp. of Remote Sensing of Environment, Nairobi, Kenya, 20.

Barrett, E.C. and G. D'Souza, 1986: The Development of an Objective Range of Dry Day Products for Drought Monitoring Over Africa, Final Report to ESA, Contract No. 6137/84/D.HS(SC), 76.

Barrett, E.C., G. D'Souza, D. Lloyd and C.H. Power, 1986: Satellite Remote Sensing: African Applications Today.

Barrett, E.C., M.G. Hamilton and C.H. Power, 1986: A Satellite Climatology of Potential Rain Clouds in the Region of the Sultanate of Oman, Final Report to the Sultanate of Oman Public Authority for Water Resources, Contract No. C/2 (1985), Remote Sensing Unit, University of Bristol, 127.

Barrett, E.C. and D.W. Martin, 1981: The Use of Satellite Data in Rainfall Monitoring, Academic Press, London, 340.

Barrett, E.C., C.H. Power, M.J. Beaumont and T.S. Richards, 1986: 'Cloud cover and rainfall in the Western Sahel', Final Report to NERC, Contract No. F6/C1/19, Remote Sensing Unit, University of Bristol, England, 85.

Cutrim, E.M.C., 1983: 'Estimating monthly rainfall from geostationary satellite imagery over Amazonia, Brazil', Ph.D. Dissertation, Dept. of Atmos. Sci., Univ. of Michigan, Ann Arbor, 107.

Clark, D. and R. Borneman, 1984: 'Satellite precipitation estimates program of the synoptic analysis branch', preprint Vol., 10th Conf. on Weather Forecasting and Analysis, June 25-29 (Clearwater Beach), Amer. Meteor. Soc., Boston, 392-399.

Del Beato, R. and S.L. Barrell, 1985: 'Rain estimation in extra-tropical cyclones using GMS imagery', Mon. Wea. Rev., 113, 747-755.

Dugdale, G. and J.R. Milford, 1985: 'Rainfall estimation over the Sahel using Meteosat TIR data', paper presented to the Conf. on Parameterisation of Land Surface Characteristics, ISLSCP, Rome, December 1985.

Garcia, O., 1981: 'A comparison of two satellite rainfall estimates for GATE', J. Appl. Meteor., 20, 430-438.

Gautier, C., T. Delacroix and D.W. Martin, 1985: 'Method for determining large-scale precipitation and preliminary results for the Indian Ocean (SMONEX)', Manuscript in preparation.

Griffiths, C.G., W.L. Woodley, P.G. Grube, D.W. Martin, J. Stout and D.N. Sikdar, 1978: 'Rain estimation from geosynchronous satellite imagery - visible and infrared studies', Mon. Wea. Rev., 106, 1153-1171.

Griffith, C.G., J.A. Augustine and W.L. Woodley, 1981: 'Satellite rain estimation in the U.S. High Plains', J. Appl. Meteor., 20, 53-66.

Heitkemper, L.J., J.F. Moses, E.S. Merritt and J.N. Cooper, 1985: 'An interactive method for estimating rainfall from satellites', preprint Vol., Int. Conf. on Interactive Information and Processing Systems for Meteorology, Oceanography and Hydrology, January 7-11 (Los Angeles), Amer. Meteor. Soc., Boston, 271-278.

Kilonsky, B.J. and C.S. Ramage, 1976: 'A technique for estimating tropical open-ocean rainfall from satellite observations', J. Appl. Meteor., 15, 972-975.

Krishnamurti, T.N., S. Cocke, R. Pasch and S. Low-Nam, 1983: 'Precipitation estimates from rain gauge and satellite observations, Summer MONEX', Report No.83-7, Dept. of Meteor., Florida State Univ., Tallahassee.

Kruger, L.R., G. Strubing and G.A. Schulz, 1982: 'Satellite data as a basis for the estimation of rainfall and runoff', Advances in Hydrometry, IAHS Pub. No.134, 273-284.

LeComte, D., 1977: 'Evaluation of satellite rainfall estimates in Northern Haiti', Unpublished notes, Environmental Data and Information Service, U.S. Dept. of Commerce, Washington, DC, 8.

Lovejoy, S. and G.L. Austin, 1979a: 'The delineation of rain areas from visible and IR satellite data for GATE and mid-latitudes', Atmosphere-Oceans, 17, 77-79.

Lovejoy, S. and G.L. Austin, 1979b: 'The sources of error in rain amount estimating schemes from GOES visible and IR satellite data', Mon. Wea. Rev., 107, 1048-1054.

Martin, D.W. and M.R. Howland, 1985: 'Grid history: a geostationary satellite technique for estimating daily rainfall in the tropics', submitted to J. of Clim. and Appl. Meteor.

Martin, D.W. and W.D. Scherer, 1973: 'Review of satellite rainfall estimation methods', Bull. of Am. Meteor. Soc., 54, 661-674.

Moses, J.F., 1985: 'Interactive global precipitation estimates', preprint Vol., Int. Conf. on Interactive Information and Processing Systems for Meteorology, Oceanography and Hydrology, January 7-11 (Los Angeles), Amer. Meteor. Soc., Boston, 206-212.

Moses, J.F. and E.C. Barrett, 1986: 'Interactive procedures for estimating precipitation from satellite imagery and surface characteristics', Proc. of the Int. Workshop on Hydrologic Applications of Space Technology, Cocoa Beach, Florida, August 18-23, 1985, IAHS, in press.

Murakami, M., 1983: 'Analysis of the deep convective activity over the Western Pacific and southeast Asia: Part I, Diurnal variation', J. Meteor. Soc., Japan, 61, 60-76.

Negri, A.J., R.F. Adler and P.J. Wetzel, 1984: 'Rain estimation from satellites. An examination of the Griffith-Woodley technique', J. Clim. and Appl. Meteor., 23, 102-116.

NERC, 1986: Monitoring of Pastureland Ecosystems in the Sahel and Mapping of Cloud Cover and Rainfall, Final Report to the Commission of the European Communities Development Directorate (DG VIII).

Robertson, F.R., 1985: 'IR precipitation in the South Pacific during FGGE', SOP-1, preprint Vol., 16th Conf. on Hurricanes and Tropical Meteorology, May 14-17 (Houston), Amer. Meteor. Soc., Boston.

Scofield, R.A. and V.J. Oliver, 1977: 'A scheme for estimating convective rainfall from satellite imagery', NOAA Tech. Memo NESS 86, Washington, DC, 47.

Van Dijk, A., 1981: 'Precipitation assessment from environmental satellite data for Northwest Libya including the Gefara Plain', Food and Agriculture Organization (RSC Series 4), Rome, 45.

Whitney, L.F. Jr. and L.D. Herman, 1979: 'A statistical approach to rainfall estimation using satellite data', Satellite Hydrology (M. Deutsch, ed.), American Water Resources Association, 139-143.